BIOLOGY AND MANAGEMENT OF *BACTROCERA* AND RELATED FRUIT FLIES

BIOLOGY AND MANAGEMENT OF *BACTROCERA* AND RELATED FRUIT FLIES

Anthony R. Clarke

School of Earth, Environmental and Biological Sciences, Queensland University of Technology (QUT), Brisbane, Australia

PUBLISHING

First published in 2019 by CABI, with ISBN 978 1 789 241 822.

CABI is a trading name of CAB International

CABI	CABI
Nosworthy Way	745 Atlantic Avenue
Wallingford	8th Floor
Oxfordshire OX10 8DE	Boston, MA 02111
UK	USA
Tel: +44 (0)1491 832111	Tel: +1 (617)682-9015
Fax: +44 (0)1491 833508	E-mail: cabi-nao@cabi.org
E-mail: info@cabi.org	
Website: www.cabi.org	

Library of Congress Cataloging-in-Publication Data

Names: Clarke, Anthony R., author.
Title: Biology and management of Bactrocera and related fruit flies / Anthony R. Clarke.
Description: Boston, MA : CABI, 2019. | Includes bibliographical references and index.
Identifiers: LCCN 2019016395| ISBN 9781789241822 (hbk : alk. paper) | ISBN
    9781789241846 (epub)
Subjects: LCSH: Tephritidae. | Tephritidae--Control.
Classification: LCC QL537.T42 C53 2019 | DDC 595.77/4--dc23 LC record available
at https://lccn.loc.gov/2019016395

Published exclusively in Australia and New Zealand, with ISBN 9781486312139, by:
CSIRO Publishing, Locked Bag 10, Clayton South VIC 3169, Australia
Telephone: +61 3 9545 8400
Email: publishing.sales@csiro.au
Website: www.publish.csiro.au

A catalogue record for this book is available from the National Library of Australia.

CSIRO Publishing publishes and distributes scientific, technical and health science books, magazines and journals from Australia to a worldwide audience and conducts these activities autonomously from the research activities of the Commonwealth Scientific and Industrial Research Organisation (CSIRO). The views expressed in this publication are those of the author(s) and do not necessarily represent those of, and should not be attributed to, the publisher or CSIRO. The copyright owner shall not be liable for technical or other errors or omissions contained herein. The reader/user accepts all risks and responsibility for losses, damages, costs and other consequences resulting directly or indirectly from using this information.

Commissioning editor: Ward Cooper
Editorial assistant: Tabitha Jay
Production editor: Ali Thompson

Typeset by SPi, Pondicherry, India
Printed and bound in the UK by Bell & Bain Ltd, Glasgow

Contents

Acknowledgements

As an academic, you are only as good as your research fellows and students. I've been extraordinarily fortunate in my career to have had the opportunity to mentor many exceptional beginning, early and mid-career researchers. I acknowledge particularly former and current research fellows of the QUT Fruit Fly Group: Drs Solomon Balagawi, Paul Cunningham, Matt Krosch, Katharina Merkel, Kumaran Nagalingam, Chandra Prabhakar, Mark Schutze, and Rehan Silva. Prof Stephen Cameron, a quiet leader in the lab before being snatched to greater things at Purdue is not forgotten. Prof David Gust, as my head-of-school when I was building up the lab, is thanked for his practical support, encouragement and advice.

Mrs Amy Carmichael kindly drew or reworked many of the illustrations in this book (Figures 2.2, 2.3, 4.1, 5.1, 5.3, 5.5, 6.4, 7.2 and 8.1): many thanks Ames.

Mr Peter Leach 'guest' authored chapter 10, *Phytosanitary Measures*. The pre-harvest and post-harvest fruit fly research communities still remain, unfortunately, largely separate despite the increasing need for them to be working closely together. My interactions with Peter over many years have greatly increased my broader understanding of fruit flies as regulatory pests, and particularly the bottom line of how they impact on growers. The correct citation format for chapter 10, acknowledging Peter as author, is: Leach, P. (2019) Phytosanitary measures. In: Clarke, A.R. *Biology and Management of Bactrocera and Related Fruit Flies*. CAB International, Wallingford, U.K., 195–225.

The greater portion of this book was written while on a six-month academic study leave in the Department of Animal Ecology and Tropical Biology at the University of Würzburg, Bavaria, Germany. I thank my host Prof Dr Ingolf Steffan-Dewenter for making facilities available to me, and to all members of the department for making me so welcome. The Plant Biosecurity

Cooperative Research Centre and the Queensland University of Technology co-funded the study-leave and I thank them for their support. The book was finalised during a short study-leave undertaken in the Royal Museum for Central Africa, Tervuren, Belgium. I similarly thank my host Dr Marc De Meyer for making me welcome.

Dedication

I dedicate this book to Linda, who has made my career possible by being loving and supporting always.

Disclaimer: Mention of a commercial product does not infer endorsement.

1 General Introduction

1.1 Scene Setting

On the 16th of February 2015 a single male insect was caught in a permanent surveillance trap in the suburb of Grey Lynn, Auckland, New Zealand. Within 24 hours it had been positively identified and within 48 hours a report had been tabled by the relevant Minister to the New Zealand Parliament and public press releases had been made. Further trapping collected another 13 adult insects and immatures. Within 72 hours a biosecurity *Controlled Area Notice* had been announced, effectively locking down the suburb, and television reporters started doing 'emergency reports from the field'. Public information was released not only in English, but also in Cantonese, Mandarin, Korean, Samoan, Tongan and Punjabi to ensure no local ethnic group missed the information. By the end of an emergency response, which lasted nearly the entire year until eradication was declared on the 4th of December 2015, over 4500 additional traps had been distributed around Auckland, 1000 bins placed in the controlled area for the disposal of potentially infected goods and emptied over 99 000 times, 530 tonnes of potentially infected goods were collected and disposed of, the public at 25 events (including those at the nation's premier sports stadium, Eden Park) were managed to ensure movement of suspect goods out of the controlled area did not occur, and NZ$36 million had been spent.

So what caused this massive emergency response? Some form of exotic mosquito carrying a life-threatening disease? A mutant housefly set to destroy the world? No, the answer is far less dramatic, but no less important. The problem was a small population of the Queensland fruit fly, *Bactrocera tryoni*. As its name suggests the Queensland fruit fly is a native of Australia and it is that country's worst insect pest of horticulture, causing direct and indirect losses estimated at over AU$100 million per year. New Zealand is one of the few countries in the world free of fruit flies and the possible establishment of the

exotic Queensland fruit fly threatened at least 90% of the different fruit and vegetable types grown in that country's multi-billion dollar horticultural industry.
Fortunately that incursion did not result in the establishment of a new pest, but
other countries have not been so lucky. In the last decade the highly successful
Californian olive industry, worth an estimated US$73 million per year, has been
threatened by the establishment of the exotic olive fly, *Bactrocera oleae*, an
insect native to the Mediterranean and Middle East. In 2003 the Oriental fruit
fly, *Bactrocera dorsalis*, was discovered infesting fruit and vegetables in Kenya.
This fly has subsequently spread through all of sub-Saharan Africa, threaten-
ing food security and destroying valuable export markets, such as mangoes to
Europe, which were a major source of income to many African rural producers.
Its estimated impact to Africa is over US$2 billion per year.

1.2   *Bactrocera* Fruit Flies

The non-expert often confuses the pest fruit flies of agriculture with the small dark
insects, often incorrectly called fruit flies, which fly around your fruit bowl when
your fruit is overripe. These latter insects are drosophilid fruit flies and should be
more correctly known as vinegar flies, as they respond to the smell of acetic acid
given off by the fermenting, rotting fruit in which they breed. These insects may
be a nuisance, but with the exception of spotted wing drosophila, *Drosophila
suzukii*, are not pests of agriculture. Agricultural pest fruit flies on the other hand,
belong to the insect family Tephritidae and are medium-sized insects that attack
sound fruit on trees: drosophilids are rarely pests, tephritids often are. Tephritid
fruit flies are the insects that, in tropical and sub-tropical regions especially, are
the most likely cause of maggots in fruit and they are found on all continents
except Antarctica. In North America and Europe the most common tephritid pests
belong to the genus *Rhagoletis* Loew, in South America to *Anastrepha* Schiner,
and in Africa and the Mediterranean to *Ceratitis* Macleay and *Dacus* Fabricius.
However, in the broad sweep of the globe from Pakistan and India in the west,
across Asia and down into Australia and the South Pacific, the pest fruit flies come
predominantly from the genera *Bactrocera* Macquart and *Zeugodacus* Hendel.
        Bactrocera is a huge genus of 461 species and in its endemic range
Bactrocera species are the dominant insect pests of fruits and fleshy vegetables.
Female *Bactrocera* lay their eggs into fruit (Fig. 1.1), where the emergent larvae
feed before leaving the fruit to pupate in soil. *Bactrocera* species cause direct
production losses of both commercial and non-commercial crops; they can
impact on food security for subsistence level producers; and due to quarantine
concerns they can be the cause for restrictions on domestic and international
fresh produce trade. While long recognised pests in countries where they naturally occur, some *Bactrocera* species are now also found outside their native
ranges, such as the already mentioned Oriental fruit fly and olive fruit fly, along
with others such as the peach fruit fly (*B. zonata*) that is devastating fruit production in the Nile delta of Egypt and currently spreading both east and south.
        While commonly thought of only as pests, *Bactrocera* species are also
important components of local ecosystems. The majority of *Bactrocera* species
remain restricted to tropical and sub-tropical rainforests, the presumed native

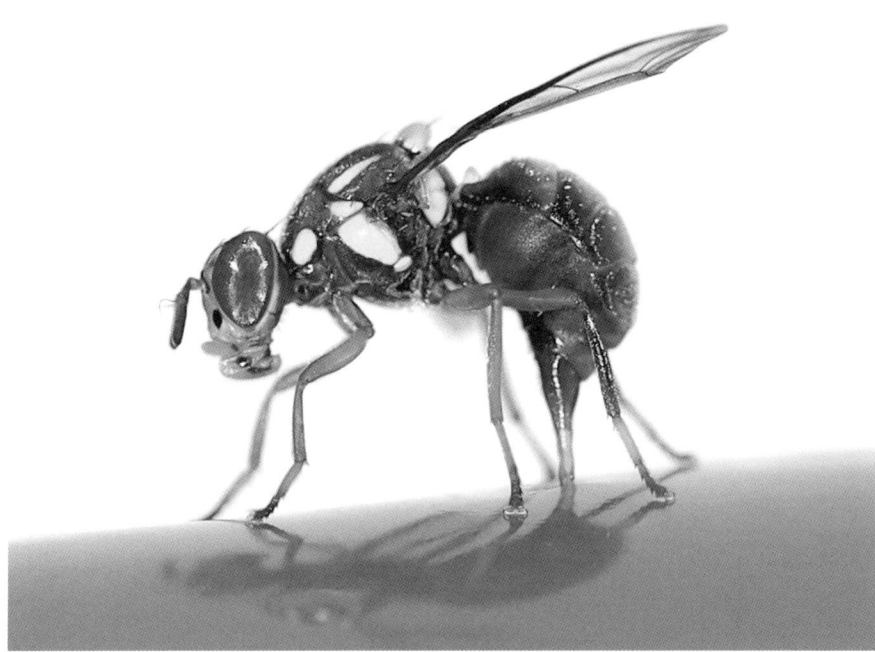

Fig. 1.1. Ovipositing *Bactrocera tryoni* (photo credit: Ms Jaye Newman).

habitat of the genus. In those rainforests they play an indirect role in aiding fruit dispersal and seed germination, and as such may be an important but largely ignored component of forest regeneration. For ecologists and evolutionary biologists the great diversity of *Bactrocera* in rainforests also makes these flies an ideal system to study questions of biogeography, speciation, herbivore–host plant relationships and community ecology.

Closely related to *Bactrocera* are two sister genera within the tribe Dacini: *Dacus* (273 species) and *Zeugodacus* (196 species). Both genera occur in the same broad geographic range as *Bactrocera*, although *Dacus* is by far the most prevalent of the three genera in Africa, where both *Bactrocera* and *Zeugodacus* are rare. While common in Africa, *Dacus* becomes rarer in Asia and the Pacific where it is replaced by *Bactrocera* and *Zeugodacus*. Only recently has *Zeugodacus* been elevated to generic level; it was previously treated as a subgenus of *Bactrocera*. While the elevation of *Zeugodacus* has not been without debate, it is now accepted by most workers. The general biology of all three genera is similar, although their host plant preferences are slightly different with *Zeugodacus* being most prevalent on Cucurbitaceae (melons, squashes) and *Dacus* on Cucurbitaceae and Asclepidaceae (milkweeds). With the exception of the major pest species melon fly, *Zeugodacus cucurbitae*, there is surprisingly little known about the biology of most species in these two genera, although their taxonomy is well documented. Where biological, ecological or pest management information is available for *Zeugodacus* or *Dacus* species, and it adds to information available for *Bactrocera*, then I include it in this book, but to focus on these two genera outside of taxonomy and systematics only highlights how little we know about them.

The remainder of this Introduction provides more information on *Bactrocera* as pests, as well as introducing their value as model organisms for broader scientific research; both issues are developed more fully in later chapters. The chapter concludes with a detailed statement of the structure and philosophy of the book.

1.2.1   *Bactrocera* as crop pests

Bactrocera fruit flies cause significant damage as direct crop pests, but this is not the only reason they are considered such significant threats to agriculture. In addition to direct crop loss, there are the additional costs associated with the task of producing uninfested fruit in a fruit fly area, or conversely the regulatory costs of maintaining and proving an area is fruit fly free. At subsistence level fruit flies may impact on food security, while at a trade level the presence of Dacini fruit flies in a production area can severely limit opportunities for market access without appropriate market access treatments. Depending on the fruit fly species, and the cropping area, all of these costs may occur.

Direct crop damage
Bactrocera are best known as pests for the direct damage they do to fruit and fleshy vegetables. The fruit fly life cycle involves the female fly depositing her eggs into a host fruit. Depending on the thickness of the peel, the eggs may be laid into or under the peel, but regardless of where they are laid the emergent larvae migrate to the fruit flesh, which they consume as they grow. Larval feeding causes direct fruit damage, while the introduction of bacteria and other decay organisms also damages the fruit. In mature but still green fruit, fruit fly infestation may lead to localised ripening around the oviposition site, and fruit fly-induced premature fruit fall is common. Any fruit fly infestation will normally make fruit unmarketable, but whether the fruit is also inedible depends on the type of fruit and the degree of infestation. In some cases for larger fruit, or where the larvae are restricted in their movement by internal structures of the fruit (as often occurs in citrus), the infested portion of fruit may be able to be cut away, leaving the remainder of the fruit perfectly sound for human or animal consumption.

The ability to attack multiple types of fruit (= polyphagy or generalism) is a key element of the pest status of many dacine species, especially within *Bactrocera*. The olive fly is unusual for a pest *Bactrocera* in that it specialises on only one host, olive. Most of the other pest *Bactrocera* species (e.g. *B. tryoni*, *B. dorsalis*, *B. zonata*) have very large host ranges, sometimes in excess of 100 different plant species across 20 or more plant families. Such large host ranges make them difficult to manage as they can breed in non-commercial host plants (e.g. wild plants), or move from one fruiting crop to another. Host ranges tend to be more restricted in the *Zeugodacus* and *Dacus*, with most limited to species of cucurbit (*Zeugodacus* and *Dacus*) or asclepiad (*Dacus* only); but there are exceptions. The melon fly, *Z. cucurbitae*, is a major pest of commercial cucurbits (i.e. melons, cucumbers, gourds), but is also a pest of long beans and chillies. The issue of host use and host range is discussed more fully in Chapter 6, but with respect to pest status a key issue to remember is that for the polyphagous species not all hosts are used equally, and that host use will change depending on the presence or absence of other hosts.

Infestation rates by dacine fruit flies in the field can be highly variable. One-hundred per cent infestation rates are known for highly susceptible crops grown in areas where pest dacines are abundant. However, much lower infestation rates occur if fly abundance is low, or if the host fruit is of low suitability. It is also not clear how the infestation rate is modified by the intensity of the agricultural system, or the density of crops, but infestation of rainforest fruits at less than 10% is lower than that generally recorded in agricultural systems. Data from Papua New Guinea (PNG) (Table 1.1) is probably reflective of infestation rates in cultivated crops when the natural environment is relatively unmodified and alternative non-crop hosts available. In this system infestation rates ranged from zero to nearly 100%, with variation depending on crop, locality and sampling occasion.

Production costs
To produce commercial crops in areas where fruit flies occur requires active control. Even backyard growers must actively control fruit flies if they want to pick any significant amount of crop. The different types of control options available are dealt with in Chapter 9, but no dacine fruit fly controls come without direct cost to producers. The costs of those controls vary greatly, but some may always be prohibitively expensive unless subsidised by the state (e.g. the sterile insect technique), while others may only be economically viable where labour costs are cheap (e.g. individual fruit bagging).

Food security
While dacine species are often implicated in food security issues, it is not clear how much of this is true. In developing countries where food security issues are real, there are few staple commodities that fruit flies attack. Fruit flies do not affect any cereal crops (rice, wheat, etc.), nor are they problems for starch crops such as taro, cassava or sweet potato. In countries or regions where bananas (which are fruit fly susceptible) are the starch staple, the local cooking varieties tend to have high resistance to tropical fruit fly and any infestation is commonly minor and cut away during food preparation. Fruit flies do cause significant losses to non-starchy fruit and vegetable crops in developing countries, but again the potential impact of this on food security is unclear. Many tropical fruit trees ripen their fruit over short periods and, if not being harvested for sale, more fruit is produced than can often be eaten, so the impact of partial losses to insects is unlikely to significantly impact local food security. However, tropical fruit flies can significantly impact food security in an indirect fashion. Subsistence growers often produce small amounts of high-value fruit or vegetables for sale, to create or supplement a cash income. If this fruit is lost to fruit fly attack, then fruit fly damage may severely impact food security by stopping or limiting an income flow, thus meaning other food purchases cannot be made and general quality of life is lowered.

Table 1.1. Infestation of different fruit types in Papua New Guinea (PNG), from Leblanc *et al.* (2001). PNG predominantly consists of subsistence level fruit and vegetable production, with large amounts of native vegetation remaining. Levels of active crop protection are minimal or absent. The infestation rates thus represent those occurring in a relatively complex agricultural setting with mixed cropping and wild-fruit breeding sites. Note the large variation between fruit types and collection regions, also that multiple fly species are commonly collected from the same fruit type.

Host	Stage of maturity	Government area (= Province)	% Mean infestation	% Range of infestation in samples	Mean number pupae/ infested fruit	Fruit fly species
Banana	Mature to ripe	Central	22.0	0–75	22.4	*B. musae*
Banana	Mature to ripe	East New Britain (ENB)	0.3	0–0.3	25.0	*B. frauenfeldi*
Banana	Mature to ripe	Morobe	17.6	0–17.6	36.6	*B. musae*
Banana	Ripe	Oro	–	10–40	–	*B. musae*
Breadfruit	Ripe	ENB	75.3	–	115.3	*B. umbrosa, B. frauenfeldi, B. curvifera*
Carambola	Ripe	ENB	13.8	0.8–38	5.1	*B. frauenfeldi*
Carambola	Ripe	Central	18.7	0–74	6.6	*B. frauenfeldi, B. dorsalis*
Carambola (Malaysian)	Ripe	Central	82.0	10–96	22.5	*B. frauenfeldi*
Cashew apple	Ripe	ENB	5.2	6–66	5.0	*B. frauenfeldi*
Guava	Ripe	Central	75.0	17–92	14.3	*B. frauenfeldi, B. trivialis*
Guava	Ripe	ENB	6.2	28–96	29.4	*B. frauenfeldi, B. obliqua*
Guava	Ripe	Morobe	61.5	59–64	16.4	*B. frauenfeldi, B. trivialis*
Guava	Ripe	ENB	74.1	52–82	13.9	*B. frauenfeldi, B. obliqua*
Mandarin	Ripe	ENB	0.6	–	3.0	*B. frauenfeldi*
Mango	Fallen	ENB	50.8	–	9.8	*B. frauenfeldi*
Orange	Ripe	Highlands	2.8	0–9	1.4	*B. trivialis*
Pumpkin	Flower	Central	25.0	–	9.5	*Z. cucurbitae, Z. atrisetosa*
Pumpkin	Mature fruit	ENB	24.0	–	36.6	*Z. cucurbitae, Z. decipiens*
Pumpkin	Mature fruit	Central	14.5	0–66	29.0	*Z. cucurbitae, Z. atrisetosa*
Tahitan chestnut	Ripe or fallen	Central	34.4	26–42	24.6	*B. frauenfeldi, B. moluccensis*
Tropical almond	Fallen	Central	33.2	22–80	8.0	*B. frauenfeldi, B. trivialis*
Watermelon	Flower	Central	31.9	31–36	6.6	*Z. cucurbitae*
Watermelon	Young fruit	Central	26.0	–	12.5	*Z. cucurbitae*

1.2.2 The indirect costs of *Bactrocera*

Dacini fruit fly species are pests of global quarantine importance. Those parts of the world without dacine fruit flies wish to keep them out, while in regions where dacine species do occur then most countries still have at least one pest species that is not shared with another country. In some large countries, such as Australia, dacine species may be endemic in one part of the country but an absent or regulated pest in another part. Dacine fruit flies can spread both naturally through the flight of adults, or through human-assisted carriage of infested fruit. The fact that fruit infested with only eggs or young larvae can appear perfectly sound from the outside, means that travellers may be unaware that fruit flies are being moved. This is a particular problem for air travel, as in the time that it takes for a clutch of eggs to hatch, fruit may be carried from one side of the world to the other. Naturally enough, plant health officials want to stop the entry and spread of dacine fruit flies and this is achieved predominantly through border quarantine and regulated market access.

Quarantine
Many countries maintain active border surveillance programmes against dacine fruit flies, particularly island nations where there are physical (as opposed to political) boundaries that can limit the entry of flies. The case of New Zealand cited at the start of this chapter is a good example. New Zealand maintains nearly 3500 permanent fruit fly traps, predominantly against *Bactrocera* and *Zeugodacus* species, but also against Mediterranean fruit fly, *Ceratitis capitata*. Early detection of an exotic entry allows for rapid response and a much greater likelihood of eradicating an incursion. The Australian Commonwealth maintains port and city surveillance for exotic fruit flies, as well as a targeted surveillance programme across its entire tropical northern coastline (the North Australian Quarantine Strategy, or NAQS). While many pests and diseases are targeted by the NAQS programme, exotic *Bactrocera* and *Zeugodacus* species are a key priority. One component of the NAQS programme is a combined activity with plant health officers of the State of Queensland, who monitor and respond to *Bactrocera* incursions from PNG into the numerous small islands of the Torres Strait, the narrow seaway between Australia's Cape York Peninsular and the PNG mainland. Every year the Torres Strait programme implements an active fruit fly management programme to stop the entry and establishment of PNG fruit flies, such as *B. dorsalis* and *B. trivialis*, into mainland Australia.

As tropical fruit flies can be moved as eggs or larvae in fruit, there is a potential that travellers may carry fruit fly if they are unaware of the quarantine risk. Border quarantine is therefore active in many countries against incoming passengers, be they on flights or in boats. Fruit carriage is a particular problem in cultures where fruit are carried and presented as traditional gifts. In an analysis of fruit carriage on PNG domestic airline flights, it was found that up to one in three passengers was carrying fresh fruit or vegetables for personal consumption or as gifts. Using basic risk analysis, it was estimated that 1% of passengers travelling from a part of PNG where banana fruit fly (*B. musae*) was endemic, to a part of PNG where banana fly was absent, was likely to be carrying an *infested* banana. Perhaps not surprisingly, banana fly now occurs in parts of PNG where it did not traditionally occur.

Market access

It is as market access problems that dacine fruit fly species come into their own as internationally important pests. *Market access* is the term used to describe the ability to sell your produce into a market. Clearly if your market is your local town or village, then market access describes the simple physical process of transporting your crop to that market. However, for most commercial producers of fresh commodities, the point of sale is removed from the point of production and may involve sending the crop to other states within your own country (domestic market access) or to other countries (international market access). In such cases the physical movement of the crop is the last and often least worrying part of market access; the most difficult part is obtaining regulatory permission to sell into those markets.

If a plant pest or disease occurs in a production area, but not in the importing area, then the movement of any fresh commodity poses a biosecurity risk to the importing region. Well established international regulations exist to allow trade to continue while minimising the plant health risk, operating under the global agreement of the International Plant Protection Convention (IPPC). The IPPC states that sovereign nations are allowed to impose 'risk-reduction measures' (i.e. quarantine or market access treatments) that reduce the plant health risk of importing a commodity from a specific exporting country to an 'acceptable level'. The details of regulatory controls and market access are discussed in Chapter 11, but in summary what this means for production areas where pest dacines are present is that access to markets without those pests is often reliant on implementing treatments that reduce the risk of accidentally exporting fruit flies within fruit to very low levels. Market access treatments include regulatory treatments, such as managing and maintaining zones officially 'free' of pests, imposing rigid pre-harvest management regimes, applying post-harvest disinfestation treatments, or integrating several such steps into a 'systems approach'. The costs of obtaining and maintaining market access can be prohibitory for some producers, which effectively means that fruit flies are causing a 100% market loss. Even where treatments have been approved and applied, significant additional costs are normally involved to meet these treatments. Additionally, the unexpected failure of approved market access protocols (e.g. due to a fruit fly outbreak in a previously recognised fruit fly free area) can see export markets closed immediately.

Research, development and extension costs

Because of their pest status, millions of dollars are spent annually around the globe on research, development and extension (RD&E) to better control and manage dacine fruit flies. Research and development covers everything from the development of fruit fly diagnostics (i.e. how to tell one fruit fly from another) for front-line quarantine officers, through to working with farmers on how to implement field controls. Estimating the total cost of this effort has never been done, but it is certainly not trivial. For example, approximately 165 scientific papers are published each year on *Bactrocera, Dacus* or *Zeugodacus*. Various sources suggest that between US$20 000 and US$50 000 worth of research is required to generate one scientific paper. Using these estimates, then between US$3.3 million and $8.2million is spent annually just on research that leads to papers. This does not cover the huge amount of the research and development that is not published, or any of the extension. Estimating that tens of millions of dollars are spent annually on dacine fruit fly RD&E is therefore realistic.

1.2.3 *Bactrocera* as model organisms for ecology, evolution and genetics

Dacine fruit flies should not simply be considered as pest organisms. In their natural ecosystems they are abundant and speciose members of the local insect fauna. They interact with their host plants in complex ways and play important ecological roles including the enhancement of seed dispersal and germination and, for certain orchid species, pollination. The rapid speciation of several lineages within the tribe Dacini make them as interesting for evolutionary study as the better known Hawaiian *Drosophila* or the South American *Heliconius* butterflies. Because of their economic importance, dacine fruit flies have also been at the forefront of genetics research and whole genomes and/or transcriptomes are available for members of all three genera. For academic purposes, particularly, dacines are ideal models as the students and researchers who work on them can simultaneously carry out scientifically important fundamental research, while also knowing that they are providing the underpinning science that is needed to support applied fruit fly pest management.

1.3 Philosophy of This Book

Within the formal structure of the book (see last section of this chapter), three philosophical elements underpin how I have approached writing: *fruit flies are a regulatory problem*; *pest management based on science*; and *process versus pattern*. Understanding these elements, which I develop below, makes the logic of the book, the chapters within it, and the sections within those chapters much easier to identify and follow.

1.3.1 Dacines as biological versus regulatory pests

Most researchers new to fruit flies think of them primarily as biological pests (i.e. they think about the direct fruit damage they cause). Fruit flies are certainly significant biological pests and in susceptible crops, with no controls, they can destroy a harvest. But if that was the only limitation caused by dacines and other frugivorous tephritids then they would be no worse than many other insect pests. Unfortunately fruit flies are not just biological pests, they are also the world's most import regulatory pests of horticulture. For this reason many established fruit fly researchers consider them more as 'regulatory pests' than as 'biological pests'.

As developed more fully elsewhere in this book, fruit flies can be easily moved as eggs and young larvae in commercial commodity trade. To reduce this risk, risk-reduction treatments (i.e. phytosanitary treatments) can be mandated by importing nations on fruit being exported from areas with endemic pest fruit flies. Carrying out such treatments may be operationally impractical and/or economically prohibitive, leading to lost market opportunities. Even where satisfactory treatments can be applied, the negotiation of market access can delay exports for years causing lost opportunity costs, and when trade is open there will still be some level of economic costs to be borne by producers and exporters. That these costs have to be met as much for crops which may be rarely attacked by pest flies as for those which are common hosts, is what makes fruit flies more than just a biological pest.

While this 'regulatory pest status' may be dismissed as simply a problem of developed countries where fruit is being grown solely for export, this is not the case. In many developing countries rural development has been underpinned by the creation of a fresh commodity export trade where exporters gather the production of many small producers to create a supply chain for international markets. For the small-scale producers, the cash return for their crops can be an important or even sole source of income. Such producers are no less affected by fruit flies as regulatory pests than are the large commercial producers.

Because of the problems fruit flies pose in international market access, this book includes chapters on phytosanitary treatments and regulations.

1.3.2   Pest management based on science

Walter (2003) wrote a detailed and elegant critique as to why pest management needs to be based on the best available science. This, he argues, is because science, including theoretical science, provides the basis for sustainable pest management. He uses the analogy of a bridge, with one end of the bridge the pest problem, the other end the successful outcome, the bridge platform the solution and the bridge piers the supporting basis for the solution. For a control solution with little underpinning science then the lack of support makes it weak and prone to failure, whereas strong underpinning science makes the control solution more robust and sustainable. As an analogy it is a good one: 'silver bullet' pest management solutions rarely succeed for long, whereas controls that do succeed in the longer term are those for which it is known why they work and what can be done to make them progressively better.

Pesticide cover sprays are perhaps the most common example given of controls that are applied without underpinning science, but this need not be the case. Pesticide applications are certainly not unique in being a pest management tool that can be applied without appropriate science. Naïvely applied pesticides can lead to significant wastage and environmental damage, but a pesticide strategy that is built upon science so that it is spatially and temporally well targeted, and mixed with other complimentary controls, can provide excellent and sustainable pest management outcomes. Similarly, the naïve use of natural enemies may make a producer feel like they are minimising environmental impacts while getting control, whereas in fact they may be achieving little because they simply do not understand the system. But a natural enemy control strategy with strong underpinning science may provide excellent, long-term control.

Because of this belief that pest management must be based on quality science, a great deal of this book provides basic scientific knowledge to underpin *Bactrocera* pest management.

1.3.3 Pattern versus process

A key element of understanding the biology, ecology and management of any organism is to understand the difference between 'pattern' and 'process'. However, these terms are often not used in the applied biology literature and their meaning may be unclear. As I use these terms in this book, I explain here what I mean by them.

Patterns in ecology are repeatable trends in the behaviour of individuals, populations or communities that can be documented by human observers. Regular, seasonal changes in the abundance of populations over time is a classic ecological pattern. Whether a herbivore species feeds on only one plant species (i.e. monophagy), or across many plant families (i.e. polyphagy) is another example of an ecological pattern. Defining patterns is important as a way of summarising often complex data, and for rapidly and concisely communicating with other scientists, growers or market regulators.

Processes are the underlying biological, behavioural and environmental drivers of patterns. For example the olive fruit fly (*B. oleae*) can be described as monophagous on olives, whereas the Oriental fruit fly (*B. dorsalis*) is polyphagous on over 300 plant species. The terms monophagous and polyphagous instantly tell a reader about the 'patterns' of host use in these two fruit fly species, but nothing about the process of host use. The process of host use, in contrast, may be that *B. oleae* responds to the specific chemical volatiles produced by olive fruit, whereas the process of host use of *B. dorsalis* may be that it responds to a suite of generic fruit ripening esters that are common across many different fruit species.

Patterns can be best considered as 'emergent properties' of underlying processes. *Bactrocera oleae* has a unique host-location mechanism by which it locates olive, the fruit it needs for oviposition: monophagy is an emergent pattern of that host-location process. *Bactrocera dorsalis* also has a unique host-location mechanism by which it can locate suitable fruit, but in this case the fruit may come from many different plant species: polyphagy is an emergent property of that particular host-location process. If trying to develop a control for olive fruit fly based on host varietal resistance, the host-location process is what needs to be understood so that it can be manipulated; the pattern is largely useless. But if communicating to a market access regulator worried about the risk posed by the importation of fresh olives to a local citrus crop, then telling that regulator that olive fruit fly is monophagous to olives (and thus by inference will not attack citrus), is instantly informative without having to explain the fine detail of how olive fruit fly finds an olive crop.

Understanding the differences between process and pattern is clear when starting with process and moving onto pattern, or when treating the two as entirely different. However, problems can arise when patterns are identified and underlying processes are inferred from them. Seasonal fluctuations in populations are a classical example of this problem. Largely because most textbook examples are based on work done in temperate climates, seasonal changes in the abundance of an insect population (a classical 'pattern') are commonly attributed to underlying seasonal changes in temperature and/or day-length (i.e. the 'processes'). But temperature is not the only driver of insect population abundance, especially in the tropics and sub-tropics. The onset of the rainy season causing rapid plant growth, fruit availability or its lack, dry season induced insect diapause, extreme heat killing populations, etc., are all other processes that may influence the patterns of insect abundance. Identifying a pattern and then assuming an underlying process (even if done formally, for example, through correlation analysis) can lead to serious errors if the correct process is not identified. For this reason, in this book I will treat pattern and process as separate states, and will only link the two when working from an initial basis of process. I will not infer process from patterns.

1.4   Structure and Purpose of This Book

The great bulk of information on the *Bactrocera, Dacus* and *Zeugodacus* exists as scientific research papers, or as chapters in specialist books resulting from scientific meetings. With the exception of taxonomic treatments there is no single work focused on these three genera, which are the dominant tephritid pest fauna of India, Asia, Australia and the Pacific, and due to the repeated introduction of highly competitive invasive species increasingly of Africa. For the dedicated fruit fly research scientist this is not a great problem and this book is not targeted at those readers. Rather, it is targeted at field entomologists, extension officers, quarantine officers and market access negotiators who need a single source that can provide a rapid, yet comprehensive introduction to the group. It is important to note that this book is not a 'tool-box' and does not provide specific recommendations for operational activities such as laying out surveillance grids, applying spray schedules or implementing area-wide management. All of these are important activities, but such guidelines already exist in easily accessed locations on the Internet, or through local departments of agriculture. Because such recommendations are often state, country or region specific, local sources of information should be pursued for the best and most relevant day-to-day operational advice. The purpose of this book is rather to support these operational resources by providing the background information that such recommendations are based upon, so as to better inform their use.

As a believer that appropriate pest management decisions should be underpinned by an understanding of the basic biology of the target pests, I divide this book into three key sections. The first section (Chapters 2 and 3) covers the evolution history of the dacine flies, including their systematic relationships, taxonomy and species level diagnosis. The second section (Chapters 4–8) covers the biology of the flies with a focus on their life history and population demography, behaviour and ecology, and natural enemies. The third and final section (Chapters 9–11) covers pre-harvest, post-harvest and regulatory controls. Each chapter concludes with a list of key monographs, papers or book chapters in the event that deeper information is sought. Some of these sources may be several decades old, but if used it will be because the information they contain has not dated and remains the best available. These lists are also not, in any way, meant to be exhaustive, but are simply an introduction to the larger literature.

1.5   Further Reading and References Cited

Aluja, M. and Norrbom, A. (2000) *Fruit Flies (Tephritidae): Phylogeny and Evolution of Behavior*. CRC Press, Boca Raton, Florida, USA.

Clarke, A.R., Armstrong, K.F., Carmichael, A.E., Milne, J.R., Raghu, S., Roderick, G.K. and Yeates, D.K. (2005) Invasive phytophagous pests arising through a recent tropical evolutionary radiation: The *Bactrocera dorsalis* complex of fruit flies. *Annual Review of Entomology*, 50, 293–319

Clarke A.R., Powell K.S., Weldon C.W. and Taylor P.W. (2011) The ecology of *Bactrocera tryoni* (Froggatt) (Diptera: Tephritidae): What do we know to assist pest management? *Annals of Applied Biology*, 158, 26–54.

Daane, K.M. and Johnson, M.W. (2010) Olive fruit fly: Managing an ancient pest in modern times. *Annual Review of Entomology*, 55, 151–169.

Dhillon, M.K., Singh, R., Naresh, J.S. and Sharma, H.C. (2005) The melon fruit fly, *Bactrocera cucurbitae*: A review of its biology and management. *Journal of Insect Science*, 5, 1–16.

Drew, R.A.I. and Romig, M.C. (2000) The biology and behaviour of flies in the Tribe Dacini (Dacinae). In: Aluja M. and Norrbom A. (eds) *Fruit Flies (Tephritidae): Phylogeny and Evolution of Behavior*. CRC Press, Boca Raton, Florida, USA, pp. 535–546.

Drew, R.A.I. and Romig, M.C. (2013) *Tropical Fruit Flies of South-East Asia (Tephritidae: Dacinae)*. CAB International, Wallingford, UK.

Ekesi, S., Mohamed, S.A. and De Meyer, M. (eds) (2016a) *Fruit Fly Research and Development in Africa. Towards a sustainable management strategy to improve horticulture*. Springer, Basel, Switzerland.

Ekesi, S., De Meyer, M., Mohamed, S.A., Virgilio, M. and Borgemeister, C. (2016b) Taxonomy, ecology, and management of native and exotic fruit fly species in Africa. *Annual Review of Entomology* 61, 219–238.

Fletcher, B.S. (1987) The biology of Dacine fruit flies. *Annual Review of Entomology*, 32, 115–144.

Koyama, J., Kakinohana, H. and Miyatake, I. (2004) Eradication of the melon fly, *Bactrocera cucurbitae*, in Japan: Importance of behaviour, ecology, genetics, and evolution. *Annual Review of Entomology*, 49, 331–349.

Leblanc, L., Balagawi, S., Mararuai, A., Putulan, D. and Clarke, A.R. (2001) Fruit Flies in Papua New Guinea. Pest Advisory Leaflet No. 37, Secretariat of the Pacific Community-Plant Protection Service, Suva, Fiji.

Shelly, T.E., Epsky, N., Jang, E.B., Reyes-Flores, J. and Vargas, R.I. (eds) (2014) *Trapping and the Detection, Control, and Regulation of Tephritid Fruit Flies: Lures, Area-Wide Programs, and Trade Implications*. Springer, Dordrecht, The Netherlands.

Walter, G.H. (2003) *Insect Pest Management and Ecological Research*. Cambridge University Press, Cambridge, UK.

White, I.M. (2000) Morphological features of the Tribe Dacini (Dacinae): Their significance to behaviour and classification. In: Aluja M. and Norrbom A. (eds) *Fruit Flies (Tephritidae): Phylogeny and Evolution of Behavior*. CRC Press, Boca Raton, Florida, USA, pp. 505–533.

White, I.M. and Elson-Harris, M.M. (1992) *Fruit Flies of Economic Significance: Their Identification and Bionomics*. CAB International and ACIAR, Wallingford, UK.

2 Systematics and Taxonomy

2.1 General Introduction

Taxonomy is the science of naming species whereas systematics involves the placement (i.e. classification) of organisms into different taxonomic groups. Traditionally, within entomology, the taxonomic groupings used are (from most exclusive to most inclusive): sub-species, species, sub-genus, genus, sub-tribe, tribe, super-tribe, sub-family, family, super-family, sub-order and order. Placement of a species within a genus is mandatory for the Linnaean binomial-nomenclature system, and thus systematics and taxonomy are effectively the same thing when describing a new species. The placing of a new genus within a family, and a new family within an order, are similarly fundamental to taxonomic practice. However, the use of other taxonomic categories within a systematic structure is much more variable, and is often dependent on historic practice within a particular research field. For example, the taxonomic community that deals with butterflies commonly uses sub-species designations, while the use of sub-species in most other fields of entomology is rare. The fruit fly literature makes no use of the sub-species designations, but uses most of the other taxonomic categories.

Bactrocera and related genera are true flies and belong to the family Tephritidae within the order Diptera. Within the Diptera, the Tephritidae belong to the super-family Tephritoidea that contains seven other related fly families. Within the family Tephritidae there are six sub-families, to which *Bactrocera* belongs in the Dacini, one of three tribes within the sub-family Dacinae. It is important to note that the higher classification of the fruit flies is treated differently by different authors. The summary below follows from the classification used in Aluja and Norrbom (2001) for the sub-families within the Tephritidae, from Wiegmann *et al.* (2011) for placement of the Tephritoidea within the Diptera, and from both sources for the families within the Tephritoidea. In this classification, the genera *Bactrocera*, *Dacus* and *Zeugodacus* constitute the tribe Dacini, within the sub-family Dacinae of the family Tephritidae. Contrary to the system used here, some authors treat *Bactrocera*, *Dacus* and *Zeugodacus* as belonging to a sub-tribe (= Dacina) within an enlarged tribe Dacini that then includes a few other closely related fruit fly groups. While of importance to insect systematists, whether *Bactrocera*, *Dacus* and *Zeugodacus* belong to the tribe Dacini, or the sub-tribe Dacina, makes no difference to their biology or pest impact, but the reader needs to be aware that their formal, higher-level systematic placement is not fully agreed upon.

2.2 Systematics

2.2.1 Higher placement of tephritids within the Diptera

Within the true flies, the Tephritidae are technically eremoneuran, cyclorrhaphan, schizophoran, acalyptrate Brachycera. That is to say they belong to the 'higher' flies that are stout bodied with short antenna (= Brachycera), they do not have an enlarged basal lobe on the wing (= Acalyptrate), they have three larval instars (= Eremoneura), they pupate in a puparium (= Cyclorrhapha), and they escape from the puparium through the use of the eversible ptilinal sac (= Schizophora).

The cyclorrhaphan Schizophora are one of the largest radiations of true flies and this radiation is thought to have occurred 65–40 million years ago in the early Tertiary period. Within this group, the super-family Tephritoidea forms a distinct clade that is sister to most other cyclorrhaphan schizophoran groups. Ageing the tephritids is difficult, as there are relatively few fossils. The earliest known fossil tephritids are from Dominican amber of mid-Miocene (15–11 MYA) to early Eocene (34–38 MYA) age, but it is probable that the tephritids go back to at least 60 MYA, and possibly further. Lack of fossils, and the ongoing debate about the use of fossils versus molecular clocks, makes dating problematic.

2.2.2 The super-family Tephritoidea

The super-family Tephritoidea consists of ten families: Lonchaeidae, Piophilidae, Pallopteridae, Eurygnathomyiidae, Ctenostylidae, Richardiidae, Ulidiidae, Platystomatidae, Pyrgotidae and Tephritidae. Some authors also include the family Tachiniscidae within the Tephritoidea, but there is debate if this constitutes a separate family or is a clade within the Tephritidae (which is how it is treated here). Both molecular and morphological analyses have shown the Tephritoidea to be a monophyletic clade (i.e. a true natural grouping). The morphological features that are common to families with the Tephritoidea are: (i) male tergum 6 strongly reduced or absent; (ii) surstylus, or medial surstylus if there are two, bearing tooth-like prensisetae; (iii) female sterna 4 to 6 with anterior rod-like apodemes; and (iv) female tergosternum 7 consisting of two portions, the anterior one that forms a tubular 'oviscape' and the posterior comprising two pairs of longitudinal taeniae (Korneyev, 2001). The tubular 'oviscape' is considered a key innovation of the Tephritoidea as it has allowed these flies to deposit their eggs inside oviposition substrates where the eggs are protected and the result-ant larvae can safely feed. Within the Tephritoidea, the families Lonchaeidae, Piophilidae, Pallopteridae and Eurygnathomyiidae create a single clade (the Piophilidae family group), while the remaining families form a second clade (the Tephritidae family group) (Fig. 2.1).

A summary description of each of the families, except for the Tephritidae, follows.

Family Lonchaeidae
This family contains 561 species in eight genera and are commonly referred to as the 'lance flies'. The flies are small but robust in shape, with blue-black or metallic bodies and are found globally except for New Zealand. The general biology of the group is diverse and some species have predacious larvae, but most species have phytophagous larvae and some, especially in the Neotropics, are pests of horticulture.

Family Piophilidae
A small family of fewer than 100 species in 23 genera, this is predominantly a Holarctic group. Most species are scavengers in animal products, carrion and

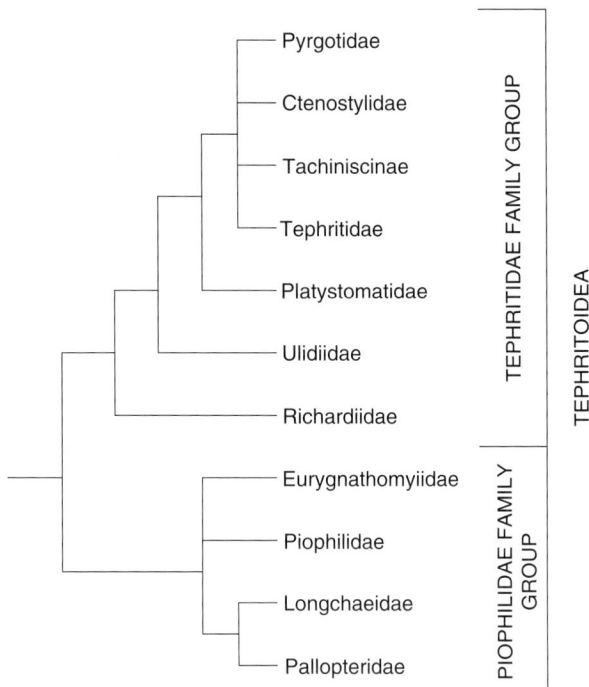

Fig. 2.1. Phylogenetic relationships of the families within the super-family Tephritoidea (redrawn from Han and Ro, 2016).

fungi, but they can also breed in cured meat, preserved fish and cheeses. In Sardinia, larvae of the cheese fly, *Piophila casei*, are deliberately introduced to pecorino cheese to make a local speciality, 'rotten cheese'.

Family Pallopteridae
Another small family of fewer than 60 described species in 15 genera known as the flutter-wing, trembling-wing or picture-wing flies. As the name suggests the wings are marked with spots or very broad costal margins and the wings are rapidly beaten in courtship. Larval records include both phytophagous and predacious species. The family is recorded predominantly from the Palearctic and Nearctic, but some species have widely disjunct distributions, for example, in New Zealand.

Family Eurygnathomyiidae
Known only from the single European species, *Eurtgnathomyia bicolor*, some authors place this fly within the family Pallopteridae.

Family Ctenostylidae
The placement of this family is uncertain. Some authors treat it as part of the Tephritoidea, others consider it unplaced within any dipteran super-family. The family contains less than 20 species placed in six genera, and all species are considered rare. It is thought all females lay larvae instead of eggs, and the larvae may be parasitoids. The adults lack ocelli and functional mouthparts, and the female antennal arista is double branched.

Family Richardiidae
Another predominantly neotropical group, the family has approximately 175 species in 30 genera. Picture wings are common and the adult body often has blue or greenish colours. Some species are also sexually dimorphic, with males having 'stalk eyes'. The larvae are plant feeders or saprophagous in decaying organic matter.

Family Ulidiidae
A large family of approximately 680 species in over 100 genera and with a worldwide distribution. Adult colouration is often bright and metallic. The use of wings in courtship is common and these flies are also referred to as 'picture-wing' flies. Larvae develop in a variety of breeding sites, including rotting fruits and vegetables; however, there is a strong tendency to attack living plant tissue. The family was previously referred to as the Otitidae and this name may sometimes still be seen, but Ulidiidae is correct.

Family Platystomatidae
The 'signal flies' are a large family of 1200 species in 199 genera. While having a worldwide distribution, platystomatids are most common in the tropics. The larvae are predominantly phytophagous or saprophagous, but others are known from decaying animal matter, including human corpses. Unlike many

other Tephritoidea, where species are commonly small, the signal flies include large species of up to 20 mm in length.

Family Pyrgotidae
An unusual family within the Tephritoidea, the pyrgotids are all endo-parasitoids of adult scarab beetles. The female fly intercepts the scarab on the wing and lays one or more eggs through the soft underside of the beetle, although only one offspring will develop per host. The adults are predominantly night active but can be caught at lights. There are some 350 species in 60 genera globally, and a second distinguishing feature of the family is the lack of ocelli in the adults.

2.2.3 The family Tephritidae

The family Tephritidae (= the true fruit flies) is one of the largest families in the Diptera with nearly 5000 described species. Tephritids are found on every continent and the family is divided into six sub-families and 29 tribes (Table 2.1). While referred to as the fruit flies, this greatly under-represents the diversity of host use within the tephritids. The largest sub-family, the Tephritinae, do not attack fruit at all but are most commonly associated with the flower heads of daisies, whereas the Tachiniscinae are, like the pyrgotids, parasitoids. The frugivorous tephritids, including all the pest species, are largely restricted to two sub-families, the Dacinae and the Trypetinae.

A summary of the sub-families, excepting for the Dacinae, follows.

Sub-family Tachiniscinae
This is a small group of 20 species placed in nine genera. Like the family Pyrgotidae, the Tachiniscinae are endoparasitoids, although in this case of the larvae of saturnid butterflies. The parasitoid lifestyle has seen the Tachiniscinae placed in their own family, the Tachiniscidae, but morphological and molecular phylogenetic analysis confirms them as belonging to the Tephritidae, where they are sister to all other tephritid sub-families (i.e. they are basal in the evolutionary lineage of the Tephritidae). Little is known about the biology of these rare and unusual flies.

Table 2.1. Sub-family and tribal classification of the dipteran family Tephritidae.

Sub-family	Tribes
Tachiniscinae	Ortalotrypetini, Tachiniscini
Blepharoneurinae	
Phytalmiinae	Acanthonevrini, Epacrocerini, Phascini, Phytalmiini
Dacinae	Ceratitidini, Gastrozonini, Dacini
Trypetinae	Acidoxanthini, Adramini, Carpomyini, Nitrariomyiini, Rivelliomimini, Toxotrypanini, Trypetini, Xarnutini, Zaceratini
Tephritinae	Acrotaeniini, Cecidocharini, Dithrycini, Eutretini, Myopitini, Noeetini, Schistopterini, Tephrellini, Tephritini, Terelliini, Xyphosiini

Sub-family Blepharoneurinae

Another basal lineage of tephritid, the Blepharoneurinae are a largely tropical group of 34 described species in five genera. However, it is estimated that there are over 200 undescribed species. The Blepharoneurinae exhibit 'typical' tephritid behaviour in that larvae develop in living plants, but their host use is highly specialised to a level not seen in other tephritids. All Blepharoneurinae use cucurbits as hosts, with different species specialising not only by cucurbit species, but also by plant part. Thus some species attack only cucurbit female flowers, others male flowers, fruit, seeds or stems. The same cucurbit species can support two or more species of Blepharoneurinae, each infesting a different plant part. The flies have highly pictured wings that are used extensively in courtship, which occurs in leks, or lek-like aggregations. The adult flies of this sub-family are also the only known tephritids that actively rasp plant leaves or other tissue and then feed on the resulting wound exudate.

Sub-family Phytalmiinae

The Phytalmiinae is a large sub-family of 95 genera and approximately 330 species, 80% of which are found in the Oriental and Australasian regions. The biology of most Phytalmiinae is unknown, although it is considered that for most species larval development occurs in decaying vegetable matter. The Phytalmiinae are better known as the 'antler flies', because many species have cuticular projections from the side of the head. In the type genus *Phytalmia*, where they are best studied, the projections can be quite dramatic and they look superficially like the antlers of deer or moose. Not all 'antlers' are so dramatic and in some species they are absent entirely, or may be represented by simple, small cuticular extensions. Antler size is also correlated with body size, and small individuals of antlered species may not show projections, or they may be reduced. The antlers are male specific and are a secondary sexual character utilised in male–male conflict. In the few cases where mating of antlered species has been documented, males guard the female oviposition resource (for example, rot spots under the bark of fallen trees) and mate with females who come to use that resource. Males assess the quality of other males, either visually or through direct combat, with the winner gaining control of the oviposition site. *Dirioxa pornia* (Walker), the Island fly, is a non-antlered Phytalmiinae that is a minor pest fruit fly in some parts of Australia.

Sub-family Trypetinae

With approximately 1000 species in 188 genera, the Trypetinae is one of the larger sub-families of the Tephritidae. It contains the economic genera *Rhagoletis* (tribe Carpomyini), *Anastrepha* and *Toxotrypana* (both tribe Toxotrypanini) and has a near global distribution, although most of the individual genera have more restricted distributions. The majority of Carpomyini and Toxotrypanini utilise fruit or immature nuts as larval habitat, and much of their biology is superficially similar to that of the Dacinae. The large tribe Trypetini is a predominantly Oriental and Palearctic group and while many are fruit infesters, this tribe also includes all the leaf-mining tephritids and some stem miners: the leaf miners are largely associated with Asteraceae. The well-developed female ovipositor, a common trait across the family Tephritidae and super-family Tephritoidea, is particularly well developed in members of the tribe Toxotrypanini and for some species may constitute 50% or more of the total female body length.

Sub-family Tephritinae

With over 1800 species in 211 genera, this is the largest sub-family of the Tephritidae. It is most diverse in the Holarctic region, but can be found in temperate or high-altitude areas in most parts of the world. With only one or two exceptions, the sub-family utilises Asteraceae (the daisy family) as its larval host. Most species lay into the developing seed heads of daisy flowers, where the larvae may consume all the developing seeds. However, the sub-family is not restricted just to flower heads, and many species also form galls in roots, stems and flowers. Because of their galling, or targeted attack on seed heads, several tephritine species have been used around the world as classical biological agents for invasive daisy weeds. Tephritine species are generally small and, like other tephritids, picture wings are common and used in mating.

2.2.4   The Tephritidae sub-family Dacinae

The sub-family Dacinae is largely restricted to Africa, Asia, Australia and the Pacific and consists of three tribes: the Ceratitidini, the Gastrozonini and the Dacini. In an alternative classification system, these three tribes are treated as three sub-tribes (they become the Ceratitidina, the Gastrozonina and the Dacina) of the tribe Dacini of the sub-family Trypetinae. Regardless of whether they are tribes or sub-tribes, they are considered to constitute true monophyletic groupings (i.e. evolved from a single common ancestor), although poor sampling of the Gastrozonini in both morphological and molecular studies means that the monophyly of that particular clade is still open to question. The Ceratitidini and the Dacini contain all the major pest fruit flies of Africa and Australasia, which are found in the genera *Ceratitis* (Ceratitidini) and *Bactrocera*, *Dacus* and *Zeugodacus* (Dacini). In contrast to the Dacini, for which the taxonomy and systematics is comprehensive, the taxonomy and systematics of the Ceratitidini is still relatively underworked, despite the fact that it contains *Ceratitis capitata*, the Mediterranean fruit fly, arguably the most damaging fruit fly in the world. Despite being closely related, the three sub-tribes have different larval host use patterns. The Ceratitidini and Dacini are either frugivorous or feed on seed capsules (which are botanically fruit, but are not considered as such by the lay person), while the Gastrozonini feed on grasses. A summary of the tribes Gastrozonini and Ceratitidini follows.

Tribe Gastrozonini

Consisting of 137 species in 27 genera, the Gastrozonini is a predominantly tropical Oriental group, although it extends west into Africa, north into the Palaearctic and east to Australia. However, at the extremes of its range it is rare; for example, there are only 14 Afrotropical species and one Australian species. The tribe's common names are the grass-breeding or bamboo-shoot fruit flies, and while the host use records are confirmed for only a relatively small number of species, it is considered probable that all species in the tribe utilise Poaceae (i.e. grasses and bamboos) as larval substrates. The tribe is not unique in the Tephritidae for utilising Poaceae, as some species of the Tribe Acanthonevrini (sub-family Phytalmiinae) are also associated with bamboos in the Oriental region. However, Gastrozonini species that infest bamboos attack damaged but living bamboo tissue and may be considered pests of commercial bamboo (harvested for building material or food), while the Acanthonevrini species that infest bamboos are associated with dead and decaying bamboos, as is typical of the host use patterns of the sub-family Phytalmiinae. Until recently, almost nothing was known about the Gastrozonini except for their taxonomy and some host records, but recently work has been done on the biology of some Asian bamboo feeding species. These studies show that penetrating the hard exterior of bamboos may be the biggest factor limiting host use, and wounds made by other insects are commonly utilised for oviposition. Larvae develop in the pithy internode tissue or young shoots, depending on species, and in the studied species pupation occurs within the bamboo internode space. Adults are most abundant during the wet season, and congregate in sheltered areas near water-ways in the dry season.

Tribe Ceratitidini

The predominantly Afrotropical Ceratitidini are closely related to the Gastrozonini and earlier researchers were unclear if the two groups were truly separate. However, recent molecular research suggests that they do represent monophyletic groups, and this is also reflected in their host use. The Gastrozonini are exclusively grass and bamboo feeders, while the Ceratitidini are, like the Dacini, predominantly frugivorous. The taxonomy of the Ceratitidini is highly confused and many of the 12 genera (*Capparimyia, Carpophthoromyia, Ceratitella, Ceratitis, Eumictoxenus, Neoceratitis, Nippia, Paraceratitella, Paratrirhithrum, Perilampsis, Trirhithrum, Xanthorrachista*) are likely paraphyletic. The largest genus within the tribe is *Ceratitis*, which holds around 80 species in six sub-genera. Despite containing the Mediterranean fruit fly, *C. capitata*, and other major pests such as *C. rosa*, the taxonomy and systematics of *Ceratitis*, and the wider tribe, is far from finalised.

2.2.5 The tribe Dacini

The Dacini is one of the largest radiations within the family Tephritidae, with 932 currently recognised species. The tribe's distribution extends from sub-Saharan Africa, across southern Asia, down into New Guinea and Australia, before ending in the western Pacific. Most Dacini utilise fruits and vegetables as oviposition and larval substrates, but it is incorrect to think of such hosts only as 'fleshy' fruits. Many of the *Dacus* utilise the seed pods of asclepiads (Apocynaceae: Asclepidoideae) as hosts, and while these are technically fruit the outer pod covering can be quite thin and sometimes woody. Some dacines also utilise the fleshy female flowers of cucurbits, while the larvae of yet other species have been recorded infesting soft, fleshy growing shoots. Despite these exceptions, fruits (in the botanical sense) are the primary plant part infested by the Dacini.

Genera

As currently recognised, the Dacini consists of four genera: *Monacrostichus* Bezzi (2 spp.), *Dacus* Fabricius (273 spp.), *Bactrocera* Macquart (461 spp.), and *Zeugodacus* Hendel (196 spp.). Until recently, *Zeugodacus* was treated as sub-genus of *Bactrocera*, and some authors retain this nomenclature. However, comprehensive molecular analysis by several independent groups shows that *Zeugodacus* forms a monophyletic group sister to *Dacus,* and is quite separate to *Bactrocera*. Some analyses show *Dacus* and *Bactrocera* as sister, with *Zeugodacus* forming a third clade sister to those two; but it never sits as sister to *Bactrocera*. Thus most fruit fly taxonomists are now recognising *Zeugodacus* as a valid genus. This name change affects some important pest species, notably *Z. cucurbitae* and *Z. tau*, which are still being referred to (incorrectly) under the names *B. cucurbitae* and *B. tau* in much applied literature.

The Asian genus *Ichneumonopsis* Hardy, containing only the one species *I. burmensis*, is still occasionally seen placed with the Dacini but is now recognised as belonging in the Gastrozonini. The larval host of the species is bamboo, and so its placement within the Gastrozonini better aligns it with the known larval feeding habitats of that tribe. The genus *Monacrostichus* consists of only two species, *M. citricola* and *M. malaysiae*, both of which are known only from South-east Asia where their larval hosts are several *Citrus* species. Little is known about the *Monacrostichus* species and all further discussion in this book refers only to *Bactrocera, Zeugodacus* and *Dacus*.

Sub-generic groupings

Dacine taxonomists have historically made extensive use of classification groupings below the level of genus. This has involved the use of sub-genera, groups of sub-genera, and within some sub-genera 'species complexes/species groups'. As an example, in his 1989 revision of the Australian and Oceanic dacine fauna, Drew identified 21 sub-genera within the genus *Bactrocera*, allocated to four 'groups of sub-genera' (the *Bactrocera, Melanodacus, Queenslandacus* and *Zeugodacus* groups). Within the sub-genus *Bactrocera* (*Bactrocera*) he further identified 21 species complexes. This is just one example of the extensive use of groupings below the genus level: *Dacus* has been similarly divided, with 10 sub-genera and 67 species groups (Hancock and Drew, 2006).

Molecular phylogenetic analyses have shown that many of these groupings below the genus level are of limited value. A species placed in one sub-genus on morphology will sit in another based on molecular data; or separate evolutionary lineages, defined using molecular data, may include species from two or more morphological sub-genera. Given this, it is now widely recognised that many of the morphological characters previously used to group Dacini sub-genera are not phylogenetically informative.

It is important to note that classifications below the level of genus are not required in scientific naming: the Linnaean binomial system requires only a genus (e.g. *Bactrocera*) and a species (e.g. *dorsalis*) name. The use of sub-generic groupings by dacine taxonomists has been a mechanism of convenience to try and deal with the large number of species found in the Dacini genera, but as molecular evidence increasingly shows that such groupings have little biological meaning then they may disappear from usage.

2.2.6 Species complexes

For anyone becoming involved in Dacini research, the term 'complex' or 'species complex' is rapidly encountered. Unfortunately, within the Dacini, the term has several confounded meanings, which can easily confuse both non-specialist and specialist alike. For this reason, the following section is an extended discussion of what the phrase 'species complex' can mean in the Dacini literature, and why it is important to understand species complexes.

What are complexes?

The term 'species complex' has several meanings in the biological literature, but as most commonly used in the dacine literature a species complex is a group of taxonomic species, within sub-genera, where members share common morphological characters. R.A.I. Drew uses complexes as a core part of his taxonomic treatment of dacines, and in doing so he follows a tradition started by the late Hawaiian taxonomist Elmo Hardy. Taxonomically there are many complexes in the tribe Dacini, with some of the better known including the *B. dorsalis* complex, the *Z. tau* complex, the *B. musae* complex and the *B. tryoni* complex. The other, confounded, use of the term 'species complex' in the Dacini literature comes from evolutionary biology, where a species complex is a group of genetically closely related species, often morphologically similar or identical.

Because of this confounding of the term between dacine taxonomics and evolutionary biologists, it is a common misconception that all dacine species within a complex are difficult to tell apart from each other: this is not the case. As used taxonomically by Drew, species within a complex share a group of common morphological attributes, but other than that they need not be particularly difficult to tell apart from each other as they may each also have different characters to each other. Ian White in his work on the African dacines uses the term 'species groups' rather than 'complex' and this is arguably a more correct terminology as it makes no inference about relatedness of species within the 'group'.

While species within a dacine species complex may be easily told apart, this is not always the case and this is where true confusion can arise. Some species within complexes can be hard to tell apart and in such cases it may be unclear if the species are truly cryptic, i.e. biologically different but morphologically identical, or if different taxonomic names have been applied to subtly different populations of the one biological species. This was shown to be the case for some species of the *B. dorsalis* complex, where the former species *B. papayae*, *B. invadens* and *B. philippinensis* were demonstrated to be populations of

B. dorsalis, while the morphologically similar *B. carambolae* was demonstrated to be a distinct biological species (see below, 'Major pest complexes, *Bactrocera dorsalis* complex', for further detail).

It is not clear why there are so many species complexes within the dacines. Part of it is simply taxonomic practice, and many of the taxonomic complexes do not accurately represent true evolutionary groups. However, some species complexes as defined by the taxonomists are being confirmed as distinct evolutionary groupings when subject to genetic analysis. This suggests that dacine fruit flies have been subject to frequent and repeated speciation events – leading to clusters of closely related and morphologically similar species. Rapid evolutionary radiation, host-driven ecological speciation and the extensive use of chemical rather than visual courtship cues may be some of the reasons why there are so many groups of morphologically similar species within the dacines.

Why is it important to understand species complexes?

When different fruit fly species look similar to each other and cause the same type of damage, it can be easy to stop trying to understand what the different species are and to treat them all as one combined problem. In some cases there may be some merit in that, for example, if the treatment is a generic pesticide that kills everything, but more commonly there are several good reasons why it is important to be able to accurately identify the different species within a complex.

Firstly, different species have different biologies that may affect their pest status. For example, of the 17 species within the *B. musae* species complex (i.e. the banana fly complex), only *B. musae* itself is pestiferous on bananas. The remaining 16 species are non-pests and do not attack commercial crops. This is the case in many of the species complexes where only one or a few species within the complex are pests, and the rest are not. If it is assumed that all flies that superficially look like banana fly are banana fly, then controls may be applied that are simply not needed.

Understanding species complexes is particularly important if attempts are being made to develop the sterile insect technique (SIT) for fruit fly control (see Chapter 9). The SIT requires mass-reared male flies to successfully mate with wild females. If the flies in mass culture are not the same biological species as the wild flies being targeted, then mating will not occur, or may occur but only at very low levels. It is particularly important to get this right in situations where the mass-reared flies may be being produced in one country and then released in another, as incorrectly identifying the different species may lead not only to failure of the SIT, but also the mass release of a new pest species where it did not previously occur.

Incorrect identification of flies within a complex may also stop the SIT being applied in situations where it may otherwise have been possible to do so. For example, when the fly invading Kenya in the early-2000s was identified as a new species, *Bactrocera invadens*, this led directly to an assumption that nothing was known about the fly and all new research needed to be carried out. However, if the fly had been correctly identified as a population of *B. dorsalis*, which we now know it is, then many existing controls already developed for that species, including the SIT, may have been able to be applied quickly.

The third major reason to understand species complexes is for trade and market access. While not a dacine, the case of *Anastrepha fraterculus* is an excellent example. The South American fruit fly, *A. fraterculus*, is a well-recognised species complex for which there are many biological species 'hidden' under this one scientific name. In Mexico, *A. fraterculus* was thought to exist and for that reason market access of Mexican oranges into the USA was severely restricted, as *A. fraterculus* is a major citrus pest. However, Mexican oranges were not damaged by *A. fraterculus* in Mexico and it was subsequently demonstrated that the Mexican population of *A. fraterculus* was a different biological species that was a non-pest. Once this was proven, market access restrictions were eased as there was no scientific justification to stop the fruit being exported. Within the dacines, the demonstration that the former *B. invadens* in Africa is a population of *B. dorsalis* means that there is no scientific justification (at least for this species) for limiting fresh commodity movements between Africa and Asia, as both continents have the same pest fly, rather than the two continents having a different pest species each.

Major pest complexes

BACTROCERA DORSALIS COMPLEX.   The *Bactrocera dorsalis* species complex (i.e. the Oriental fruit fly species complex) is the best known and most pestiferous of the dacine species complexes. *Bactrocera dorsalis sensu stricto*, the Oriental fruit fly, is the most pestiferous dacine fruit fly, with an extraordinarily large commercial and native host list. The fly is native to Asia, but it is also a highly aggressive invasive species and it now occurs widely outside this endemic range, including all through sub-Saharan Africa and on various islands of the Pacific. The species complex contains some 80+ species, with the greatest diversity in the Indonesian Archipelago. As defined morphologically, members of the complex are as follows:

> Species of *Bactrocera* (*Bactrocera*) with distinct fascial spots; scutum black, lateral postsutural yellow vittae present, medial postsutural yellow vitta absent; wing colourless except for a narrow costal band (never confluent with R_{4+5}) and anal streak, cells bc and c colourless or, at most, with an extremely pale tint, without dense microtrichia covering cells bc and c; femora entirely or mostly fulvous but may possess dark patterns particularly on or around apices; scutellum yellow with a dark basal band and never with other dark patterns; abdominal terga III–V with a dark 'T' pattern and with, in some species, variable lateral dark patterns' (Drew and Romig, 2013).

However, the complex is not a natural grouping, with molecular work demonstrating that the complex is paraphyletic, with species currently placed in the complex appearing at many locations across the larger *Bactrocera* phylogeny. Further, the restriction that flies have an entirely, or nearly entirely, black scutum (i.e. the top of the thorax) is entirely artificial as some species, such as *B. dorsalis* itself, are highly variable in their colour patterns, with scutum colour ranging from red-brown through to black.

Despite its reputation, only a few species within the complex are considered pestiferous, these being: *B. dorsalis, B. carambolae, B. kandiensis,*

B. caryeae, B. occipitalis, B. pyrifoliae and *B. trivialis*, and of these *B. dorsalis* and *B. carambolae* appear to be the most significant. What were regarded as other major pest species with in the group have now been synonymised with *B. dorsalis*, these being the former species *B. philippinensis* (the Philippines fruit fly), *B. papayae* (the Asian papaya fruit fly) and *B. invadens* (the Invasive fruit fly).

Bactrocera papayae and *B. philippinensis* (along with *B. carambolae*) were described in a taxonomic revision in 1994, while *B. invadens* was described from an invasive population in Kenya in 2005. These three species were all considered major pests, with *B. papayae* having been referred to as 'the most pestiferous species in the complex', and *B. invadens* a highly ingressive invader occupying nearly all of sub-Saharan Africa between 2005 and 2015. However, this group of five species (including *B. dorsalis* and *B. carambolae*) proved to be particularly difficult to tell apart and over time evidence accumulated from difference sources that they may not be separate species at all, but different names for one, or at most two biological species.

In 2000, an internationally coordinated, 5-year research project was initiated by the Joint Food and Agricultural Organisation/International Atomic Energy Agency (FAO/IAEA) Programme on Nuclear Techniques in Food and Agriculture. Through a collaborative, multinational effort, the project demonstrated that based on multiple molecular approaches, cytogenetics, mating tests, morphometrics, morphology and pheromone analysis, that *B. papayae, B. philippinensis* and *B. invadens* were the same biological species as *B. dorsalis*. Subsequently, these three former species have been formally synonymised with *B. dorsalis*. Synonymisation is a formal taxonomic action, which means that these three names are no longer scientifically valid. Attempts to use these names for regulatory purposes (for example, as part of market access negotiations) would not be making a case based on the current, best available science. In contrast, *B. carambolae* was demonstrated to be a unique biological species, and its name remains current and valid.

BACTROCERA TRYONI COMPLEX.   *Bactrocera tryoni*, the Queensland fruit fly, is Australia's major pest fruit fly and occurs along the Australian east coast and across the 'top end' to Darwin. It belongs to a species complex with three other species; *B. neohumeralis* (= lesser Queensland fruit fly*), B. aquilonis* and *B. melas*. All of these species are sympatric with each other for all or part of their geographic ranges with the exception of *B. aquilonis*, which occurs separately from the others in north-western Australia. There has not been a comprehensive systematic analysis of the complex, so the relationships of species within the complex are unknown. It is also not known if the complex is monophyletic, or if additional species currently not placed within the complex belong there. As defined morphologically, members of the complex are as follows:

> Species of *Bactrocera* (*Bactrocera*) with clear wing membrane except for a narrow costal band (not confluent with R_{4+5}) and anal streak; costal cells fulvous or fuscous and generally covered with microtrichia; lateral postsutural vittae present, medial postsutural vitta absent; scutellum yellow with a narrow dark basal band; mesonotum red-brown (with or without dark colour patterns); abdominal terga generally red-brown with variable dark colour patterns, males attracted to cue lure (Drew, 1989).

The species status of flies within the *B. tryoni* complex is not well understood. Significant population genetic work has been done on *B. tryoni* and there is no evidence of unrecognised, cryptic species within that taxon. While morphologically and genetically near identical to *B. tryoni*, *B. neohumeralis* is clearly a distinct biological species to *B. tryoni* based on differences in mating behaviour: *B. tryoni* mates over a narrow time window at dusk, whereas *B. neohumeralis* mates over several hours in the middle of the day. However, the species status of the two other species in the complex, *B. aquilonis* and *B. melas*, is less clear.

Bactrocera aquilonis, the third member of the *B. tryoni* complex, was described from material collected around Darwin, Northern Territory, Australia in 1961. While morphologically similar to *B. tryoni*, two subsequent papers supported the validity of this species, while a third using microsatellite analysis failed to distinguish between *B. tryoni* and *B. aquilonis*. The uncertainty of *B. aquilonis*' species status became an issue in the late 1980s when this previously non-pest species expanded its known host range from four commercial crops to 40. At the current time *B. aquilonis* is recognised as a distinct species, but there is active research to resolve the issue.

Like *B. aquilonis*, the species status of the fourth member of the complex, *B. melas*, is unclear. *Bactrocera melas* was described in 1949 from material collected in southern Queensland, but R.A.I. Drew subsequently discussed the likelihood that *B. melas* was simply a melanic form of *B. tryoni* and it is treated as such by most Australian workers. Nevertheless, *B. melas* continues to exist as a valid taxonomic species as it has never been formally synonymised with *B. tryoni*.

The *B. tryoni* species complex is an excellent example of why understanding the species within a complex is important. Despite being similar genetically, *B. tryoni* and *B. neohumeralis* have different pest status. *Bactrocera tryoni* is a major pest and known invasive species, having spread to south-eastern Australia and some Pacific Island nations. In contrast *B. neohumeralis* is, at worst, a pest in northern parts of Queensland and appears never to have extended its pre-European settlement native range. *Bactrocera melas*, despite being widely regarded in Australia as a melanic form of *B. tryoni*, remains on Australia's list of fruit fly species and as such some Australian trading partners are requesting pest risk information about that fly. If firm evidence was gathered that justified synonymising *B. melas* with *B. tryoni*, then that would be one less pest for Australian market access negotiators to have to deal with. Finally, there is the question of *B. aquilonis* in Western Australia. Currently, that Australian state is recognised as free of *B. tryoni* and this gives it significant market access opportunities. But conversely, if *B. aquilonis* was demonstrated to be a western population of *B. tryoni* (as some authors suggest), then market access may be negatively influenced. But conversely, the SIT that is currently being developed for *B. tryoni* could then be applied against *B. aquilonis* populations with a good likelihood of success. All these examples illustrate why, even in this small complex of only four species, the need to accurately determine the biological species within the complex is of importance for pest management and market access.

ZEUGODACUS TAU COMPLEX. The *Z. tau* complex is entirely Asian in distribution, with its greatest diversity in the Thai/Malay peninsula. *Zeugodacus tau*, in the strict sense, is a major pest of cucurbits and has a distribution that extends from South-east Asia across to India, and northwards into southern China. There are currently 21 species formally described within the *Z. tau* complex. As defined morphologically, members of the complex are as follows: Species of *Zeugodacus*

> scutum black or without large areas of red-brown, lateral and medial postsutural yellow vittae present; wing with a narrow costal band generally overlapping R_{2+3} and expanding into a spot at apex and a distinct anal streak, occasionally with infuscation on dm-cu crossvein, cells bc and c colourless or with pale tint, dense microtrichia in outer corner of cell c only; femora fulvous with or without dark apical spots; scutellum yellow with or without dark patterns, scutellum with 4 sc. setae (rarely with 2 sc. setae); abdominal terga III–V with a dark 'T' pattern and variable lateral dark patterns' (Drew and Romig 2013) [Note, Drew and Romig treat *Zeugodacus* as a subgenus of *Bactrocera*].

The morphological description of the *Z. tau* complex above is worth noting with respect to the variation encompassed: of the 14 character states referred to, eight are variable. This type of description is extremely problematic for morphological identification, but this is actually not the most problematic attribute of the complex. Thai researchers, over a period of some 20 years, have identified multiple 'forms' of *Z. tau*, utilising cytogenetics, cytochrome c oxidase subunit I (COI) haplotype variation, other neutral genes, and wing-shape variation: the most recent addition to these identified forms is 'form J'. This body of work clearly identifies that flies morphologically similar to *Z. tau* are likely to represent different biological species, and as such is consistent with the work of the taxonomists who have described new species within the complex. Unfortunately, however, none of the described species of the *Z. tau* complex can be associated with any of the 'forms' (with the exception of 'form A' which represents *Z. tau* in the strict sense). The *Z. tau* complex thus continues to be 'difficult' in the extreme. It is also worth noting that, like the *B. dorsalis* complex, the *Z. tau* complex is not a natural grouping when analysed using molecular phylogenetics.

2.3 Taxonomy and Diagnostics

Taxonomy and diagnostics, while often used interchangeably, are not the same thing. Taxonomy involves the formal naming of new species, and the revision of existing names. In contrast, diagnostics is about placing an unknown specimen into a taxonomic 'basket', be that basket a species, a genus, or some other taxonomic rank. Much of this section focuses on the latter activity, i.e. diagnostics, rather than taxonomy. This is because true taxonomic practice for dacines is a specialised field, with only a few researchers worldwide. In contrast, nearly all fruit fly workers need to understand the basics of identifying the fruit flies that are already named.

   The ability to accurately identify (or 'diagnose') *Bactrocera* species is a primary element of nearly all quarantine and pest management activities against these flies. Thanks to the work of a series of dacine taxonomists, most notably Elmo Hardy, R.A.I. (Dick) Drew and David Hancock, the morphological taxonomy of *Bactrocera* is well documented. The same can also be said for

Zeugodacus (which Drew and Hancock treat as a sub-genus of *Bactrocera*) and the African *Dacus* (thanks to taxonomists including Hugh Munro, Ian White, Marc De Myer and Hancock again). This strong morphological taxonomy base makes some aspects of identifying *Bactrocera* relatively easy, but in other situations it can still be difficult to make accurate identifications.

Identifying *Bactrocera* can be easy when adult flies are available and the fly is most likely to be part of a well-documented regional fauna. For example, the *Bactrocera* fauna of the Micronesian and Polynesian islands of the South Pacific Islands is relatively small and well documented. The adult flies are, for the most part, quite distinct from each other and building up sufficient expertise to identify captured flies in local traps is relatively easy. However, in stark contrast to this example is the problem of identifying an intercepted maggot in a fruit importation, or a single adult fly in a port surveillance trap. In such cases adult features may be entirely lacking, or even if an adult fly is available (in a trap, or following larval rearing) it may still be impossible to identify due to lack of appropriate keys. The following section introduces the key elements of identifying *Bactrocera*, and the strengths and pitfalls of different approaches.

2.3.1 Morphological character states

Full taxonomic descriptions of *Bactrocera* species use a large number of character states, but many of these are repetitive between species and so not particularly useful in separating or diagnosing a species. Nearly all Drew's species' descriptions and/or revisions provide a 'species diagnosis' statement, and knowing where to locate/how to identify this subset of characters provides the basis for fruit fly taxonomy. The following section provides as example of such a diagnosis, and crosslinks the key terms of that diagnosis with a plain language description of each term in Table 2.2 and illustrated in Fig. 2.2.

Diagnosis (for *B. dorsalis* from Drew and Romig, 2013):

> Face **fulvous** with a pair of medium-sized circular **black spots**; **postpronotal lobes** and **notopleura** yellow; **scutum** black with extensive areas of red-brown to brown below and behind **lateral postsutural vittae**, around **notopleura suture**, between **postpronotal lobes** and **notopleura**, inside postpronotal lobes; broad parallel-sided **lateral postsutural vittae** ending behind **ia. seta**; **medial postsutural vitta** absent; **mesoplural stripe** reaching midway between anterior margin of **notopleuron** and **anterior npl. seta** dorsally; **scutellum** yellow; legs with **femora** entirely fulvous; **fore tibiae** pale **fuscous** and **hind tibiae** fuscous, **mid tibia** fulvous with a small area of dark fuscous basally; wings with **cells bc and c** colourless, **microtrichia** in outer corner of cell c only; a narrow fuscous **costal band** confluent with R_{2+3} and remaining very narrow around apex of wing (occasionally there can be a very slight swelling around apex of R_{4+5}); a narrow pale fuscous **anal streak**; **supernumerary lobe** of medium development; abdominal **terga III–V** exhibits a range of colour patterns but normally possess a basic pattern of a clack 'T' consisting of a narrow **transverse black band** across anterior margin of tergite III, a narrow **medial longitudinal black band** over all three tergites, narrow **anterolateral** fuscous to dark fuscous **corners** on terga IV and V; a pair of oval orange-brown to pale fuscous **shining spots on tergum V**; **abdominal sterna** dark coloured.

Table 2.2. Terms used in the diagnoses of Dacini fruit flies.

	Term (and alternative name[s])	Plain language description
Colours	Fulvous	Reddish-yellow
	Fuscous	Dark brown, approaching black
Head	Spots	A pair of dark spots occur on the front of the head, below the antenna. Size of the spot (i.e. small, medium, large) is sometimes provided in a species description, but these categories are variable and difficult to tell apart
Thorax	Postpronotal lobes (= humeral calli)	The 'shoulder pads', one on each side on the top front edge of the thorax. Commonly bright yellow, but can also be black, or the base colour of the rest of the thorax
	Notopleuron (= notopleura)	A second pair of commonly yellow, roughly triangular shaped markings located on the upper, outer edges of the thorax just in front of the wing attachments, approximately one-third of the way along the thorax
	Notopleural suture	An indentation across the top of the thorax, approximately one-third of the way back, starting on each side at the notopleurons. The indentations from each side rarely go all the way across, but leave a gap in the middle
	Scutum	The top surface of the thorax
	Lateral postsutural vittae (= lateral vittae)	A pair of (normally) yellow stripes which run along the top outer edges of a fly's thorax. The stripes vary in shape (broad, narrow, uniform, tapering) and length
	intra-alar or ia seta	These are a pair of bristles, one located at the rear end of each of the lateral vittae. The vitta can finish before the bristle, at the bristle, or after the bristle
	Medial postsutural vitta (= medial vitta)	A yellow strip which runs along the midline of the top of a fly's thorax. It may be present (e.g. in many *Zeugodacus*) or absent (e.g. in most *Bactrocera*)
	Mesopleural stripe	A pair of normally yellow markings which run, slightly diagonally, down each side of thorax from below the notopleuron to just above the middle leg. They can be broad or narrow judged in comparison to the size of the notopleuron
	Anterior notopleural (npl) seta	The npl setae consist of a pair of bristles on either side of the thorax. The posterior npl bristle sits within the notopleuron, the anterior npl bristle sits approximately halfway between the postpronotal lobe and the notopleuron
	Scutellum	The commonly yellow, large projection from the back of the thorax
Legs	Fore, mid and hind legs	The pairs of legs, beginning from the front, are the fore (= front), mid (= middle) and hind legs (= back legs)
	Femora and tibiae	These are the two longest segments of the fly's leg. The femur is the section closest to the body, the tibia closest to the foot (= tarsus)

Continued

Table 2.2. Continued.

	Term (and alternative name[s])	Plain language description
Wing	R_1, R_{2+3}, R_{4+5}	These are the three wing veins which occur in the front-half of a fruit fly wing. The R_1 vein terminates approximately halfway along the leading edge of the wing, the R_{2+3} vein about two-thirds of the way along, and the R_{4+5} vein terminates just before the wing tip. The M vein terminates just after the wing tip
	Basi-costal (bc) and costal (c) cells	The bc and c cells occur at the base of the wing, just behind the leading wing edge. The two cells are separated from each other by a small cross vein, but are otherwise contiguous with each other. The bc cell is the one closest to the body. The cells are commonly clear, but shading can occur of one or the other (most commonly the costal cell) and this is an important character
	Microtrichia	Very small hairs
	Costal band	The dark marking which runs along the leading edge of the wing. Width of the band varies and is measured with respect to the R_{2+3} and R_{4+5} veins. In some species the costal band can swell near the wing tip
	Anal streak	The dark marking on the wing which runs backwards from near the base of the wing, to finish on the hind margin of the wing, approximately one-third of the way along the wing
	Supernumerary lobe	An extension or enlargement of the hind margin of the wing, occurring immediately after the termination point of the anal streak. The size of the lobe is variable and can range from there being no obvious extension of the wing's hind margin, to there being an obvious expansion such that there becomes a distinct indentation of the wing at the point where the anal streak terminates
Abdomen	Abdominal terga III–V	There are five external segments to a fruit fly abdomen; segment one being the one closest to the thorax. The terga are the upper surfaces of each abdominal segment
	Transverse black band	A black marking that runs along the front half of terga three
	Medial longitudinal black band	A black marking that runs along the midline of terga three, four and five
	Anterolateral corners	The outer edges
	Shining spots (= ceromata)	As named, these are a pair of large, shiny spots on the very back end of fly. If present they are obvious when rotating a specimen under light
	Abdominal sterna	The underside of the abdomen

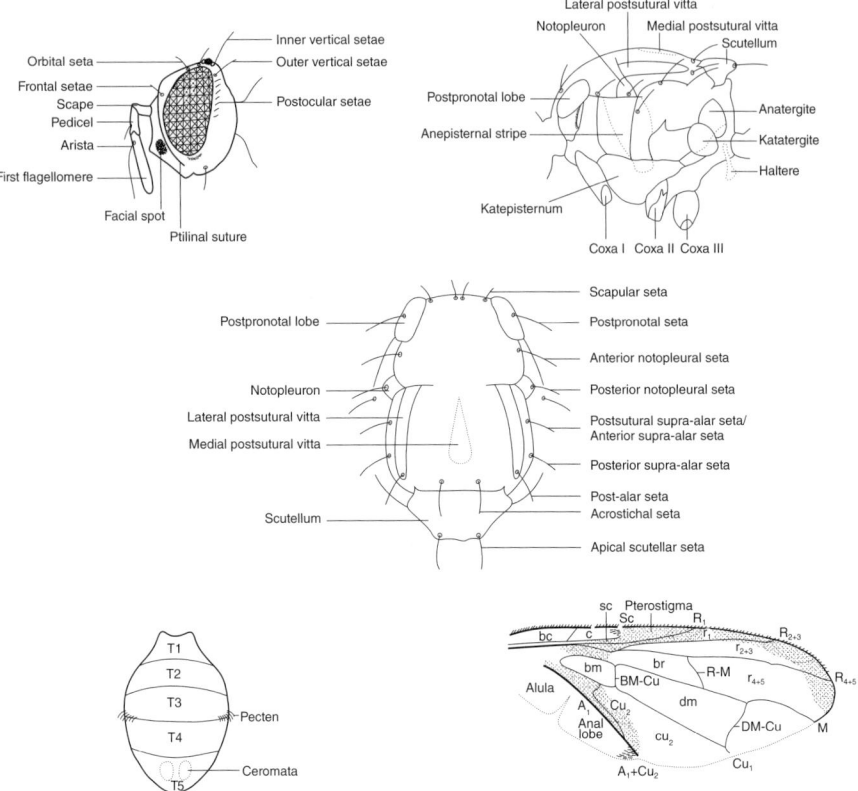

Fig. 2.2. Morphological features of a typical *Bactrocera*. These are also shown comprehensively on a 3D animation at http://fruitflyidentification.org.au/360-fruit-fly/.

2.3.2 Variability of morphological character states

One huge problem of *Bactrocera* taxonomy and diagnostics, and the main reason why so few become experts at it, is because many of the important taxonomic character states can be highly variable within a species. Within the one biological species, colour markings on the legs can be present or absent, bristles may be absent or over represented (e.g. two pairs where there should be one), or the medial band on the abdomen may vary from being almost entirely absent to distinct and well formed. Even basic body colour can vary, with the thorax of *B. dorsalis* varying from almost entirely brown to entirely black. Increasingly, morphological taxonomy of dacines is documenting and reporting this variation while also linking the variable morphology with underlying genetic data. However, while formal species descriptions often recognise this variability within a species, the variation is often poorly captured in accompanying figures and dichotomous keys. Morphological variability is particularly problematic when identifications have to be made from only one, or a limited number of specimens. The rise of molecular markers is one important mechanism to overcome the problem of morphological variation.

2.3.3 Molecular characters

Because of the need to identify larvae that have few morphological features, and because of a loss of morphological taxonomic expertise, significant effort has gone into the development of molecular markers for dacine species-level identification. The 'barcode' region of the COI gene (i.e. the mitochondrial cytochrome c oxidase I gene) is the molecular marker that has been most intensively studied and COI barcodes are publicly available for many pest and numerous non-pest tephritids. The consensus of international data is that COI has, with some important caveats, good resolving power for many pest tephritids. High-throughput systems for processing samples are being developed and will become increasingly available in the future. It is important to note that COI is certainly not the only neutral marker for which multiple dacines have been scored – others include *cytochrome oxidase II, cytochrome b, white-eye, EF-1alpha* and *16S*.

What are the limitations and exceptions of COI for dacines? COI appears to have poor or no resolution capacity for closely related species. For example, it cannot separate *B. dorsalis* from *B. carambolae* and several other members of the dorsalis species complex, nor can it separate *B. tryoni* from its sibling species *B. neohumeralis*. The reasons for this relate to: (i) trying to accurately identify the COI gene, as it has been shown in some *Bactrocera* species that nuclear copies (i.e. numt pseudogenes) of the barcode region of the gene occur; and (ii) because the apparently rapid speciation of some *Bactrocera* clades has simply not resulted in significant divergence of this (and other) neutral genes. For this reason many papers reporting COI diagnostics for dacine fruit flies will often stop at the species complex level (e.g. dorsalis complex or tryoni complex). If all members of a complex are pests this may not be a problem, but it is a problem if a complex contains both pest and non-pest species and the appropriate biosecurity management response depends on getting the identification correct.

The issue of molecular diagnostic markers for dacines (and other pest groups) is an area of ongoing work that will make anything written in this book rapidly obsolete. Current research areas include optimising COI to avoid misinformation derived from numts; the use of 'mini barcodes' to make available museum specimens that commonly have fractured, short-length DNA; identification of COI alternatives that are more informative for tephritids based on whole-genome screening; and rapid processing techniques for both individual flies and bulk samples. It is highly likely that in the near future molecular diagnostic tools will exist that allow not only positive species-level discrimination, but also highly accurate placement of the source population.

2.3.4 Other characters

Morphometrics
Morphometrics involves the use of linear measurements of body parts, or the ratios of two such measurements, as character states. Examples might include the length of the male hind tibia, or the ratio of the width to the length of the thorax.

Morphometrics is little used in Dacini taxonomy, although there are exceptions. Most notable of these is the use of male aedeagal lengths, and/or the ratio of the aedeagal length with a second body size measure (e.g. hind tibia length) to separate taxa within the *B. dorsalis* species complex. While differences in the male aedeagal length was originally the character used to separate *B. dorsalis* from *B. papayae*, later research showed that the character state was continuously variable and thus not reliable as a taxonomic character.

Geometric morphometrics

Geometric morphometrics involves the analysis of shape variation between individuals from two or more populations. Unlike traditional morphometrics, which uses linear distances, geometric morphometrics uses *x,y,z* coordinate data for multiple landmarks and after mathematical correction for variation due to translation, rotation and scaling (i.e. Procrustes analysis), what is left is variation in shape.

For dacine fruit flies, geometric morphometrics has been applied widely for analysis of wing-shape variation between populations and species. Wings are first slide mounted for convenience of handling and for removal of variation in the '*z*' coordinate (i.e. they are all flat). Repeatable landmarks are then identified, commonly vein junctions or where veins terminate at the wing edge. The *x,y* coordinates are then gathered digitally for each specimen and analysis undertaken. The output data includes numerical and graphical presentations of variation in total wing size and wing shape. There are several papers that describe the process and analysis (e.g. see Schutze *et al.*, 2012, in the reading list).

While popular, geometric morphometrics should be used with caution, as the approach is almost too sensitive. Several studies have shown that statistically significant wing-shape variation can be detected despite only the most subtle variation between populations. For example, variation in larval rearing diet and rearing temperature is sufficient to create significantly different adult wing shape in *B. tryoni* populations. Variation in wing shape between two or more populations may be evidence that those populations have some deeper genetic differentiation, but it may simply reflect some form of within species population variation, or even host plant effects. Having used geometric morphometrics widely in our group, I recommend it only as part of larger integrative study, used in conjunction with other tools such as genetics and morphology, and to be done in collaboration (at least initially) with someone who knows both the strengths and weaknesses of the approach.

Host records

Host fruit rearing records can be informative as a diagnostic tool in some cases. Where a fly species is known only from one host fruit, or a small group of closely related hosts, then a positive rearing record may quickly help with species identification. However, in regions where highly polyphagous fruit flies occur (e.g. *B. dorsalis*, *B. tryoni*), there may be few fruits that are not potential hosts of these species. Thus identifications should always be checked to make sure the unknown fly is not one of the common pest species, even if the fruit is not a normal host of the fly.

Trace elements and stable isotopes

Research has been undertaken on the use of naturally occurring, regional-level variation in trace elements and stable isotopes (δ^2H, δ^{18}O, δ^{13}C) as geolocation tools for dacine species of biosecurity threat. Essentially a case of 'you are what you eat', researchers have investigated the use of trace elements and stable isotopes to generate regional 'fingerprints' that are created when larvae eat fruit grown in a particular location, and which is then carried through to the adult. The 'fingerprint' in the adult can be used as a means of tracing that individual back to its region of origin. The approach has application for trying to identify the source location of flies of unknown origin in quarantine and emergency response management. While shown to be technically possible, this approach is unlikely to be operationalised with current technologies due to significant logistic constraints.

2.3.5   Diagnostic tools

Diagnostics involves taking an unknown individual and assigning it to some taxonomic rank – commonly identification to species level. Operationally, diagnostics is an area most commonly associated with quarantine and plant biosecurity. Diagnosticians need not be fully qualified taxonomists, and may commonly work with groups for which they have only limited expertise. The tools available for diagnosticians vary from image libraries and paper-based keys, through to high-throughput molecular processing. All such tools have strengths and weaknesses, but the most serious weakness for all of these diagnostic tools is incomplete coverage. All available keys and molecular tools for the Dacini currently deal with faunal sub-sets (e.g. keys to regional, but not complete faunas; molecular tools that target only quarantine pests), and so any positive identification or molecular 'hit' needs to be interpreted within the context of 'what other species might it be that my diagnostic tool does not include'. A summary of available diagnostic tools follows.

Image libraries

There are an increasing number of online image libraries that support dacine researchers. Images alone are normally only useful if a preliminary diagnosis has been made through some other tool (e.g. a key) or expert opinion, but they greatly increase confidence that an accurate diagnosis has been made. URLs to a number of online image libraries are provided at the end of this chapter.

Dichotomous keys

Dichotomous (i.e. couplet), paper-based keys to adults are the most common diagnostic tool available for the dacine fruit flies and these exist for regions (e.g. South-east Asia) (Drew and Romig, 2016), individual countries (e.g. India, Australia), or for selected groups of species (e.g. the *B. dorsalis* complex). If a fly is morphologically distinctive and likely to come from a fauna for which a modern key is written, then dichotomous keys can work well; however, there are also problems with them. Most importantly, dichotomous keys suffer from the 'unanswerable couplet' problem. If a couplet question requires the answer to a question that your specimen cannot provide, for example if you have a male specimen and the couplet asks about a female character state, then you can proceed no further. Paper-based documents also have limited capacity for illustration, and the structure of couplet questions (is it this or is it that) makes dealing within morphological variation and overlap in character states difficult.

Multi-access keys

A now well-established alternative to the limitations of dichotomous keys are computer-based, multi-access (i.e. multi-entry) keys. These keys operate in a fundamentally different way to dichotomous keys. In a multi-access key, all species being treated are scored for all character states in a background data matrix. When the key is opened, the diagnostician can answer any question that she or he wishes, in any order. The key automatically retains those species that have the character, and discards those that do not. The keys can handle variation within character states, and can be image intensive to aid the user. An identification is routinely reached by answering less than half the questions required for a dichotomous key, and because any question can be answered in any order, the unanswerable couplet question becomes much less problematic.

Good multi-access keys, such as those that exist for the African frugivorous tephritids, are an excellent diagnostic tool, but they are also not without problems. As for any type of key, they can only be accurate for the species included in the key. They are also time consuming and resource intensive to make. Developing a full multi-access key, with a properly and accurately scored matrix, a complete set of illustrations and supporting text, is a project in its own right. For this reason few are available.

Genetic tools

Gene-based diagnostic tools utilise unique genetic markers (i.e. informative loci) to identify unknown specimens. DNA is extracted and amplified from one or more specimens using targeted primers, and then the DNA is analysed based on its size (as inferred through gel electrophoresis: tests such as RFLP, AFLP and RAPD) or sequence variation. Sequence variation diagnostics involves the generation of sequence from the unknown specimen(s), which is then matched against existing sequence from known specimens; a >98% match of sequence is regarded as a positive match. With the increasing ease and reduced costs of genetic sequencing, and the greater ability to deal with unknown specimens within a sequence-based phylogenetic framework, it is likely that sequencing-based diagnostic tools will continue to be developed and preferred by diagnosticians. Of the sequence-based tests, utilisation of the COI barcode is the best known and most widely applied despite known limitations in separating closely related sibling taxa (e.g. *B. carambolae* from *B. dorsalis*, *B. tryoni* from *B. neohumeralis*).

The advantages of molecular tools for diagnosticians are that once protocols are developed, the diagnostician need have no understanding on the taxonomy of the group. Molecular diagnostics does not need adult specimens for identification, but can work equally well with larvae, which are the most often intercepted life stage in quarantine. Ongoing development of different molecular tools is also allowing for greatly increased rapidity of gaining results, the bulk processing of large samples, and even in-field processing of samples. Novel loci, being explicitly developed for tephritids, are also likely to overcome the current limitations of COI.

Despite these advantages, molecular diagnostics are not a universal panacea for the global decline in taxonomic expertise. Naïve application of molecular tools can lead to erroneous identifications just as easily as any other approach, and if specimens are destroyed during molecular processing then there are no vouchers for confirmation. Molecular diagnostics should not be done in isolation from traditional morphological taxonomy, any more than morphological taxonomy can be done in isolation of genetic tools.

2.4   Further Reading and References Cited

Aluja, M. and Norrbom A.L. (eds.) (2001) *Fruit Flies (Tephritidae): Phylogeny and Evolution of Behaviour*. CRC Press, Boca Raton, Florida, USA.

Anon. (n.d.) Pest and Disease Image Library (PaDIL). Available at www.padil.gov.au (accessed 17 May 2018).

Anon. (2018) Fruit Fly Identification Australia. Available at http://fruitflyidentification.org.au/ (accessed 17 May 2018).

Clarke, A.R. and Schutze, M.K. (2014) The complexities of knowing what it is you are trapping. In: Shelly T., Epsky N., Jang E.B., Reyes-Flores J. and Vargas R. (eds.) *Trapping and the Detection, Control, and Regulation of Tephritid Fruit Flies*. Springer, Dordrecht, The Netherlands, pp. 611–632.

David, K.J., Hancock, D.L., Singh, S.K., Ramani, S., Behere, G.T. and Salini, S. (2017) New species, new records and updated subgeneric key of *Bactrocera* Macquart (Diptera: Tephritidae: Dacinae: Dacini) from India. *Zootaxa*, 4272, 386–400.

De Meyer, M. and White, I. (2016) True fruit flies (Diptera, Tephritidae) of the Afrotropical Region. Available at http://projects.bebif.be/fruitfly/index.html (accessed 17 May 2018).

De Meyer, M., Clarke, A.R., Vera, M.T. and Hendrichs, J. (eds.) (2015) *Resolution of cryptic species complexes of tephritid pests to enhance SIT application and facilitate international trade. ZooKeys*, 540, 1–557.

Doorenweer, C., Leblanc, L., Norrbom, A.L., San Jose, M. and Rubinoff, D. (2018) A global checklist of the 932 fruit fly species in the tribe Dacini (Diptera, Tephrtidae). *ZooKeys*, 730, 17–54.

Drew, R.A.I. (1989) The tropical fruit flies (Diptera: Tephritidae: Dacinae) of the Australasian and Oceanian regions. *Memoirs of the Queensland Museum*, 26, 1–521.

Drew, R.A.I. and Romig, M.C. (2013) *Tropical Fruit Flies of South-East Asia (Tephritidae: Dacinae)*. CAB International, Wallingford, UK.

Drew, R.A.I. and Romig, M. (2016). *Keys to the Tropical Fruit Flies of South-East Asia (Tephritidae: Dacinae)*. CAB International, Wallingford, UK.

Freidberg, A., Kovac, D. and Shiao, S. (2017) A revision of *Ichneumonopsis* Hardy, 1973 (Diptera: Tephritidae: Dacinae: Gastrozonini), Oriental bamboo-shoot fruit flies. *European Journal of Taxonomy*, 317, 1–23.

Han, H.-Y. and Ro, K.-E. (2016) Molecular phylogeny of the superfamily Tephritoidea (Insecta: Diptera) reanalysed based on expanded taxon sampling and sequencing. *Journal of Zoological Systematics and Evolutionary Research*, 54, 276–288.

Hancock, D.L. and Drew, R.A.I. (2006) A revised classification of the subgenera and species groups in *Dacus* Fabricius (Diptera, Tephritidae). *Instrumenta Biodiversitatis*, VII, 167–205.

Korneyev, V.A. (2001) Phylogenetic relationships among the families of the superfamily Tephritoidea. In: Aluja, M. and Norrbom A.L. (eds.), *Fruit Flies (Tephritidae): Phylogeny and Evolution of Behaviour*. CRC Press, Boca Raton, Florida, USA, pp. 3–22.

Leblanc, L. (n.d.) Dacine fruit flies of Asia-Pacific. Available at https://www.flickr.com/photos/uhmuseum/sets/72157629625747182/ (accessed 17 May 2018).

Munro, H.K. (1984) A taxonomic treatise on the Dacidae (Tephritoidea, Diptera) of Africa. *Entomology Memoirs, Department of Agriculture and Water Supply, Republic of South Africa*, 61, 1–313.

San Jose, M., Doorenweerd, C., Leblanc, L., Barr, N., Geib, S.M. and Rubinoff, D. (2018) Incongruence between molecules and morphology: A seven-gene phylogeny of Dacini fruit flies paves the way for reclassification (Diptera: Tephritidae). *Molecular Phylogenetics and Evolution*, 121, 139–149.

Schutze, M.K., Krosch, M.N., Armstrong, K.F., Chapman, T.A., Englezou, A., Chomič, A., Cameron, S.L., Hailstones, D.L. and Clarke, A.R. (2012) Population structure of *Bactrocera dorsalis* s.s., *B. papayae* and *B. philippinensis* (Diptera: Tephritidae) in southeast Asia: evidence for a single species hypothesis using mitochondrial DNA and wing-shape data. *BMC Evolutionary Biology*, 12, 130.

Schutze, M.K., Aketarawong, N., Amornsak, W., Armstrong, K., Augustinos, A., Barr, N., Bo, W., Bourtzis, K., Boykin, L., Cáceres, C., Cameron, S., Chapman, T., Chinvinijkul, S., Chomič, A. De Meyer, M., Drosopoulou, E., Englezou, A., Ekesi, S., Gariou-Papalexiou, A., Geib, S., Hailstones, D., Hasanuzzaman, H., Haymer, D., Hee, A., Hendrichs, J., Jessup, A., Ji, Q., Khamis, F., Krosch, M., Leblanc, L., Mahmood, K., Malacrida, A., Mavragani-Tsipidou, P., Mwatawala, M., Nishida, R., Ono, H., Reyes, J., Rubinoff, D., San Jose, M., Shelly, T., Srikachar, S., Tan K., Thanaphum, S., Ul Haq, I., Vijaysegaran, S., Wee, S., Yesmin, F., Zacharopoulou, A. and Clarke, A.R. (2015) Synonymization of key pest species within the *Bactrocera dorsalis* species complex (Diptera: Tephritidae): Taxonomic changes based on a review of 20 years of integrative morphological, molecular, cytogenetic, behavioural, and chemoecological data. *Systematic Entomology*, 40, 456–471.

Virgilio, M. (2016) Identification tools for African frugivorous fruit flies (Diptera: Tephritidae). In: Ekesi, S., Mohamed, S.A. and De Meyer, M. (eds.) *Fruit Fly Research and Development in Africa. Towards a sustainable management strategy to improve horticulture*. Springer, Basel, Switzerland, pp. 19–34.

Virgilio, M., White, I. and De Meyer, M. (2014) A set of multi-entry identification keys to African frugivorous flies (Diptera, Tephritidae). *ZooKeys*, 428, 97–108.

Virgilio, M., Jordaens, K., Verwimp, C., White, I.M. and De Meyer, M. (2015) Higher phylogeny of frugivorous flies (Diptera, Tephritidae, Dacini): Localised partition conflicts and a novel generic classification. *Molecular Phylogenetics and Evolution*, 85, 171–179.

White, I.M. (2006) Taxonomy of the Dacina (Diptera: Tephritidae) of Africa and the Middle East. *African Entomology Memoir*, 2, 1–156.

White, I.M. and Goodger, K. (2009) African *Dacus* (Diptera: Tephritidae): New species and data, with particular reference to the Tel Aviv University collection. *Zootaxa*, 2127, 1–49.

Wiegmann, B.M, Trautwein, M.D., Winkler, I.S., Barr, N.B., Kim, J.-W., Lambkin, C., Bertone, M.A., Cassel, B.K., Bayless, K.M., Heimberg, A.M., Wheeler, B.M., Peterson, K.J., Pape, T., Sinclair, B.J., Skevington, J.H., Blagoderov, V., Caravas, J., Sujatha Narayanan Kutty, Schmidt-Ott, U., Kampmeier, G.E., Thompson, F.C., Grimaldi, D.A., Beckenbach, A.T., Courtney, G.W., Friedrich, M., Meier, R. and Yeates, D.K. (2011) Episodic radiations in the fly tree of life. *Proceedings of the National Academy of Sciences*, 108, 5690–5695.

Zaelor, J. and Kitthawee, S. (2018) Geometric morphometric and molecular evidence suggest a new fruit fly species in *Bactrocera* (*Zeugodacus*) *tau* complex (Diptera: Tephritidae). *Zoological Systematics*, 43, 27–36.

3 Evolutionary Biogeography and Biodiversity

3.1 General Introduction

Consistent genetic evidence shows that the four genera within the Dacini form a single monophyletic group – that is, they all arose from a single common ancestor. Dating the evolution of the Dacini is difficult because of a lack of fossil evidence and conflicting dates in the literature, ranging from approximately 85 million to 55–36 million years ago. The answer probably lies somewhere in between or, if skewed one way or other, then towards the lower end based on the well-documented radiation of the schizophoran flies (of which the tephritids are members). Independent of actual times, what is now clear in the evolution of the Dacini is that there are both deep evolutionary divergence, for example, between *Bactrocera* and *Zeugodacus/Dacus*, and recent, rapid and extensive species radiations, such as the *Bactrocera dorsalis* species complex. What has driven these divergences and radiations (i.e. evolutionary biogeography), and the resulting regional and local faunas (i.e. biodiversity) are the topics for the first half of this of this chapter.

The second half of the chapter focuses on what we know about *Bactrocera* in rainforests. Rainforests are considered the endemic home of *Bactrocera*, and possibly *Zeugodacus*, and most species are still restricted to that environment. Fruit flies are poorly researched in their endemic habitat, but what we do know suggests they play important ecological roles as part of the larger biodiversity of the old-world rainforests. They play a direct ecological role in orchid pollination, and indirect roles in seed dispersal and germination of their rainforest host plants. This part of the chapter does not deal with the African *Dacus* because knowledge of these flies in their endemic habitats is almost entirely unknown.

3.2   Biogeography

The Dacini is dramatically split between Africa and the Asia-Pacific, with the majority of *Dacus* being in Africa, and nearly all *Bactrocera* and *Zeugodacus* being in the Asia-Pacific region (Table 3.1.). Drew and Hancock proposed that the Dacini arose on the rafting Indian plate after it had detached from the southern super-continent of Gondwana. After docking with Laurasia, the early *Dacus* moved west into Africa, whereas the *Bactrocera* and *Zeugodacus* rapidly diversified in the rainforests of South-east Asia, moving eastward into Australia and the western Pacific. Krosch and colleagues subsequently modified this hypothesis slightly, by proposing that *Dacus* moved into Africa via Madagascar prior to the Indian docking. This latter hypothesis is based on the older evolutionary dates proposed for the Dacini (i.e. 85 MYA), but would need to be reconsidered for younger evolutionary dates. If the younger dates are correct, then divergence and speciation of both the African and Asian faunas after Indian docking is more likely, and is a case strongly and persuasively argued for by Ian White (2006). The rapid evolutionary expansion of faunal and floral groups post-docking of India with Laurasia is well documented in the biogeography literature and is known as the 'out-of-India' hypothesis.

3.3   Local Faunas

3.3.1   South-east Asia

This region is a centre of diversity for the Dacini. As treated by Drew and Romig (2013), the Dacini fauna of the region encompasses 423 species: *Bactrocera* 369 species, *Dacus* 51 species, *Monacrostichus* 2 species and *Ichneumonopsis* 1 species. Accounting for the elevation of *Zeugodacus* and related sub-genera to genus level, and the transfer of *Ichneumonopsis* to the Gastrozonini, the Dacini fauna of this region currently encompasses 422 species: *Bactrocera* 220 species, *Zeugodacus* 149 species, *Dacus* 51 species and *Monacrostichus* 2 species. As discussed above, India may possibly be the region of origin of the dacines, but it is not currently the major centre of diversity within the Asian region. The richest species diversity now occurs in the wet, tropical nations

Table 3.1. The number of Dacini species in different regions and their percentage as part of the total fauna. The worldwide numbers are slightly lower than the summed Africa + Asia-Pacific figures because of a small number of species that occur across both regions (from Doorenweerd *et al.*, 2018).

	Worldwide	Africa	Asia-Pacific
Dacini	932	207 (22%)	730 (78%)
Bactrocera	461	13 (3%)	451 (97%)
Zeugodacus	196	1 (1%)	195 (99%)
Dacus	273	193 (71%)	81 (29%)
Monacrostichus	2	0 (0%)	2 (100%)

of South-east Asia (e.g. Malaysia, Thailand, Indonesia), with species diversity declining westward into India and northward into China.

The historical biogeography of the region is complex, and there is good reason to believe this complexity has helped drive dacine speciation. Geologically, the region consists of three major elements: the northward floating Indian plate; the Asian continental plate; and the various archipelagos formed by the largely submerged Sunda shelf (Indonesia, Borneo) and the Philippine islands. Changing sea levels have repeatedly exposed and then submerged substantial land-mass areas around the South China Sea, causing repeated isolation and then expansion of plant and animal populations. Land-mass movement, through floating terrains and upheaval (in the Himalayas) can also lead to population isolation: such isolation is commonly associated with plant and animal speciation and this is presumed also to have happened to the flies.

Within South-east Asia there are some unique faunal "sub-elements". The Western Ghats region of south-western India and Sri Lanka both have locally rich faunas (21 and 28 species, respectively) with high percentages of local endemism. Bhutan has a small endemic Dacini fauna restricted to mid-altitude Himalayan valleys, and is also part of the endemic range of *Bactrocera minax* (Enderlein), one of the few Dacini to be an obligate univoltine species with a pupal diapause over winter. Whether this Bhutanese fauna extends westward along the Himalayas into Nepal is not currently known. Major radiation events are associated with wet tropical South-east Asia. For example the *Z. tau* species complex, currently recognised at 21 described species, is found almost entirely in Thailand and Peninsula Malaysia. Similarly the *B. dorsalis* species complex shows its greatest richness in the same area.

Population genetic studies on widely distributed individual species within South-east Asia have most often shown only weak, or no signals of population structuring within the region, particularly with respect to what are recognised biogeographic barriers. Neither *B. dorsalis* nor *Z. cucurbitae* showed any significant population structuring with respect to the Isthmus of Kra (the narrowest point of the Thai/Malay Peninsula), the recognised biogeographic transition zone between the Indochinese and Sundaic biotas.

3.3.2 Wallacea and the Asian-Australian faunal boundary

The English natural historian, Alfred Russel Wallace, was the first to recognise in the late 1800s the disjunction between Asian- and Australian-derived faunas that occurs in the eastern Indonesian archipelago: a biogeographic region now known as Wallacea. Subsequently known as Wallace's Line, Wallace identified the east–west disjunction as occurring across an imaginary south-to-north line lying to the west of the islands of Lombok and Sulawesi and terminating at the southern Philippines. While Wallace's Line is the best known, other lines of faunal and floral disjunction within Wallacea have since been identified. Most importantly for the Dacini, the eastern edge of Wallacea, close to the island of New Guinea, is bounded by Weber's and Lydekker's Lines. These two lines run on the same path between Australia and Timor, but diverge as they run north past the Molucca's: Weber's Line runs to the west of the island group and Lydekker's Line to the east.

Lydekker's Line appears to be a 'hard' biogeographic boundary for the Dacini. Only six Dacini species are found on both sides of this biogeographic barrier (excluding invasive pests such as *B. dorsalis* and *Z. cucurbitae*), so that the faunas of South-east Asia, and Australia and Western Pacific, are highly distinct. The breadfruit fly, *B. umbrosa*, is one of the few dacines whose natural distribution spreads from the western Pacific (Solomon Islands), west through Wallacea, into South-east Asia (Malaysia). While the fly now occurs across the transition zone, population genetic data shows strong genetic structuring either side of Lydekker's Line, and it is possible the fly was originally from the western Pacific (where its genetic and morphological diversity is highest), and was transported by human movement of breadfruit, *Artocarpus altilis*, in pre-history.

3.3.3 Australia and Pacific

Australia and the Pacific have a large and unique *Bactrocera* fauna, but a highly reduced *Dacus* and *Zeugodacus* fauna. The regional fauna is well-known taxonomically, although there are an estimated 50 species collected but currently undescribed for Papua New Guinea (PNG). Despite the large number of undescribed species, PNG still has the largest described fauna within the region, with approximately 180 species, approximately three-quarters of them being *Bactrocera*. The western half of the island of New Guinea (i.e. the Indonesian province of West Papua) is under-explored with respect to its local fruit fly fauna and it is unclear how many additional species occur there which are not already known from PNG. Australia's fauna consists of 93 *Bactrocera* species, 13 *Dacus* and 7 *Zeugodacus*. Nearly all the *Bactrocera* are placed within the sub-genus *Bactrocera* (*Bactrocera*). In the Pacific Islands the fauna is still relatively rich in the Melanesian islands of the Solomons, Vanuatu and New Caledonia (76 species), but declines to seven or less species in any of the islands of Polynesia and Micronesia.

The biogeographical work on the Australia and Pacific fauna is limited and none has been done using molecular phylogenies. The existing hypotheses are that the fauna is derived originally from Asia, possibly rafting down on land masses that now make up the 'accreted terrains' which constitute the geological northern half of PNG (i.e. north of the Owen Stanley Ranges). The islands of New Britain, New Ireland and the Solomons are geologically of the same material as the accreted terrains, but they have not docked with the Australian craton (which makes up the southern half of PNG) and retain largely unique fruit fly faunas.

The Australian fruit fly fauna is thought to have derived from the New Guinea fauna, but shares few species with it. Within Australia the fruit flies have a classically Torresian distribution: that is they spread west to east across northern Australia and then down the east coast to stop at the southern edge of the subtropical zone, in far northern New South Wales. The greatest majority of species (approximately 80%) are restricted to far northern Queensland.

The biogeographic barrier known as the Burdekin Gap, located just south of Townsville in northern Queensland, seems to be a highly significant natural barrier to the southward movement of many Australian *Bactrocera* species.

3.3.4 Africa and the Middle East

Dacus reaches its greatest diversity in Africa and the Middle East, with 193 species found in the combined region. The sub-generic classification of *Dacus* is not stable, with conflict between the work of White, and Hancock and Drew. White places the African *Dacus* into six sub-genera (*Dacus, Didacus, Leptoxyda, Lophodacus, Neodacus, Psilodacus*), while Hancock and Drew place it into eight (the six of White [redefined] plus *Mictodacus* and *Metidacus*). *Dacus* (*Leptoxyda*), *D.* (*Lophodacus*) and *D.* (*Neodacus*) (as defined by Hancock and Drew) all have species found in both the African and Indo-Australian regions, while in Africa *D.* (*Neodacus*) is restricted to Madagascar. The other sub-genera are Afrotropical restricted. The sub-genera as currently recognised do not align well with molecular analysis, and as for *Bactrocera*, significantly more work needs to be done to align molecular data with sub-generic classifications.

The African *Dacus* have restricted host associations, being found predominantly on fruits and seed capsules of species from the families Cucurbitaceae, Passifloraceae and Apocynaceae. There is significant debate about which of these are the ancestral hosts, but this depends largely on the sub-generic classification accepted. Molecular analysis shows that species feeding on each of the three plant families form monophyletic clades, and it may be that further systematics work aligns the sub-genera fully with feeding habitat (which it currently only partly does).

Within the Afrotropical region, species of the sub-genera *D.* (*Dacus*) and *D.* (*Mictodacus*) are found most commonly in forests and moist woodlands, while all other sub-genera are more common in dry woodlands and savannah. In a detailed study of East African frugivorous fruit flies, 51 *Dacus* species were represented in all the ecoregions sampled (west Guinean lowland rorest, east Guinean forest, Nigerian lowland forest, cross-Nigerian transition forest, west Sudanian savannah, Guinean forest savannah mosaic, Jos plateau forest grassland mosaic, central African mangroves), but were most prevalent in the moist, broadleaf east Guinean forest (25 spp.), the drier Nigerian Lowland Forest (26 spp.), and the drier again Guinean forest savannah mosaic (20 spp.) and the west Sudanian savannah (26 species).

The *Bactrocera* and *Zeugodacus* of Africa and the Middle East are restricted to a small endemic fauna, and an even smaller, but economically impactful, invasive fauna (*B. dorsalis, B. zonata*, and *Z. cucurbitae*). The endemic African *Bactrocera* consists of nine species in the sub-genus *B.* (*Daculus*) and two in the sub-genus *B.* (*Gymnodacus*): *B.* (*Daculus*) is entirely restricted to Africa, while a third *B.* (*Gymnodacus*) species occurs in PNG. This division of the sub-genus *B.* (*Gymnodacus*) between Africa and PNG strongly suggests that the sub-genus does not represent a monophyletic group. With three exceptions, *B.* (*Daculus*) is distributed in eastern and southern sub-Saharan Africa. The exceptions are

B. montyanus, *B. menanus* and *B. nesiotes*, which are known from the Indian Ocean islands of Mauritius and La Réunion (*B. montyanus*) and Madagascar (*B. menanus* and *B. nesiotes*). *Bactrocera* (*Daculus*) *oleae* has been present in the Mediterranean since pre-history, but its endemic distribution is also considered sub-Saharan Africa. The *B.* (*Gymnodacus*) occur in both East and West Africa.

3.4   Likely Drivers of Speciation

The mechanisms of speciation in Dacini have not been directly tested, but at least two models of speciation may be involved. What is known about speciation in the group is that radiation of some lineages has been extreme, and in some cases rapid. Trying to tease apart speciation mechanisms, and how frequently one speciation mechanism may occur versus another, requires deeper understanding of the evolutionary phylogeny of the Dacini than we have so far.

3.4.1   Allopatric speciation

This is the traditional model of speciation, where a daughter population becomes geographically isolated from the parent population. Due to random genetic drift over time, or directional selection following adaptation to the new environment, the two lineages become isolated from each other. That 'isolation' was originally considered as purely associated with mating behaviour, but it is now recognised that other behavioural/physiological and/or genomic changes can lead to lineage separation.

There is strong likelihood that allopatric speciation has played a major role in Dacini diversification. For example, over the last 50 000 years South-east Asia has been subject to repeated rises and falls in sea height of as much as 100 m. During glacial maxima the oceans have fallen, exposing the large land mass of Sundaland that underlies most of modern Indonesia. This created large areas of rainforest and land bridges for population expansion. During glacial minima the oceans have risen, isolating land masses and islands from each other and creating situations ideal for allopatric speciation.

Sea-level changes are not the only factor that will have isolated populations. Significant climate changes that co-occur with glacial cycles can cause major changes in vegetation patterns, which may link or divide populations. Along the east coast of Australia major biogeographic barriers are associated with the climate-driven contraction and expansion of the east coast rainforests and this has driven speciation in both plants and animals, including insects. An argument for significant co-speciation, that is, simultaneous speciation of both host plant and the fruit fly utilising it, has been made for *Bactrocera*. This may be possible as there is huge diversity of flies and their host plants in the Asian-Pacific rainforests. However, such an argument can only be used for the host specialist species, and to date has not been formally tested through appropriate comparative phylogenetic approaches: this is research waiting to be done.

Over geological time, regional plate movements will also have caused isolation between populations, almost certainly leading to speciation. This is best demonstrated by the fruit fly fauna of PNG. Occupying the eastern half of the island of New Guinea and several smaller off-shore islands, PNG has three main geological elements. The southern mainland component is part of the northward moving Australian craton, the northern mainland component is made up originally of southward moving 'accreted terrains' that have docked with the Australian carton. The off-shore islands (e.g. Manus Island, New Ireland, New Britain) constitute the same terrains that have yet to dock and so remain isolated. *Bactrocera* fauna associated with off-shore islands show high levels of endemism to the land mass they are associated with, while the mainland fauna shows low levels of local endemism but high levels of endemism for the country as a whole. As also argued for birds and several plant groups, the inference from this pattern is that there were high levels of isolation and subsequent speciation on the individual terrains that floated south from Asia, but following docking with the Australian craton the species on those terrains have subsequently spread over the island of New Guinea. This is a likely explanation for the extraordinarily high endemic species diversity of *Bactrocera* in PNG, which at well over 150 species is the highest for any single nation.

3.4.2 Sympatric or ecological host-driven speciation

There is increasing evidence that speciation need not be driven only through physical isolation, but can also occur when two populations occur in the same geographic area (i.e. sympatry). The argument for sympatric speciation in such conditions is that there are behavioural, ecological and/or physiological attributes of two populations such that individuals of those populations rarely breed and so the populations diverge over time. The 'textbook' example of sympatric speciation is the tephritid *Rhagoletis pomonella*, where significant population divergence has occurred in the last 200 years between populations that breed predominantly on apple and those that breed on hawthorn. Originally thought just to be linked to host preference, the *R. pomonella* story is now recognised as being much more complex and is additionally linked to population level differences in seasonal time of mating and pupal diapause length.

Advances in understanding of genetic processes has substantially altered how biologists view sympatric speciation. As little as 20 years ago it was still dismissed by most biologists as it was not clear how divergence could occur between two populations in the presence of probable gene flow (i.e. mating) between those populations. New molecular advances, and changing theory to match those advances, now recognise that subtle mutations in response to different ecological drivers (such as preferential host plant use in herbivores) can occur between populations that are geographically close to each other, and that such changes over time can lead to increasing population divergence. Having said that, confirmed examples of sympatric speciation are still rare and no cases have been suggested, tested or confirmed in the Dacini.

3.5   *Bactrocera* in Rainforest

The *Bactrocera* are widely regarded as rainforest specialists, and certainly their greatest species diversity occurs in the old-world rainforests of Asia, New Guinea and Australia. Despite this, there are few studies on *Bactrocera* in rainforest ecosystems, beyond one-off fruit collections. This stems from the fact that rainforests are logistically difficult to work in, and that most *Bactrocera* studies focus on a restricted number of pest species in human modified ecosystems. Much greater work needs to be done on *Bactrocera* within rainforests and it needs to be recognised that the following summary comes from a small number of papers.

3.5.1   Abundance in rainforest

While regarded as rainforest specialists, within rainforests *Bactrocera* species are rare. In a study in lowland rainforest of PNG the median infestation rate was 1 (range 0–12) fruit fly per kilogram of fruit and 1 (range 0–17) fruit fly per 100 fruits (Novotny *et al.*, 2005). In contrast to the pest species occurring outside the rainforest, many of the non-pest, rainforest-restricted species are host specialists to one, or a small number of closely related plant species. The low density of flies per fruit piece means that large numbers of samples need to be taken if accurate assessments are to be made of the local fruit fly fauna based on fruit-rearing data.

Why are fruit flies rare in the rainforest? There are two probable reasons. Firstly, there is the availability of fruit in which to breed. Contrary to popular opinion, rainforests do not fruit continuously. Commonly, most tropical rainforests fruit over a relatively short period, strongly linked to the monsoon driven wet and dry seasons. For example, the rainforests of far north Queensland fruit dominantly over a period of approximately 2 months in October/November and have limited fruit at other times. Further, many rainforest tree species 'mast' fruit, which means they set heavy flower (and subsequent fruit) loads only once every 2 years, or even only once every 3 to 4 years. In the intervening years, flowering and fruiting will be greatly reduced or absent. Additionally, while tropical rainforests have high plant species richness, individual tree species may be rare at the local level. All of this means that fruit is temporally and spatially patchy within a rainforest; being rare or absent most of the time, with occasional periods of abundance. Fly populations will obviously build up during the periods of abundance, but decline during the periods of fruit absence.

The second reason *Bactrocera* are rare in the rainforest is that they appear to suffer high levels of mortality from generalist fruit predators. Fleshy fruits, particularly in rainforest species, are 'designed' to assist in seed dispersal. Some 70% of old-world rainforest plants have fleshy fruit that is evolved to be eaten and the seed subsequently dispersed via the droppings of bats, birds, rodents and other small mammals. Several authors have shown the fruit fly eggs and larvae suffer high levels of mortality because the fruit they are in is eaten by these vertebrate frugivores. Indeed, it has been speculated (but without evidence) that predation of fruit on the rainforest floor is the reason why pre-pupal *Bactrocera* maggots so quickly leave a fruit piece once it has fallen from the canopy. In contrast with many other Tephritidae groups where pupation occurs inside the host, *Bactrocera* rarely pupate in a host unless physically constrained (for example, by hard, dried out peel) from leaving the host. Interestingly, Chinese citrus fruit fly, *B. minax*, which breeds in citrus that is not consumed by native animals after falling to the ground, can have larval development periods of several weeks in fallen fruit. This suggests that larvae rapidly leaving fruit is not a character that is common across all *Bactrocera*, and it may be a trait that evolutionary selection has acted upon.

3.5.2 Location within the rainforest

Although poorly studied, and largely determined by inference, it is most likely that *Bactrocera* are canopy species. When using male traps, catches increase the higher the trap is placed within a forest and, as most rainforest fruit occurs in the canopy, it is reasonable to assume that the flies also spend most, if not all, of their time in the canopy. In the only study to have examined the question of spatial distribution within a rainforest, *Bactrocera* species had aggregated distributions within the forest, but what caused this spatial patchiness is unknown.

3.5.3 Species diversity in rainforest

Only ever assessed for the northern coastal lowland rainforests of PNG, it is considered that rainforest *Bactrocera* have high alpha diversity but low beta diversity (Novotny *et al.*, 2007). This means that within any given patch of rainforest there will be a lot of different *Bactrocera* species, but as you move to a new patch of rainforest that group of species will not change much. For this particular example the coastal belt of rainforest poses no obvious physical barriers to fly movement and it appears all local species are spread throughout that forest.

3.5.4 Impact on host fruit and seed

Insects and vertebrates that feed on plants are almost invariably considered to be damaging to those plants. There are exceptions, for example, specialist vertebrate frugivores that disperse seed, but herbivores are almost invariably considered to damage plants by reducing their size and vigour and then, either directly or indirectly, reducing their seed producing capacity.

In contrast, dacine fruit flies are unlikely to cause fitness damage to their larval host plants, and it may be argued that, to at least some plant species, they are beneficial. Why is this? Dacini fruit flies most commonly attack fruit when it is reproductively mature, that is, at the ripening stage of colour break or later. By this time seeds within a fruit are mature and fully set. With one or two rare examples, dacine larvae do no consume seeds and so they have no direct fitness cost to the host plant in terms of decreased seed production. However, while having no direct fitness costs to the larval host plant, fruit fly larval infestation may have indirect fitness benefits to the host.

Several authors have shown that fruit fly-infested fruit become more attractive to vertebrate frugivores. For birds, particularly during their egg-laying period, it may be because infested fruit is higher in protein, both because of the little protein 'bundles' that are fruit fly larvae, and also because larval infestation increases amino acid levels in fruit. For one group of small mammals in north Queensland rainforest, the preference seemed to be driven by the softening of fruit caused by larval infestation, rather than the presence or absence of the larvae themselves. Only one group of vertebrate frugivores appear not to prefer larval infested fruit, and these are the primates of South-east Asia: in a study of their preferences, they rejected infested fruit (as do most humans). But increased attraction of vertebrate frugivores to infested fruit is indirectly beneficial to the host plant, as it increases the chance of their fruit being dispersed.

The second effect of fruit fly larvae on hosts, with likely indirect beneficial action, is to aid the rapid breakdown of fruit flesh, and so promote germination. Most seed has to have the fruit flesh removed before germination can occur. Indeed some fruit, such as tomatoes, have a germination inhibiting layer around the seed that has to be entirely broken down before seed germination can occur. Research has shown that *Bactrocera* larval infestation significantly increases the breakdown of firm (but not soft) fleshed fruit and increases subsequent seed germination rates. The two-way effect of attracting vertebrate seed dispersers and increasing the breakdown rate of fruit not consumed by vertebrates, means that rainforest fruit flies probably play an important but unrecognised role in rainforest seed ecology.

3.5.5 Orchid pollination

A second important role that dacines play in the old-world rainforests is as pollinators of *Bulbophyllum* orchids (Fig. 3.1). It is likely that they are the sole pollinators of *Bulbophyllum* species in the section *Sestochilus* (Brada) Benth. and Hooker f., as well as *Bulbophyllum* species from other sections of the genera. The flies are attracted to the orchids by secondary plant chemicals produced by the orchid flowers (discussed in depth in the section on 'male lures', see Section 5.7). Depending on the orchid species only one or a small number of dacine species are attracted, but for other orchids large numbers of species can be attracted. In Australia, a relatively short-term survey identified 18 *Bactrocera* species (approximately 20% of the Australian fauna) attracted to the flowers of *Bulbophyllum baileyi*, the only Australian fruit fly-pollinated orchid.

Different orchid species have different mechanisms by which pollen transfer occurs. *Bulbophyllum baileyi*, as one example, has a trigger lip such that when a fly lands on the lip mechanism it triggers, rapidly flicking the fly over on its back pushing the top of the fly's thorax against adhesive pads of the awaiting pollinia (i.e. the pollen bundles of orchids). The caught fly then struggles out of the flower, carrying with it the pollinia (Fig. 3.2), which it may ultimately transfer to a second flower to achieve cross-pollination.

Fig. 3.1. *Bactrocera tryoni* adult males feeding on the petals of *Bulbophyllum baileyi* (upper); and *B. tryoni* male with the paired pollen bundles of *B. baileyi* attached to its thorax (lower) (photo credit: Dr Jacinta Zalucki).

The published work on orchid pollination by *Bactrocera* is largely restricted to systems in Malaysia, due to the extensive work of Professors Hong Keng Tan and Ritsuo Nishida and their students. Nevertheless, orchid fruit fly pollination systems are found all the way from South-east Asia, down through the Indonesian Archipelago and the Island of New Guinea, to northern Australia. As such they are a widespread pollination interaction of much importance to the survival of indigenous orchids native to this region.

3.6   *Bactrocera* in Rainforest Fringe and Other Non-rainforest Habitats

While considered rainforest insects, studies that have simultaneously trapped *Bactrocera* in rainforest and non-rainforest environments have frequently caught equal numbers of species in both environments. Sometimes individual species will be restricted to rainforest or non-rainforest environments, and in

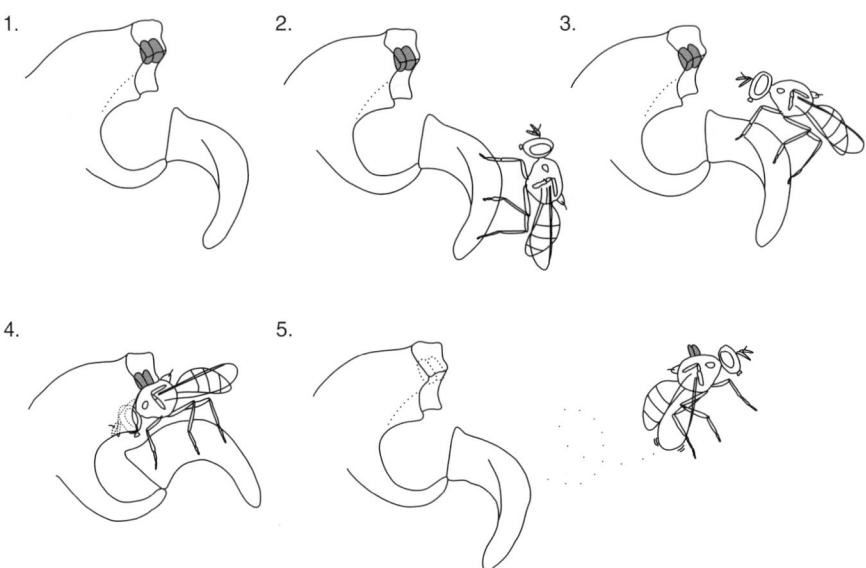

Fig. 3.2. Pollen removal from *Bulbophyllum baileyi* by *Bactrocera tryoni*.

others a species may be equally abundant in both habitat types. This suggests that rainforests need not be considered the 'primary habitat' of the genus, but just one important habitat.

This should not be surprising. Even for *Bactrocera* species whose host plants are rainforest restricted, the larval host is just one resource (i.e. the oviposition substrate) needed by one sex of the mobile adult flies. Other resources include sheltering sites, protein, water, sugar, phenylpropanoids (for males) and mates. These resources may be as equally likely to be found on rainforest margins or outside the rainforest, as they are inside the rainforest. Importantly, native larval hosts, and not just introduced exotics, also occur outside rainforest environments. For example, the Pacific almond, *Terminalia catappa*, has been postulated as the most preferred native host of Oriental fruit fly through much of its endemic and introduced range. This plant species has a specific habitat, growing on the coastal fringe of the Indian Ocean and western Pacific. Only rarely do rainforests come down to the sea, so for Oriental fruit fly to be using this host it must be seeking the larval resource in different habitat types.

It is probable that the preferred habitat requirements strongly impact on the likelihood of a fly becoming pestiferous or not. In three independent studies, *B. tryoni* has been found to be rare in rainforests, while hugely abundant in nearby non-rainforest vegetation or urban environments. Similarly, Oriental fruit fly never entered the north Queensland rainforests when it invaded Australia in the mid-1990s, despite being in huge numbers in other agroecosystems. Similarly in PNG and Asia, Oriental fruit fly is rare in rainforest but common in open agricultural environments. Thus it seems that some dacine fly species do not necessarily like the closed canopy ecosystems of rainforests and maybe flies that are more adapted to open ecosystems are more likely to be those which have become pests.

3.7 *Dacus* in their Natural Environment

In the Afrotropical region, abundance of *Dacus* is presumed to be linked to the suitability of habitat for host plants. Hence the *Dacus* species that specialise on the dryer environment Apocynaceae are more common in the savannah and drier woodlands, whereas those associated with the Passifloraceae and Cucurbitaceae are more commonly found in forests and moist woodland. Through observations of Apocynaceae pollen seen on the tarsi of some African *Dacus* species inference has been made that these flies are pollinators of those plants, but this has not been confirmed.

3.8 Further Reading and References Cited

Clarke, A.R., Balagawi, S., Clifford, B., Drew, R.A.I., Leblanc, L., Mararuai, A., McGuire, D., Putulan, D., Sar, S. and Tenakani, D. (2002) Evidence of orchid visitation by *Bactrocera* species (Diptera: Tephritidae) in Papua New Guinea. *Journal of Tropical Ecology*, 18, 441–448.

De Meyer, M., White, I.M. and Goodger, K.F.M. (2013) Notes on the frugivorous fruit fly (Diptera: Tephritidae) fauna of western Africa, with description of a new *Dacus* species. *European Journal of Taxonomy*, 50, 1–17.

Doorenweerd, C., Leblanc, L., Norrbom, A.L., San Jose, M. and Rubinoff, D. (2018) A global checklist of the 932 fruit fly species in the tribe Dacini (Diptera, Tephrtidae). *ZooKeys*, 730, 17–54.

Drew, R.A.I. (1987) Reduction in fruit fly (Tephritidae: Dacinae) populations in their endemic rainforest habitat by frugivorous vertebrates. *Australian Journal of Zoology*, 35, 283–288.

Drew, R.A.I. (2004) Biogeography and speciation in the Dacini (Diptera: Tephritidae: Dacinae). *Bishop Museum Bulletin in Entomology* (Evenhuis, N.I. and Kaneshiro, K.Y. (eds.) D. Elmo Hardy Memorial Volume. Contributions to the Systematics and Evolution of Diptera), 12, 165–178.

Drew, R.A.I. and Hancock, D.L. (2000) Phylogeny of the tribe Dacini (Dacinae) based on morphological, distributional, and biological data. In: Aluja, M. and Norrbom, A.L. (eds.) *Fruit Flies (Tephritidae): Phylogeny and Evolution of Behavior*. CRC Press, Boca Raton, Florida, USA, pp. 491–504.

Drew, R.A.I. and Romig, M. (2013) *Tropical Fruit Flies of South-East Asia (Tephritidae: Dacinae)*. CAB International, Wallingford, UK.

Drew, R.A.I., Zalucki, M.P. and Hooper, G.H.S. (1984) Ecological studies of Eastern Australian fruit flies (Diptera: Tephritidae) in their endemic habitat. *Oecologia*, 64, 267.

Hancock, D.L. and Drew, R.A.I. (2006) A revised classification of the subgenera and species groups in *Dacus* Fabricius (Diptera, Tephritidae). *Instrumenta Biodiversitatis*, VII, 167–205.

Krosch, M.N., Schutze, M.K., Armstrong, K.F., Graham, G.C., Yeates, D.K. and Clarke, A.R. (2012) A molecular phylogeny for the Tribe Dacini (Diptera: Tephritidae): Systematic and biogeographic implications. *Molecular Phylogenetics and Evolution*, 64, 513–523.

Novotny, V., Clarke, A.R., Drew, R.A.I., Balagawi, S. and Clifford, B. (2005) Host specialization and species richness of fruit flies (Diptera: Tephritidae) in a New Guinea rainforest. *Journal of Tropical Ecology*, 21, 67–77.

Novotny, V., Miller, S.E., Hulcr, J., Drew, R.A.I., Basset, Y., Janda, M., Setliff, G.P., Darrow, K., Stewart, A.J.A., Auga, J., Isua, B., Molem, K., Manumbor, M., Tamtiai, E., Mogia, M. and Weiblen, G.D. (2007) Low beta diversity of herbivorous insects in tropical forests. *Nature*, 448, 692–695.

Raghu, S., Clarke, A.R., Drew, R.A.I. and Hulsman, K. (2000) Impact of habitat modification on the distribution and abundance of fruit flies (Diptera: Tephritidae) in South-east Queensland. *Population Ecology*, 42, 153–160.

Tan, K.H., Nishida, R.J. and Toong, Y.-H. (2002) Floral synome of a wild orchid, *Bulbophyllum cheiri*, lures *Bactrocera* fruit flies for pollination. *Journal of Chemical Ecology*, 28, 1161–1172.

Tan, K.H., Tan, L.T. and Nishida, R.J. (2006) Floral phenylpropanoid cocktail and architecture of *Bulbophyllum vinaceum* orchid in attracting fruit flies for pollination. *Journal of Chemical Ecology*, 32, 2429–2441.

White, I.M. (2006) Taxonomy of the Dacina (Diptera: Tephritidae) of Africa and the Middle East. *African Entomology Memoir*, 2, 1–156.

Wilson, A.J., Schutze, M.K., Elmouttie, D. and Clarke, A.R. (2012) Are insect frugivores always plant pests? The impact of fruit fly (Diptera: Tephritidae) larvae on host plant fitness. *Arthropod-Plant Interactions*, 6, 635–647.

Zalucki, M.P., Drew, R.A.I. and Hooper, G.H.S. (1984) Ecological studies of eastern Australian fruit flies (Diptera: Tephritidae) in their endemic habitat. II. The spatial pattern of abundance. *Oecologia*, 64, 273–279.

4   Basic Biology and Demographic Ecology

This chapter introduces the basics of Dacini fruit fly biology. The first part of the chapter covers their life cycle and life stages, and identifies the resources needed by those different life stages. The second part of the chapter introduces the flies' demographic ecology, which covers population dynamics, landscape ecology and community ecology. While these are three large research areas, surprisingly little is known about the landscape and community ecology of dacines and so they can be included in the one chapter. Absent from this chapter, except to briefly introduce, are two critical areas of dacine biology: reproduction and host use. These two areas are dealt with in detail in Chapters 5 and 6, respectively.

4.1 Dacini Life Cycle

All Dacini have the same basic life cycle, as illustrated in Fig. 4.1. The adult males and females are free flying in the environment and need to locate each other to mate. Females mate only once or a limited number of times in their lives, whereas the males can potentially mate daily if females are available. The inseminated female then seeks fruiting host plants where she lays her eggs under or into (depending on thickness and density) the skin of the fruit. The fruit is generally at the mature, colour-change stage or later, but there are numerous exceptions to this rule. The emergent larvae feed within the fruit, most likely on a diet consisting of a mixture of the fruit flesh and bacteria that are inoculated into the fruit by the female ovipositor during egg laying. The larvae undergo three instars before leaving the fruit to pupate. Pupation most commonly occurs in the soil, but may also occur within the fruit for some fly species, especially in *Dacus*. Most species are multivoltine (i.e. multiple generations per year) and pupation occurs without a developmental interruption; a small number of dacine species are

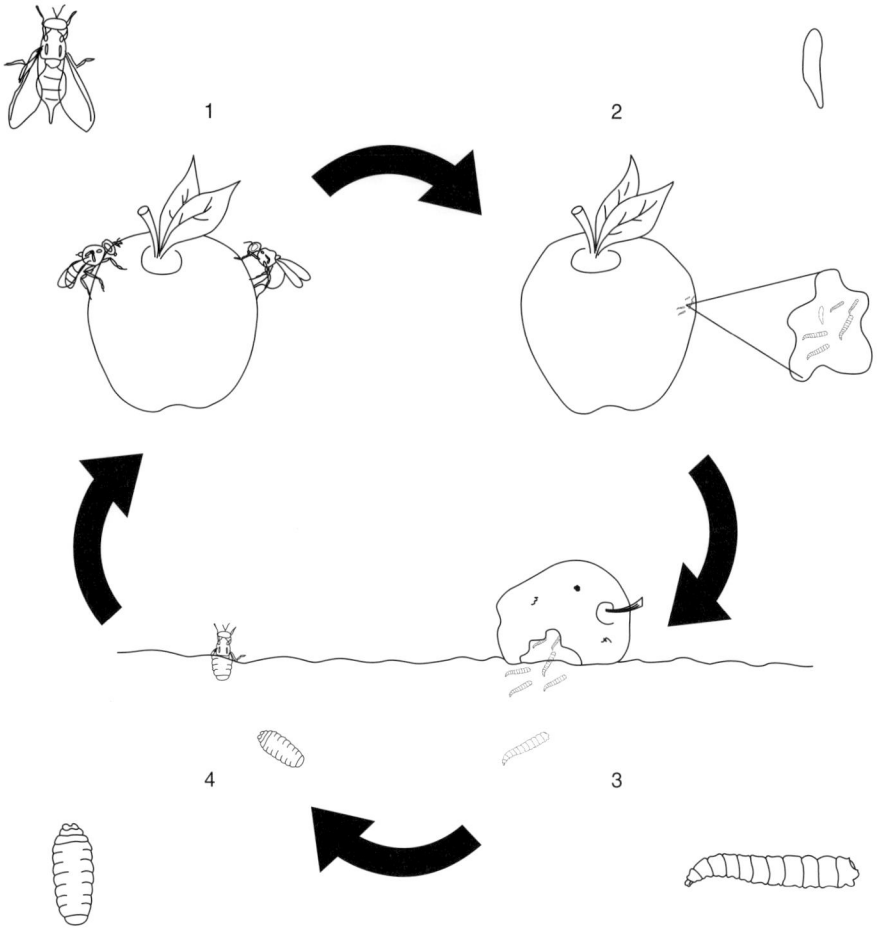

Fig. 4.1. The basic life-cycle of a Dacini fruit fly. (1) The sexually mature, gravid adult female fly lays eggs into fruit; (2) the eggs hatch within the fruit where the resultant larvae feed, passing through three instars; (3) larval feeding commonly causes premature fruit to drop and the now mature larvae leave the fruit; (4) the pupal stage occurs in the soil, from which the new generation adults emerge. The new adults are sexually immature and need to mature and mate before the cycle continues.

known to be univoltine (i.e. one generation per year) and have an overwintering pupal diapause. Teneral adults emerging from the soil take 24–48 hours for the cuticle to harden and the body colouring to develop. Flies at this stage are sexually immature, and sexual maturation can take from 6 to 30 days depending on species and environmental conditions. Once sexually mature, mating can occur and the life cycle repeats. The following sections expand on this general overview.

4.2 Immatures

4.2.1 Morphology and development of immature stages

Bactrocera eggs are creamy-white in colour, approximately 1 mm long and 'banana shaped'. Their size and volume can vary significantly between species (e.g. 0.93 mm long in *B. krausii*, 1.23 mm in *B. jarvisi*), but is not correlated to adult female body size. Hatch time under field conditions is from 1–2 days. Clutch size is highly variable within and across fly species from 1–50 eggs: drivers of clutch size variation are discussed in detail in Section 6.2.3.

There are three larval instars, ranging in length from approximately 1 mm at hatching through to approximately 10 mm in the third instar. As for most higher dipteran maggots, the larvae are typified by a general reduction of morphological traits and have a soft, unsclerotised body. The most obvious external characters are associated with the head and include the antennal sensilla, oral plates and mouth hooks, or with the last abdominal segment where the paired posterior spiracles are located. Various morphological features of the larvae are diagnostic at the species level, including the mouth hooks and the larger cephalopharyngeal skeleton, the oral plates and the posterior spiracles. While keys to species are available using larval characters, most diagnostics using larvae now relies on molecular rather than morphological characteristics.

The development time of larvae is variable between species, but runs at around eight to 12 days, although larval duration times of less than six days have been recorded for *Z. tau* held at a constant 27°C. Development time can be affected by fruit type, quality and crowding conditions, with longer development time in poorer quality hosts. For example, in *Z. tau* larval duration in sponge gourd is significantly shorter than larval duration in bitter gourd, guava and tangerine, but not significantly different to development time in cucumber. Larval crowding in small fruit, or insufficient artificial diet, can lead to resource depletion that results in shortened larval development times and smaller pre-pupal larvae. Larval survival can also be directly affected by host, but because of the difficulty of teasing apart survival within a fruit at the egg or larval stages it is often not clear if the host affect is influencing the egg, neonate larvae, more mature larvae, or some combination of the three. In a study of *B. tryoni* immature survival in different citrus species and varieties, differential survival to the pupal stage was eventually attributed largely to neonate (i.e. just hatched) larval mortality.

Third instar larvae cease feeding at the 'pre-pupal' larval stage. The pre-pupal larval stage can last from 1–2 days and involves the cessation of feeding, voiding of gut, location of the pupal site and initial physiological and morphological changes before pupation proper. The larvae of most of the pestiferous dacine species leave the fruit and pupate in the soil, with the pre-pupal larvae actively seeking suitable pupation sites (see Section 4.2.3 for more detail), but many dacines, especially among the *Dacus*, also pupate within the fruit. Within-fruit pupation can be a species normal behaviour, but it may also occur in a facultative fashion for larvae that normally pupate external to the fruit if the fruit peel has toughened and so the larvae cannot escape the fruit. In such situations it is likely that the subsequent adults would also fail to leave the fruit and so subsequently die.

Bactrocera pupae are the typical 'drum' like, highly sclerotised pupae of the higher Diptera with few external features. Development time is from 7–14 days, but it can be greatly extended to several months in the few *Bactrocera* species (such as *B. minax*) that have a pupal diapause. When the pharate adult emerges from the pupa it breaks open the end of the pupal shell (in a fashion similar to taking the top off a boiled egg) by the expansion of the ptilinum.

4.2.2 Larval diet

Dacini larval diet under natural conditions is extremely poorly researched. Fruit are low in protein and it is assumed that larvae gain the protein they need by feeding on bacteria that are introduced into the fruit, along with the eggs, by the female ovipositor. The larval mouth hooks are used to rasp and break down the fruit flesh and it is presumed that the bacteria multiply in this pulp, creating a 'bacterial soup'. Dacine maggots are considered to be liquid feeders and this bacterial/fruit pulp soup is ingested through the simple oral cavity. The relative importance of bacteria versus fruit to the nutrition of the larvae is an area of active research, with results showing that it may be variable depending on bacterial species and fruit type.

Contrary to larval feeding in fruit, there is a large body of literature on artificial larval diets for fruit fly rearing. Artificial larval diets fall into two broad categories: bulked diets and liquid diets. Both types of diets contain approximately the same nutritional constituents, but vary in the mechanical support offered to larvae. Bulked diets are those traditionally used by fruit fly researchers and are bulked by agents such as wheatgerm, dried carrot, dried alfalfa or sugarcane bagasse. Some nutrients may be provided by the bulking agents, but that is not their primary purpose in most cases. Liquid diets do away with the bulking agents and instead use artificial substrates (organic sponge cloth, cotton cloth, etc.) to provide substrate support for the larvae. Liquid diets are reported to have advantages over bulk diets through space saving, better utilisation of diet and significantly less waste after the diet is spent. However, for sterile insect technique (SIT)-level production of flies the upscaling of liquid diets to factory scale has proven difficult. An improved modification of liquid diets are 'gel diets', where the 'liquid' diet is solidified using a natural gelling agent. Gel diets appear to have all the advantages of liquid diets, while providing their own substrate for larvae to feed upon. Properly managed, a spent gel diet has no solid waste, just the spent diet. Gel diets have been developed in Australia for use in *B. tryoni* mass rearing and are currently being incorporated into factory-scale use.

Artificial larval diets contain a relatively small number of ingredients that perform different purposes within the diet (Table 4.1). Nutrients are generally provided in the form of yeast and, in bulked diets, sometimes through the bulking agent, whereas carbohydrates are provided as sugar. In liquid diets, plant

Table 4.1. Artificial larval rearing diets for selected Dacini fruit flies. Figures for each ingredient are a percentage of the total diet unless otherwise noted. For some papers multiple diets were tested and reported; if so the most successful are listed below.

Ingredient	Z. cucurbitae bulk diet	Z. cucurbitae liquid diet	B. tryoni bulk diet I	B. tryoni bulk diet II	B. zonata bulk diet	B. zonata liquid diet	B. dorsalis bulk diet	B. dorsalis liquid diet I	B. dorsalis liquid diet II	B. oleae bulk diet I	B. oleae agar diet II
Preservative/ antimicrobial	Sodium benzoate 0.11%; Nipagen, 0.11%	Sodium benzoate, 0.09%; Nipagen, 0.09%	Sodium benzoate, 0.20%; Nipagin, 0.15%	Sodium benzoate, 0.18%	Nipagin, 0.1%	Nipagin, 0.15%	Sodium benzoate 0.10%; Nipagen, 0.12%	Sodium benzoate, 0.15%; Nipagen, 0.15%	Sodium benzoate, 0.15%; Nipagen, 0.15%	Tween 80 2.81 ml; Nipagen 0.75 g; Potassium sorbate 0.18 g	Tween 80 2.81 ml; Nipagen 0.75 g; Potassium sorbate 0.18 g
Carbohydrate (sugar)	7.35%	5.96%	5.03%	0.0	11.0%	8.0%	12.45%	8.99%	8.99%	7.5 g	7.5 g
Nutrients	Torula yeast, 3.55%	Brewer's yeast, 11.51%	Torula yeast, 2.51%	Torula yeast, 2.18%	Ground maize 6.0%; Waste brewer's yeast 6.0%; Wheat bran 6.0%	Waste brewer's yeast 15.0%; Corn oil 0.30%	Torula yeast, 3.47%	Brewer's yeast, 15.06%; Wheat germ oil 0.15%	Lallamand yeast, 15.06%; Wheat germ oil 0.15%; or Corn oil 0.15%; or Soybean oil 0.15%	Soy hydrolysate, 11.25 g; Brewer's yeast 28 g; Olive oil 7.5 ml	Brewer's yeast 28 g; Casein, 7.5 g; Wheat germ, 3.75 g; Olive oil 7.5 ml
Bulking agent	Mill feed, 31.19%	–	Lucerne chaff, 62.89%	Dried granulated carrot, 24.24%	Sugarcane bagasse, 6.0%	–	Mill feed, 27.49%	–	–	Cellulose powder 115.5 g	Agar 2.52 g
Support matrix	NA	Sponge cloth	–	–	NA	Several trialled	–	Sponge cloth	Sponge cloth	–	–
Water	57.70%	81.08%	23.93%	72.73%	64.8%	75.0%	56.38%	73.81%	73.81%	206.3 ml	206.3 ml
Acid	–	Citric acid, 1.26%	Citric acid, 0.26%	Citric acid, 0.65%	Benzoic acid, 0.1%	Ascorbic acid, 1.40%; Benzoic acid 0.15%	–	Citric acid, 1.70%	Citric acid, 1.70%	HCl 2N, 11.25 ml	HCl 2N, 11.25 ml
Source	Chang et al. (2004)	Chang et al. (2004)	Khan (2013)	Khan (2013)	Sookar et al. (2014)	Sookar et al. (2014)	Chang et al. (2006)	Chang et al. (2006)	Ekesi et al. (2014)	Tsitsipis (1975)	Hanife (2008)

oils are added to replace nutrients that the larvae may originally have derived from the bulking agent. Wheatgerm oil was the original recommendation for larval diets, but subsequently different researchers have shown equivalent effects can be obtained with maize oil, soybean oil, rice bran oil and mixed vegetable oil. A food preservative and/or broad-acting antimicrobial is added to stop the diet putrefying and/or becoming mouldy. The addition of general antimicrobials to artificial diet is one of operational necessity, but may be a limitation if larvae obtain proteins from bacteria as has been suggested in natural diets. Some form of acid is added to moderate pH, with diets often being quite acidic (pH 3.3–3.8). The diet is finished with water and, if relevant, the bulking agent. Diets for specialist fruit flies (e.g. olive fruit fly) often contain some component derived from the host fruit to supply essential dietary components that are lacking from 'general purpose' diets developed for polyphagous flies. What role these specialist components play in the artificial diets of specialist flies is not well understood, but it is known they are important and cultures will not thrive without them.

Experience has shown that larvae do not automatically adapt to artificial diets, even for the polyphagous flies. One study of *B. dorsalis* in Africa showed that it took three to five generations for egg-to-adult recovery on a high-quality artificial diet to match that obtained on mango. This suggests, as is well documented for adults, that there is strong natural selection on larvae in artificial rearing conditions.

4.2.3 Larval behaviour

There is little information on the behaviour of larvae within fruit. Many reviews claim that larvae migrate into the middle of fruit (or at least until reaching an internal seed), but what these claims are based upon is unclear. Experience shows that larval cohorts tend to remain together and it is not unlikely that there is facilitation of feeding, as larger numbers of maggots can rasp a larger feeding site, while the generated metabolic heat might increase development rates; but this has not been tested. Migration within fruit appears limited. In a study of *B. dorsalis* in mango, larval cohorts did not migrate from the bottom to the top (stalk end) of the fruit, despite the top being riper and of better nutritional value. Internal divisions within a fruit, such as the segments of citrus, seem sufficient to stop larval movement in many species. Competition between larvae of the same species has been reported within fruit, but only in experimental conditions and at larval densities greatly exceeding that normally seen in the field. Larval competition between larvae of different species is well documented (see Section 7.1.5)

4.2.4 Pupae

Location
Most *Bactrocera* pupate in the soil below the host tree. Mature third instar larvae that leave the fruit enter a wandering phase, where they search for an appropriate pupation site. The wandering phase is brief, from a few minutes to

approximately 2 hours. In one laboratory study, the mean length of 'wandering' was only 20 cm. Wandering larvae make active choices of pupation site and can discriminate between soil that is too wet or too dry, in the shade versus light, and with an open pore structure versus a dense pore structure. This can lead to a patchy distribution of pupae within an orchard. Depth of pupation appears to be variable and dependent largely on the soil conditions and type. In friable, moist soils larvae may bury themselves to 5 cm depth, but in compacted soils the pupation depth may be only 1–2 cm, or in extreme cases even on the soil surface under leaf material. Across studies, a pupal burial depth of 3–4 cm seems most common for *Bactrocera*.

Pupation in the host fruit is not common in *Bactrocera*, but in species where it does occur then often soil pupation also occurs; for example, in olive fruit fly both forms of pupation seem normal. Species that normally pupate in the soil may occasionally be seen to pupate in the fruit if the peel has dried and the larvae cannot exit the fruit easily. Within-fruit pupation is commonly seen in *Zeugodacus* and *Dacus* species where the fruits often have hard peel or 'shells'. It has been suggested for *Dacus*, that pupation within fruit is a way of avoiding their often arid environments.

Pupal survival
Dry soils have been shown to be detrimental to dacine pupal survival and emergence. In a laboratory study of *B. tryoni*, 100% pupal mortality was recorded in fully dry loamy-sand and sandy-clay. Fully flooded soil may also kill pupae or emerging teneral adults, but the influence appears less damaging than 100% dry soil. In contrast, pupal mortality of *B. oleae* pupae was high in flooded soil or very moist soil. In *B. oleae*, mortality of teneral adults increases with increasing pupal depth. The population level consequences of dry topsoil have not been examined for *Bactrocera*, but environmental dryness is negatively correlated with *Bactrocera* abundance or environmental suitability.

4.3 Adults

4.3.1 Adult demographics

Sexual maturation
All dacines emerge from the pupal stage as sexually immature adults. The post-emergence teneral stage, during which the wings expand and the body takes on full colouration takes 24–48 hours, but sexual maturation can take from 5 to 30 days depending on species. A sexual maturation period of approximately 10 days is often assumed for *Bactrocera* species, but this may largely be the result of laboratory selection for early maturation as an increasing number of species are being shown to have significantly longer sexual maturation periods. For example, *B. carambolae* has been shown to have a maturation period of 20 to 30 days, *B. pyrifoliae* has a 21-day maturation period, while *Z. tau* can take anywhere from 18 to 45 days until first mating. How common these longer maturation periods are in other *Bactrocera* species is largely unknown.

Laboratory mass rearing in *Bactrocera* and other tephritids can markedly reduce the maturation period, down to as little as 5 to 7 days. *Bactrocera tryoni* has been reported to become sexually mature from anywhere from 5 to 31 days with the reported variation due to laboratory rearing history, constant versus fluctuating day length, light intensity and constant versus fluctuating temperature regimes. During maturation flies need access to sugar, water and protein, and a positive protein foraging response is strongest during this period (see dietary requirements).

Egg production

Trying to provide definitive figures on the number of eggs produced by a female for any species appears largely meaningless, as comparisons across studies show that egg production can be affected by culturing history (wildish versus long-term cultured flies), the ratio of males to females, female diet, when females had access to protein (e.g. directly after eclosion, later in life, or continuously), female access to oviposition substrates, and type of oviposition substrate. For example, *B. pyrifoliae* produced an average of 216 or 496 eggs per female depending on the ratio of males to females in the rearing cages (216 eggs when sex ratio was 1:1, 496 eggs when there were three times as many females as males). *Bactrocera dorsalis* has been reported to produce anywhere from 250 to 3000 eggs per female over an adult life-span of 75–86 days, with egg production strongly influenced by the oviposition substrate (different fruit types and artificial diet). In broad terms, potential egg production does vary across species. For example, *B. tryoni* laid four times more eggs than *B. jarvisi* by 28 days post-eclosion (500–570 eggs/female versus 116–135 eggs/female, respectively), due to *B. tryoni*'s greater number of ovarioles. Life-table studies show that peak egg production occurs early in a fly's life, and tails-off as the flies pass the age of 4 to 6 weeks.

Adult longevity

In the laboratory, most studies show adult flies surviving for a maximum of around 80 days, with population mortality commencing from around 5 weeks. However, *B. dorsalis* has been reported to survive in excess of 200 days in the laboratory, whereas in field cages in temperate Australia, *B. tryoni* has been reported to survive from 7 to 8 months in the period from autumn through winter and into spring, although most flies died substantially earlier, at around 4 to 5 months of age. Dietary inhibition, notably a lack of protein, can increase longevity, as do cooler temperatures.

Mating and mating propensity

This topic is covered separately in Chapter 5.

4.3.2 Adult diet

Adult fruit flies actively forage for nutritional and water resources. Nutritionally, adult flies need carbohydrate, protein in the form of free amino acids, minerals,

B-complex vitamins and water. Teneral adults carry only limited resources through from the larval stage and will die quickly in the absence of carbo-hydrate and water; protein is required for sexual maturation. The response to proteins and carbohydrates (as sugar) is highest in immature flies and such flies will seek these resources over fruit or a male-lure source. In sexually mature *B. cacuminata*, diurnal foraging patterns for sugar and protein were fairly con-stant during the day, although slightly higher in the morning, while immature flies continued to aggregate at these resources throughout the day. Females are considered to have a greater lifetime need for protein than males, due to the ongoing requirements for egg production. In *B. tryoni*, it appears that much of a male's protein need may be carried through from the larval stage, and only one, or a limited number of protein feeds early in the life of the male is sufficient. However, females require protein on an ongoing basis, although the specific amount of protein required varies between fly species.

In nature, it is considered that most nutritional resources for adult flies come from leaf-surface leachates (which contain amino acids, organic acids and sugars), extrafloral nectaries and bacteria (developed below). Feeding on bird droppings is rare, but has been observed, as has incidental feeding on liquid seeping from ripe fruit. Like their larvae, adult Dacini are fluid feeders and will regurgitate droplets from the crop onto a dry feeding sur-face, which they then return to and re-imbibe. The proboscis is sponge-like, with a fine, sieve-like structure. The droplet/regurgitation process is a mech-anism for taking up micro-particulate material from the dry substrate (e.g. leachates or bacteria) that could not otherwise be consumed. It may also be a mechanism for concentrating crop contents, if the regurgitated content evaporates slightly before being re-ingested. This has been suggested but not confirmed.

In a field study of *B. tryoni*, it was demonstrated that sexually mature adult males and females were never hungry enough, as measured by internal pro-tein and carbohydrate levels, to reach threshold levels that physiologically 'prompted' active protein and carbohydrate forging. This strongly suggests that, at least for this species, the flies are well adapted to meeting their nutritional needs and those needs are readily available in their environment. In contrast, free water from which flies need to drink can be limiting, particularly in open cropping environments. Low relative humidity environments are detrimental to the survival of adult flies and, within both large and small scale landscapes, flies will locate themselves into areas of high humidity. At an individual tree level, this may explain why trees pruned to open, vase-like canopies, or pruned hard to trellis-like canopies, are reported to suffer less fruit fly damage than orchards with traditional heavy canopies.

4.3.3 Bacterial interactions

The interaction between Dacini and gut microbiota, predominantly bacte-ria, is arguably the 'hottest' area of current fruit fly research: it is also one of the oldest. In 1910, the German microbiologist Julius Petri (credited with the invention of the Petri dish) identified the bacteria *Pseudomonas savastanoi* in association with the olive fruit fly. While this particular bacteria is not now considered part of the normal gut fauna of olive fruit fly, the pioneering work of Petri did initiate a field of study that has occupied the attention of many fruit fly workers.

In summary, *Bactrocera* and related fruit flies are routinely associated with bacteria, and less frequently yeasts, as part of their gut microbial fauna in both the adult and larval stage. Bacteria can be picked up by foraging flies or larvae directly from their environment, or they can be passed inter-generationally via the eggs or through inoculation of the fruit via the ovipositor of the parental female. From as early as the 1930s it was recognised that at least some component of the microbial fauna was beneficial to the flies as a nutritional source. Through the 1970s to 1990s, using traditional microbial plating techniques and antimicrobial knockdown, it was further demonstrated that at least some of the microbes could be regarded as true symbionts, with reduced larval growth rates and adult fitness parameters if the microbial community was absent. The bacteria most commonly associated with *Bactrocera* belonged to the family Enterobacteriaceae, in the genera *Enterobacter, Klebsiella, Citrobacter, Erwinia, Escherichia, Pantoea, Pseudomonas* and *Serratia*. Literature from this period shows conflict between studies and unresolved questions. Using the same fly species, one study might show a positive relationship between microbial fauna and fruit fly nutrition, while a second study might not. Some authors believed the fruit fly/microbial relationship was a closely co-evolved symbiotic relationship, while others, based on the often high variability in the microbial communities recorded, that the relationship was not co-evolved but rather flies picked up bacteria randomly as part of their daily foraging.

The development in the last decade of sensitive molecular techniques for screening and identifying microbial communities has helped answer many of these earlier questions. As for most science, the answers lie between the two extremes of tight co-evolution and random ecological interactions.

As currently understood, there are clear examples within the Dacini of mutualistic relationships between flies and their gut microbial community. In the olive fruit fly, *Erwinia dacicola* is considered to form an inseparable, essential part of the fly's nutritional ecology by allowing the fly to utilise a food resource, the olive, which has low nitrogen levels. In this case the bacteria directly contribute essential amino acids and metabolise urea from bird droppings upon which the fly feeds. Similar beneficial relationships have been reported in *B. dorsalis* and *Z. cucurbitae*. Additionally, symbiotic bacteria have been shown to produce volatile metabolites that are attractive to male and female flies (Table 4.2).

However, also as shown in Table 4.2, some experiments have shown no fitness benefits of particular bacteria, or even negative effects. This is not unexpected given the huge diversity of gut microbial fauna being discovered in flies with molecular tests of increasing sophistication, such as pyrosequencing of entire microbial communities. Such studies are showing differences in microbial communities across all life stages of a single species,

Table 4.2. Examples of Dacini–microbial interactions from the recent literature.

Microbe	Fly species	Interaction	Reference
Enterobacter spp.	*Z. cucurbitae*	Bacteria added as a probiotic to larval diet significantly increased several fitness parameters	Yao *et al.* (2017)
Klebsiella oxytoca	*B. zonata*	Metabolites from the symbiont attracted male flies	Naaz *et al.* (2016)
Candidatus erwinia dacicola	*B. oleae*	Females whose bacterial community was suppressed had reduced fitness	Ben-Yosef *et al.* (2014)
Candidatus erwinia dacicola	*B. oleae*	Presence of the symbiont neither increased nor decreased mating probability	Estes *et al.* (2014)
Candidatus erwinia dacicola	*B. oleae*	Bacteria helped the fly larvae feed in unripe fruit by counteracting the inhibitory effect of oleuropein	Ben-Yosef *et al.* (2018)
Enterococcus phoeniculicola; *Lactobacillus lactis*	*B. dorsalis*	Diets enriched with *E.p.* positively influenced fly immature development; diets enriched with *L.l.* negatively influenced development	Khaeso *et al.* (2018)
Klebsiella oxytoca	*B. dorsalis*	Females with *K.o.* received more matings than females without	Damodaram *et al.* (2016)
Bacillus cereus, *Enterococcus faecalis*	*B. dorsalis*	Metabolites from the symbiont attracted male and female flies	Wang *et al.* (2014)
Citrobacter spp.	*B. dorsalis*	The symbiont increases pesticide resistance	Cheng *et al.* (2017)

between different parts of the gut of an individual, between different body parts (e.g. gonads versus gut), between different populations of a single species, and across species.

More detailed experiments, in both tephritids and other taxa, are also giving more detail about the mechanisms of the mutualistic relationships. In *B. minax*, for example, bacteria have significant fitness impacts through diet, but only when the adult diet is of poor quality. In this case bacteria themselves do not provide a protein source, but metabolise dietary components to provide essential amino acids. If the diet is not lacking this effect is not observed in fitness trials, but if the diet is poor the bacterial effect is significant. Such mechanistic based approaches to fruit fly/bacteria interactions help answer much of the conflict in the early literature.

4.3.4 Sheltering requirements

Bactrocera roost, predominantly inactive, for up to 80% of the day. Plants preferentially used for roosting have been identified for several species in different agro-ecosystems. In Hawaii castor bean (*Ricinus communis*), Christmas berry (*Schinus terebinthefolius*), ti (*Crodyline fruticose*), maize (*Zea mays*) and cassava (*Maniohot esculenta*) are some of the preferred roosting plants for *Z. cucurbitae*; while maize, cassava, panax (*Polyscias guilfoylei*) and papaya (*Carica papaya*) are some known roosting plants for *B. dorsalis*. In Réunion, *Z. cucurbitae*, *D. demmerezi* and *D. ciliates* will roost in maize, and the former two species prefer maize over Napier grass (*Pennisetum purpureum*) as a roosting site. The work in both Hawaii and Réunion has been driven largely by the application of protein bait spray for control of cucurbit pests (see Section 9.3.2), as these flies (particularly) are known to enter the crop largely for oviposition, and will roost, forage for protein and mate in surrounding non-crop vegetation. By manipulating border plantings so that they provide preferred roosting sites, while also fitting into the production landscape, protein bait spray can be applied more effectively.

   Although roosting is such an important part of a fly's daily activity, and important to know for pest management purposes, relatively little is known about what drives a fly's preference for different roosting sites: as such the search for preferred roosting plants is largely through trial and error. Some plants provide nutrients to the flies through extrafloral nectaries (e.g. castor bean), while maize pollen is a demonstrated source of protein and a likely attractant to maize border plantings. *Bactrocera tryoni* prefers to roost in tall plants over short plants; while roosting of *B. cacuminata* in wild tobacco is positively influenced by leaf density.

4.3.5 Host plant as centre of activity

The phrase 'host plant as centre of activity' has been used by R.A.I. Drew to describe the relationship between dacines and their larval host plant, and is commonly seen in older literature. Drew and colleagues postulated that all life activities (feeding, sexual maturation, mating, egg laying and larval development) of dacines happened on the larval host plant, an interaction not just mediated by the fruit, but also bacteria. Specifically, the theory proposed that gravid female flies would first arrive at a fruiting host plant to oviposit. While there they would inoculate the plant with leaf phylloplane bacteria upon which they would feed. The presence of bacteria would subsequently attract immature flies, which would also feed on the bacteria and sexually mature, after which mating would follow. The cycle could then be repeated on the same plant if it were still fruiting, or the now mature and gravid females could disperse to find a new plant.

   Experiments designed to explicitly test the 'centre of activity' hypothesis have met with mixed results. In field cage trials gravid females were more likely to respond to fruit cues than bacterial cues, but protein hungry females did respond to bacterial odours. In contrast, the level of mating and feeding observed on the host plant varies greatly between studies, and experimental tests have demonstrated that the presence of fruit on a plant is not a perquisite for mating in *B. tryoni* (the species upon which Drew developed his theory). Wild tobacco fly (the second species with which Drew worked) may mate exclusively on the host plant, but observations of feeding on wild tobacco plant are rare and plants used for mating are patchy within a wild tobacco stand, suggesting there is more to mate location than simply being on 'any' host plant.

The most significant weakness with the 'centre of activity' hypothesis has been failure to look for behaviours in other places. When in the field, fruit fly workers almost universally seek fruit flies on fruiting hosts, because we know we can find flies there. This does not mean they do not occur in other locations within the environment, but we are not sure where those are. There is no doubt that some feeding and mating will occur on the host plant, but these activities may also occur elsewhere. Phylloplane bacteria, for example, are not the only source of protein used by dacines, nor are they exclusively found on fruiting plants, and so feeding is likely to occur in other locations. The melon-infesting *Zeugodacus* are well documented to only use the host for oviposition, and carry out most other activities on non-host plants. Within a fruiting orchard the sex ratio of flies is female biased, but dacines have a 1:1 sex ratio – so what are all the other males doing? The 'centre of activity' hypothesis has not been widely adopted by the dacine and larger tephritid community and this is a reasonable view. While the role of bacteria in dacine biology is important, it does not follow that the presence of bacteria on a fruiting host negates the need for other resources that are elsewhere available in the environment.

4.4 Population Dynamics

4.4.1 Seasonal dynamics and effects of weather

In general, the *Bactrocera* are regarded as multivoltine insects, having multiple generations per year. *Bactrocera* (*Tetradacus*) *minax*, endemic to the Himalayas and southern China, comes from areas where cold winters are normal and it is a rare obligatory, univoltine *Bactrocera* species (see Section 4.4.2). It is unknown if other dacine species that are endemic to mid-altitude Himalayas (e.g. *D. fletcheri* and *D. dorjii*) are also univoltine, or if the characteristic is restricted to the subgenus *Tetradacus*.

For nearly all other *Bactrocera* species, however, multiple overlapping generations per year are the norm, with the number of generations observed being based on host availability (see Section 4.4.3) and temperature. To illustrate the role of temperature, and assuming hosts are always available, temperature-based development models for *B. tryoni* predict only one generation per year in temperate Tasmania, three or four generations in northern Victoria/southern New South Wales, and up to 12–15 generations in far northern Queensland. In most tropical and subtropical regions temperature is not considered limiting for *Bactrocera* species and an average of at least 12 generations per year is theoretically possible. Even for *B. oleae* in Greece, where cool to cold winters occur, four to five generations per year have been recorded.

Temperature is not the only weather variable to influence *Bactrocera* populations. Large-scale predictive climate models for several *Bactrocera* pest species reinforce the importance of rainfall and/or irrigation on the abundance and persistence of *Bactrocera* populations, with areas of low moisture being poor for *Bactrocera*. The mechanisms by which moisture influences *Bactrocera* population dynamics are not well understood, but likely include both direct and indirect effects. Direct effects will include the desiccation of adults, mortality of pupae through water loss (in dry soils) and drowning (in saturated soils), and increased mortality of emerging teneral adults in dry soil. Adult desiccation is poorly studied, but observations of *Bactrocera* adults locating themselves in local areas of high humidity suggest that excessive water loss is a problem for them. Indirect effects of low moisture will include reduced vegetation for sheltering, fewer fruiting host plants, and lower number and quality of fruits on those plants.

4.4.2 Overwintering

In most parts of the endemic ranges of *Bactrocera* and other dacines, how the flies 'overwinter' (i.e. survive winter months that have temperatures below activity thresholds) is irrelevant, just as winter itself is a largely irrelevant term (in the tropics, wet and dry seasons are more relevant). Nevertheless, at the northern and/or southern limits of some species' endemic or invasive ranges, winter temperatures do fall below activity thresholds and flies need to have behavioural and or physiological strategies to overcome this seasonally unfavourable period.

Overwintering as diapausing pupae seems uncommon in *Bactrocera*, although it is confirmed and characterised in *B. minax* and *B. oleae*. In the obligatory univoltine *B. minax*, pupal diapause is probably initiated early in larval development and, without appropriate chilling exposure, the pupae die without emerging if held at a constant 25°C: field chilling exposure of less than 30 days also results in high pupal mortality. With increasing duration of field chilling exposure, the development time and mortality of pupae returned to the laboratory and held under a constant 25°C decreased. Accumulated cold chilling is a diapause termination mechanism recorded in several temperate tephritids, and across many temperate insects in general.

While documented in *B. minax* and *B. oleae*, pupal diapause as an overwintering mechanism is otherwise rare or absent in the Dacini. Where studied, overwintering of *Bactrocera* is thought to occur in the adult stage. There are reports of *Bactrocera* populations aggregating to overwinter in sheltered areas of dense foliage, although how much of this is true aggregation (e.g. as driven by an aggregation pheromone or similar) versus many individuals locating and utilising the same suitable habitat is unclear. Similarly unclear is how regularly such behaviours occur, as the documented reports of such are rare. One report of overwintering *B. tryoni* recorded adults foraging for food on warm days, but otherwise the flies simply roosted throughout the winter.

For both *B. oleae* and *B. tryoni* it has been documented that females will undergo a reproductive diapause during winter. Again, how much of this is a true physiological diapause, versus a simple slowing of degree day accumulation is unclear. *Bactrocera tryoni* females are known to resorb already mature eggs during winter, which suggests a true physiological response to cold. The ability to survive temperatures well below normal activity thresholds in *Bactrocera* is related to cold acclimation. Flies exposed to gradually cooling temperatures (i.e. as occurs normally with the onset of winter) have a much greater cold hardiness and low temperature survival ability than flies that are rapidly exposed to a very cold temperature (i.e. as might be experienced through an early season frost).

4.4.3 'Bottom-up drivers' of population dynamics

In areas where temperatures and moisture are not limiting for *Bactrocera* (which are most endemic regions) host fruit availability is likely the single most important driver of population dynamics and abundance. In a study of *B. dorsalis* in India, 78% of variation in trap catch could be explained by a single host variable – the amount of small, immature guava fruit. Applying a large range of other host-related and weather variables in linear regression models did not increase this explanatory power. In Africa, seasonal abundance of *B. dorsalis* is closely linked to the fruiting of its two major commercial hosts, mango and guava.

Host plant abundance is strongly correlated with fruit fly abundance at the landscape level. In a study based in and around the city of Brisbane in Australia, *B. tryoni* abundance was significantly greater at suburban trapping sites than in sites increasingly distant from suburbia. This was despite the fact that these distant sites included rainforest, the supposed endemic habitat of the fly. When available host plants around the traps were quantified, it was shown that there were significantly more hosts available in urban areas than at other sites, and that the fruiting times of these hosts gave near year-round breeding resource for *B. tryoni*.

Another excellent example of the role of multiple hosts in maintaining flies within the larger landscape comes from Benin in West Africa. In a 5-year study of fruit flies in an agroecosystem dominated by cashew and mango, the size of yearly seasonal peaks in *B. dorsalis* numbers were positively correlated with 'high' and 'low' mango production years, illustrating the importance of that one crop on the fly. However, the entire system is much more complex than just 'mango'. The mango itself consisted of 15 varieties, which extended the crop's fruiting season over 5 months. Outside the mango season, the presence of the fly in the region was maintained by large numbers of lesser crops and wild hosts that provided continuous hosts, even if only infrequently used, throughout the rest of the year. Cashew apple then played another role in helping build up the local *B. dorsalis* population just prior to the mango crop. This example highlights the importance of primary and secondary crops, as well as non-crop plants, in driving and sustaining the population dynamics of polyphagous dacines.

4.4.4 'Top-down drivers' of population dynamics

In population dynamics parlance, top-down factors are considered to be predators and other natural enemies. The natural enemies of *Bactrocera* are covered in Chapter 8, and are only briefly introduced here within a population dynamics framework. The main top-down mortality agents of *Bactrocera* are hymenopteran parasitoids of eggs, larvae and pupae; vertebrate frugivores (birds, bats, rodents, etc.) that are indirect predators of larvae within fruit; and generalist invertebrate predators, especially ants, that prey on the pre-pupal maggots and emerging teneral adults.

For the most part these top-down regulators of *Bactrocera* population abundance are quite poorly studied, especially in agroecosystems where species are endemic and not invasive. Several of the parasitoid species have been deliberately introduced and released as classical biological control agents of invasive fruit flies (most notably in Hawaii which pioneered such work for the biocontrol of *B. dorsalis*, *Z. cucurbitae* and *C. capitata*) and the effects of these agents are reasonably well documented and in some cases they have achieved notable population reductions. In areas where *Bactrocera* are endemic (Asia, Australia, etc.) data on parasitism rates in both agricultural and native ecosystems are extraordinarily sparse. Those that are available show huge variance in the parasitism data, from lows of 10% parasitism in a sample to up to 80% in others.

Direct predation by ants, and indirect predation by vertebrate frugivores, may play an important role in regulating *Bactrocera* numbers, with some studies recording between 50% and 80% mortality by such agents. But again these studies are highly restricted and the inferences that can be drawn from them are limited for this reason.

In summary, the available information (covered more fully in Chapter 8) suggests that the top-down drivers of *Bactrocera* population dynamics may be important in reducing population numbers, but with the exception of a small number of classical biological control agents they are too poorly studied for their effects to be properly understood. Given this, the author is not aware of any previous or current effort to incorporate natural enemies into *Bactrocera* population dynamics models.

4.4.5 Immigration and emigration

Immigration and emigration feature heavily as drivers of population dynamics in the older *Bactrocera* literature, but increasingly the importance of these traditional population attributes is being questioned as usage of the terminology changes over time. At an individual orchard level flies will move in and out with fruiting cycle, and at this scale immigration or emigration plays a major role in population dynamics: this is the scale at which fruit fly population dynamics research in the 1960s and 1970s operated. However, modern usage of the terms immigration or emigration tends to focus more on long-distance movement, for example, the migration cycles of birds or whales. Local level movement, between different resource patches, is now captured as landscape dynamics and interpreted as local resource foraging rather than migration. It is not documented that any *Bactrocera* have regular long-distance immigration or emigration (in the sense of a seasonal migration pattern), and even the existence of a dedicated dispersal phase (notably the widely used concept of post-teneral dispersal) is debatable. Rather, at the local landscape level, immigration and emigration are unlikely to play a major role, if any role, in changing population numbers.

While dismissing the role of immigration or emigration in local population dynamics, movement is a key attribute of *Bactrocera* and understanding it is critical to their management as both endemic pests and off-shore threats. For this reason Chapter 7 contains a large section dedicated to *Bactrocera* movement.

4.4.6   Influence of climate change and global-scale climate variables

Despite their economic importance little has been done on understanding how global warming may change *Bactrocera* species distribution and abundance in the long term, or how global-scale climate variables (such as the El Nino/ La Nina pattern) may do the same in the medium term. Several studies have used the 'climate change' feature in distributional modelling programmes such as CLIMEX and MaxEnt, and these invariably show fruit flies extending their ranges either northwards or southwards of the equator in response to increasing average temperatures in cooler areas. However, this is a limited assessment of climate change response, which also needs to assess changes in host availability and phenology, how elevated CO_2 levels may change fruit quality, and the capacity of predominantly tropical species to cope with increased average temperatures.

Two studies have focused on how climate change may simultaneously influence the distribution and abundance of a fruit fly and its host, both using commercial olive (*Olea europaea*) and olive fruit fly as the models. One study focused on fine scale effects on the island of Sardinia, the second study (by the same authors) looked at larger scale geographic effects for the whole of Italy and California: I will focus here on the Sardinia study. In this relatively simple two-component system, predicted climate change scenarios were still complex. Increasing temperatures were predicted to see both olive and olive fruit fly become more common in the higher altitude mountain districts of Sardinia, but current high-risk areas for olive fruit fly on the coastal plains were predicted to become lower risk because rising temperatures would likely become partially lethal for olive fruit fly, but still be suitable for olive (because of differences in their temperature tolerances). Potential desertification on the southern coast of Sardinia is also likely to more dramatically impact on olive fruit fly than olive – although it could potentially affect both significantly. Further, a predicted earlier blooming time and reduced season length for olives may also negatively impact on olive fruit fly as it has a high-temperature induced reproductive dormancy. This high-temperature dormancy currently helps olive fruit fly match its main oviposition period with the olive crop phenology, but in the future it may put it out of sync with olive fruiting phenology as the crop will mature while the fly remains in dormancy. This one example well illustrates the complexity of interactions that other *Bactrocera* workers need to consider when making climate change predictions.

Only one *Bactrocera* study, another olive fruit fly study, has incorporated a global climatic indicator into long-term population models. In this case the North Atlantic Oscillation (NAO) was fitted as a model parameter for explaining population dynamics of olive fruit fly in Israel. Only in one site was the NAO a significant explanatory variable, but this was also the site with the longest continuous sampling period (5 years, versus 2–3 years for the other sites) and the shorter periods of sampling are likely to have been simply too short to pick up this longer term effect. Both the NAO and Southern Oscillation Index have been implicated in the population dynamics of South American *Anastrepha* species and more attention needs to be paid to them in analysing long-term *Bactrocera* data.

4.4.7   Unexplained dynamics

Despite all of the above, there is much we still do not know about the phenology of pest *Bactrocera*. For example, several polyphagous *Bactrocera* species (including *B. dorsalis* and *B. tryoni*) show a decline in abundance in 'winter months' (May to August in the southern hemisphere, December to March in the northern hemisphere), which appears to have nothing to do with cold-related population reduction. Similarly *B. oleae* has a peak of activity in the spring long before olive fruit is available that makes little biological sense. Clearly, much more needs to be done to understand *Bactrocera* phenology patterns.

4.5   Dacine Landscape and Community Ecology

4.5.1   Dacine communities

Despite the abundance and diversity of dacine species, little research has been done on dacine community ecology. In contiguous lowland rainforest in Papua New Guinea, *Bactrocera* communities have been identified as having high alpha diversity (i.e. high diversity in a local area), but low beta diversity (i.e. low changeover in species between sites). This pattern has also been reported over a large area of mainland South-east Asia, covering Cambodia, Thailand, Laos and China. This pattern was not seen, however, in Central Africa, where sampling of frugivorous tephritids (more than just dacines) from four locations along the Congo River were identified as having relatively low alpha diversity (only 29 species across seven genera collected in total), but high beta diversity between locations. In the relatively small geographic area sampled, this was considered surprising at it illustrated that despite the large amount of human trade traffic along the Congo for hundreds, if not thousands of years, these frugivorous insects had not been widely moved. In East Africa (Uganda) the composition of frugivorous fruit fly communities also differed significantly between three sampled regions, but in this case the sampled flies were nearly all pest species, and the differences could be correlated to the abundance of different crops and cropping styles in the three regions.

In both Asia and Australia human modified sites have been found to have high dacine abundance, but low diversity. Towns and agricultural systems exhibit a dominance of a small number of pest species that may be extremely abundant, while undisturbed native forests have much higher species diversity but most species are rarer, non-pest endemics. Studies in Asia, Australia and Papua New Guinea have all shown that these rarer fruit flies, which are commonly host specific to a small group of closely related rainforest taxa, rarely leave the rainforest and so there is commonly a distinct separation of dacine communities within a landscape based on rainforest host specialists, or pestiferous host generalists. Interestingly, the reverse also holds. Again reported several times from across *Bactrocera*'s native range, pest species are rarely trapped within rainforest. This has been shown both with endemic pest species (e.g. *B. dorsalis* and *Z. tau* in central South-east Asia, *B. tryoni* in Australia) and with these species in invasive parts of their range. Thus *B. dorsalis* was not collected in lowland rainforest in Papua New Guinea or Australia, despite being hugely abundant only a few kilometres away in cleared agricultural county or in nearby urban areas such as Madang (PNG) and Cairns (Australia). Why host specialist flies may not leave the rainforest that is their only source of larval host(s) is relatively easy to understand; why polyphagous pest species will not enter rainforests where potential hosts are available is less obvious.

The pattern of urban–forest divide was not seen in the Congo study, where the four locations each consisted of paired sites: one site in undisturbed rainforest and the second in village agroecosystems. Only in two of the locations was there a difference between the disturbed and undisturbed sites.

The ability to rapidly collect many dacine species in simple and inexpensive traps has been identified as a feature that makes dacines highly suitable for assessing ecosystem health over time and space. In environmentally stable native forest in Papua New Guinea, male-lure catches in modified Steiner traps made several years apart had a high degree of similarity, reinforcing the value of *Bactrocera* as organisms for studying both long and short-term environmental changes. While as fruit fly workers we commonly focus on the few pest species and ignore the ecological value of dacine flies, it is important to recognise that analysis of dacine communities has played a role in questions as important as 'how many species are there on the planet?'.

4.5.2 Landscape ecology of individual species

The field of landscape ecology examines where and why individual species, or communities of species, are distributed in the environment. Fundamentally this is something that ecologists have always attempted to research and explain, but the field has become much more formalised over the last 20 years as new spatially explicit ecological theory has developed and computer-assisted mapping tools have become available. The big change in landscape ecology has been the increasing focus on real landscapes, mapped into geographic information systems (GIS), with research outcomes used directly for regional management. In entomology such management can include biodiversity conservation, manipulating resources for crop pollinators to enhance ecosystem pollination services, or identifying how an agricultural landscape might be modified to disadvantage pests while aiding natural enemies. With the exception of recent work with olive fruit fly, little 'modern' landscape ecology has been applied to the dacines. This is surprising as landscape-level approaches are essential for area-wide management and the SIT. For these reasons, I expect landscape ecology to be a growth field of research for dacines over the next decade, and several large research projects are currently underway.

This section is arbitrarily divided below into two sections, 'old' and 'new' landscape ecology. The 'old' section discusses generic landscape attributes that are known to influence dacines, whereas the 'new' section focuses on recent olive fruit fly work to illustrate what can be achieved with advances in this field. As for much of this book, the discussion below focuses almost exclusively on the pest species because there is essentially no data available on how locally endemic, non-pest dacines are located in the environment, other than being restricted to rainforests or not.

'Old landscape ecology'

At large spatial scales, the polyphagous dacine flies occur widely over landscapes. Lure trapping can capture pest flies in nearly all habitat types within a landscape, including what is known as landscape 'matrix', which are habitats that offer no resources (e.g. food or sheltering sites) for the organism under study. Grasslands between forest or agricultural production areas are a typical matrix habitat for fruit flies. The capture of flies within matrix may represent captures of dispersing flies, or simply reflect the strong pulling power of the male lures.

Typically, flies are most abundant in areas of high larval resources. Urban environments, be they large cities in Australia or small villages in Malaysia, have been shown to have higher numbers of pest fruit flies than surrounding agricultural or forest areas because of the higher density of host plants, which are often not protected from pests as they would be in a commercial production setting. High densities of *B. tryoni* in rural towns of inland Australia have also been attributed not just to fruit availability, but also to higher local humidity and the availability of free water due to local irrigation (i.e. people watering their gardens). Within a larger landscape, dacines are reported to be more likely to be associated with waterways versus open, less sheltered areas, but there is actually little hard data to support this often repeated statement. In a study in subtropical Australia, large numbers of dacines were captured during mid-summer in remnant eucalypt forest within a horticultural production landscape. It is assumed that these flies were sheltering in the forest to avoid the high summer heat.

Dispersion from a local resource area (i.e. a fruiting tree or orchard) to other localities in the environment may be associated with an innate dispersal of newly emerged adults, with the search for new fruiting plants, or in temperate areas with the seeking of long-term roosting sites for overwintering, but there is a high degree of uncertainty in all of these behaviours. For example, it is unclear if there is an innate dispersive stage of new adults and how far insects might forage to find a new fruiting host plant. In Hawaii, released *Z. cucurbitae* moved only short distances (<2 km) within an agricultural landscape so long as oviposition resources and adult roosting sites were available. A detailed discussion on *Bactrocera* movement is given in Section 7.1.3.

At fine spatial scales, evidence suggests that individual dacines are more likely to shelter in dense canopies at height over low-lying or open canopies, and this may be correlated with their dispersion and local abundance within an individual crop or orchard. For example, *B. tryoni* was found to be evenly dispersed within a full canopied apple orchard, but edge effects were detected in strawberry fields surrounded by taller vegetation: it was assumed that flies were roosting in the edge vegetation. Cucurbit infesting *Dacus* and *Zeugodacus* species show strong edge effects, with roosting, resource foraging and mating occurring away from the crop, and entry into the crop largely restricted to ovipositing females. Within a tree, flies will spend more time in trees with fruit than without and they have a behavioural foraging pattern that sees them eventually move to the tops of trees. At dusk it is assumed that most dacines aggregate to mate, and this local aggregation will modify their fine-scale distribution and abundance: being able to predict aggregation sites would be immensely useful for SIT releases but we do not yet have that knowledge.

'New landscape ecology'

Several groups in the Mediterranean region have applied modern landscape ecology tools to understanding the distribution and abundance of olive fruit fly. The fly data being inputted is of a type available to all fruit fly researchers (i.e. trap catch and fruit infestation levels, location of trap and date of trap), but the analysis of this data in conjunction within GIS surface-layer data (e.g. altitude, aspect, amount of olive groves with a certain radius of the trap, forest area, habitat fragmentation, etc.) provides much greater information than can trap catch alone.

In northern Greece, spatial analysis of trap catch over time in one intensively managed olive district of approximately 3300 ha, with the only additional input being altitude of each trap site, allowed visualisation of the spatial changes of fly populations over the management season and identification of 'at risk' areas. Altitude was shown to play a significant role in the temporal and spatial dynamics of olive fruit fly, with significant aggregations of the flies during summer occurring at mid- to high-altitude sites, and low numbers of flies found in valleys. In autumn the pattern reversed, with most flies found at lower altitudes. This information helped inform the design of spatially and temporally 'clever' monitoring programmes that identified what parts of the region were at most risk and needed to maximise sampling effort at different times of the season.

Spatial analysis and GIS tools have also been applied to olive fruit fly in Liguria, a region in northern Italy that runs in an arc along the Mediterranean coast. This region is topographically diverse, and in a distance of only 34 km from the coast rapidly rises from coastal areas with hot and humid summers, to the Ligurian Alps that at over 2000 m elevation have true alpine winters: olive growing extends from sea-level to 300 m elevation. Attempts to develop accurate, cumulative day degree models for predicting olive fruit fly flight period, oviposition and overwintering emergence are greatly hampered by this great variation in topography. The inclusion of three GIS surface-layer data sets (elevation, distance from sea, aspect), with appropriate validation, allowed the development of a day degree model that allows site-specific predictions of the targeted olive fruit fly behavioural traits at an error level of <10%, sufficient to allow site-specific pest management recommendations.

GIS tools are now routinely applied as part of surveillance and monitoring programmes against invasive tephritids in many parts of the world, or as part of SIT programmes. Nevertheless, the application of spatial analysis tools, both software and hardware, to understand the landscape ecology of dacines is still in its infancy and must be a research growth area of the future.

4.6   Further Reading and References Cited

Alyokhin, A.V., Mille, C., Messing, R.H. and Duan, J.J. (2001) Selection of pupation habitats by Oriental fruit fly larvae in the laboratory. *Journal of Insect Behavior*, 14, 57–67.

Atiama-Nurbel, T., Deguine, J.-P. and Quilici, S. (2012) Maize more attractive than Napier grass as non-host plants for *Bactrocera cucurbitae* and *Dacus demmerezi*. *Arthropod-Plant Interactions*, 6, 395–403.

Balagawi, S., Jackson, K. and Clarke, A.R. (2014) Resting sites, edge effects and dispersal of a polyphagous *Bactrocera* fruit fly within crops of different architecture. *Journal of Applied Entomology*, 138, 510–518.

Ben-Yosef, M., Pasternak, Z., Jurkevitch, E. and Yuval, B. (2014) Symbiotic bacteria enable olive flies (*Bactrocera oleae*) to exploit intractable sources of nitrogen. *Journal of Evolutionary Biology*, 24, 2695–2705.

Ben-Yosef, M., Pasternak, Z., Jurkevitch, E. and Yuval, B. (2018) Symbiotic bacteria enable olive fly larvae to overcome plant defences. *Royal Society Open Sciences*, 2, 150–170.

Castrignanò, A., Boccaccio, L., Cohen, Y., Nestel, D., Kounatidis, I., Papadopoulos, N.T., De Benedetto, D. and Mavragani-Tsipidou, P. (2012) Spatio-temporal population dynamics and area-wide delineation of *Bactrocera oleae* monitoring zones using multi-variate geo-statistics. *Precision Agriculture*, 13, 421–441.

Chang, C.L., Caceres, C. and Jang, E.B. (2004) A novel liquid larval diet and its rearing system for Melon fly, *Bactrocera cucurbitae* (Diptera: Tephritidae). *Annals of the Entomological Society of America*, 97, 524–528.

Chang, C.L., Vargas, R.I., Caceres, C., Jang, E. and Cho, I.K. (2006) Development and assessment of a liquid diet for *Bactrocera dorsalis* (Diptera: Tephritidae). *Annals of the Entomological Society of America*, 99, 1191–1198.

Cheng, D., Guo, Z., Riegler, M., Xi, Z., Liang, G. and Xu, Y. (2017) Gut symbiont enhances insecticide resistance in a significant pest, the oriental fruit fly *Bactrocera dorsalis* (Hendel). *Microbiome*, 5, 13.

Clarke, A.R., Merkel, K., Hulthen, A.D. and Schwarzmueller, F. (2019) *Bactrocera tryoni* (Froggatt) (Diptera: Tephritidae) overwintering: an overview. *Austral Entomology*, 58, 3–8.

Damodaram, K.J.P., Ayyasamy, A. and Kempraj, V. (2016) Commensal bacteria aid mate-selection in the fruit fly, *Bactrocera dorsalis. Microbial Ecology*, 72, 725–729.

Dimou, I., Koutsikopoulos, C., Economopoulos, A.P. and Lykakis, J. (2003) Depth of pupation of the wild olive fruit fly, *Bactrocera* (*Dacus*) *oleae* (Gmel.) (Dipt., Tephritidae), as affected by soil abiotic factors. *Journal of Applied Entomology*, 127, 12–17.

Dong, Y.C., Wang, Z.J., Clarke, A.R., Pereira, R., Desneux, N. and Niu, C.-Y. (2013) Pupal diapause development and termination is driven by low temperature chilling in *Bactrocera minax. Journal of Pest Science*, 86, 429–436.

Dong, Y., Wan, L., Pereira, R., Desneux, N. and Niu, C.-Y. (2014) Feeding and mating behaviour of Chinese citrus fly *Bactrocera minax* (Diptera, Tephritidae) in the field. *Journal of Pest Science*, 87, 647–657.

Drew, R.A.I. (1987) Reduction in fruit fly (Tephritidae: Dacinae) populations in their endemic rainforest habitat by frugivorous vertebrates. *Australian Journal of Zoology*, 35, 283–288.

Ekesi, S., Nderitu, P.W. and Rwomushana, I. (2006) Field infestation, life history and demographic parameters of the fruit fly *Bactrocera invadens* (Diptera: Tephritidae) in Africa. *Bulletin of Entomological Research*, 96, 379–386.

Ekesi, S., Mohamed, S.A. and Chang, C.L. (2014) A larval liquid diet for rearing *Bactrocera invadens* and *Ceratitis fasciventris* (Diptera: Tephritidae). *International Journal of Tropical Insect Science*, 34 (No. S1), S90–S98.

Estes, A.M., Segura, D.F., Jessup, A., Wornoayporn, V. and Pierson, E.A. (2014) Effect of the symbiont *Candidatus* Erwinia dacicola on mating success of the olive fly *Bactrocera* oleae (Diptera: Tephritidae). *International Journal of Tropical Insect Science*, 34, S123–S131.

Fitt, G.P. (1990) Variation in ovariole number and egg size of species of *Dacus* (Diptera; Tephritidae) and their relation to host specialization. *Ecological Entomology*, 15, 255–264.

Fletcher, B.S. (1973) The ecology of a natural population of the Queensland fruit fly, *Dacus tryoni IV*. The immigration and emigration of adults. *Australian Journal of Zoology*, 21, 541–565.

Fletcher, B.S. (1975) Temperature-regulated changes in the ovaries of overwintering females of the Queensland fruit fly, *Dacus tryoni. Australian Journal of Zoology*, 23, 91–102.

Fletcher, B.S. (1979) The overwintering survival of adults of the Queensland fruit fly, *Dacus tryoni*, under natural conditions. *Australian Journal of Zoology*, 27, 403–411.

Fletcher, B.S. (1989) Temperature–development rate relationships of the immature stages and adults of tephritid fruit flies. In: AS Robinson and G Hooper (eds.) *World Crop Pests. Fruit Flies, Their Biology, Natural Enemies and Control Vol. 3A*. Elsevier, Amsterdam, The Netherlands, pp. 273–289.

Han, P., Wang, X., Niu, C.-Y., Dong, Y.-C., Zhu, J.-Q. and Desneux, N. 2011. Population dynamics, phenology, and overwintering of *Bactrocera dorsalis* (Diptera: Tephritidae) in Hubei Province, China. *Journal of Pest Science*, 84, 289–295.

Hanife, G. (2008) Modified agar-based diet for small scale laboratory rearing of Olive fruit fly, *Bactrocera oleae* (Diptera: Tephritidae). *Florida Entomologist*, 91, 651–656.

Harwood, J.F., Chen, K., Mueller, H.-G., Wang, J.-L., Vargas, R. and Carey, J.R. (2013) Effects of diet and host access on fecundity and lifespan in two fruit fly species with different life-history patterns. *Physiological Entomology*, 38, 81–88.

Huang, K. Y.-B., Atlihan, R., Gökçe, A., Huang, J.Y.-B. and Chi, H. (2016) Demographic analysis of sex ratio on population growth of *Bactrocera dorsalis* (Diptera: Tephritidae) with discussion of control efficacy using male annihilation. *Journal of Economic Entomology*, 109, 2249–2258.

Jayanthi, P.D.K. and Verghese, A. (2011) Host-plant phenology and weather based forecasting models for population prediction of the oriental fruit fly, *Bactrocera dorsalis* Hendel. *Crop Protection*, 30, 1557–1562.

Khaeso, K., Andongma, A.A., Akami, M., Souliyanonh, B., Zhu, J., Krutmuang, P. and Niu, C.-Y. (2018) Assessing the effects of gut bacteria manipulation on the development of the oriental fruit fly, *Bactrocera dorsalis* (Diptera; Tephritidae). *Symbiosis*, 74, 97–105.

Khan, M. (2013) Potential of liquid larval diets for mass rearing of Queensland fruit fly, *Bactrocera tryoni* (Froggatt) (Diptera: Tephritidae). *Australian Journal of Entomology*, 52, 268–276.

Khanh, L.D., Hien, N.T.T., Trang, V.T., Toan, T.T. and Rull, J. (2014) Basic biology and artificial rearing of *Bactrocera pyrifoliae* (Diptera: Tephritidae), a pest of peaches and plums in northern Vietnam. *International Journal of Tropical Insect Science*, 34, S148–S153.

Kounatidis, I., Papadopoulos, N.T., Mavragani-Tsipidou, P., Cohen, Y., Tertivanidis, K., Nomikou, M. and Nestel, D. (2008) Effect of elevation on spatio-temporal patterns of olive fly (*Bactrocera oleae*) populations in northern Greece. *Journal of Applied Entomology*, 132, 722–733.

Leblanc, L., San Jose, M., Wright, M.G. and Rubinoff, D. (2016) Declines in biodiversity and the abundance of pest species across land use gradients in Southeast Asia. *Landscape Ecology*, 31, 505–516.

McQuate, G.T. and Vargas, R.I. (2007) Assessment of attractiveness of plants as roosting sites for the melon fly, *Bactrocera cucurbitae,* and oriental fruit fly, *Bactrocera dorsalis*. *Journal of Insect Science*, 7, 57.

Mwatawala, M.W., De Meyer, M., Makundi, R.H. and Maerere, A.P. (2006) Seasonality and host utilization of the invasive fruit fly, *Bactrocera invadens* (Dipt., Tephritidae) in central Tanzania. *Journal of Applied Entomology*, 130, 530–537.

Naaz, N., Choudhary, J.S., Prabhakar, C.S., Moanaro and Maurya, S. (2016) Identification and evaluation of cultivable gut bacteria associated with peach fruit fly, *Bactrocera zonata* (Diptera: Tephritidae). *Phytoparasiticam* 44, 165–176.

Novotny, V., Miller, S.E., Hulcr, J., Drew, R.A.I., Basset, Y., Janda, M., Setliff, G.P., Darrow, K., Stewart, A.J.A., Auga, J., Isua, B., Molem, K., Manumbor, M., Tamtiai, E., Mogia, M. and Weiblen, G.D. (2007) Low beta diversity of herbivorous insects in tropical forests. *Nature*, 448, 692–697.

Ordano, M., Engelhard, I., Rempoulakis, P., Nemny-Lavy, E., Blum, M., Yasin, S., Lensky, I.M., Papadopoulos, N.T. and Nestel, D. (2015) Olive fruit fly (*Bactrocera oleae*) population dynamics in the Eastern Mediterranean: influence of exogenous uncertainty on a monophagous frugivorous insect. *PLoS ONE*, 10(5), e0127798.

Petacchi, R., Marchi, S., Federici, S. and Ragaglini, G. (2015) Large-scale simulation of temperature-dependent phenology in wintering populations of *Bactrocera oleae* (Rossi). *Journal of Applied Entomology*, 139, 496–509.

Ponti, L., Cossu, Q.A. and Gutierrez, A.P. (2009) Climate warming effects on the *Olea europaea–Bactrocera oleae* system in Mediterranean islands: Sardinia as an example. *Global Change Biology*, 15, 2874–2884.

Prabhu, V., Perez-Staples, D. and Taylor, P.W. (2008) Protein: Carbohydrate ratios promoting sexual activity and longevity of male Queensland fruit flies. *Journal of Applied Entomology*, 132, 575–582.

Raghu, S., Clarke, A.R., Drew, R.A.I. and Hulsman, K. (2000) Impact of habitat modification on the distribution and abundance of fruit flies (Diptera: Tephritidae) in South-east Queensland. *Population Ecology*, 42, 153–160.

Sookar, P., Alleck, M., Ahseek, N., Permalloo, S., Bhagwant, S. and Chang, C.L. (2014) Artificial rearing of the peach fruit fly *Bactrocera zonata* (Diptera: Tephritidae). *International Journal of Tropical Insect Science*, 34 (No. S1), S99–S107.

Tsitsipis, J.A. (1975) An improved method for the mass rearing of the olive fruit fly *Dacus oleae* (Gmel.) (Diptera, Tephritidae). *Zeitschrift fur Angewandte Entomologie*, 83, 419–426.

Vargas, R.I. and Nishida, T. (1985) Life history and demographic parameters of *Dacus latifrons* (Diptera: Tephritidae). *Journal of Economic Entomology*, 78, 1242–1244.

Vargas, R.I., Miyashita, D., Nishida and T. (1984) Life history and demographic parameters of three laboratory-reared tephritids (Diptera: Tephritidae). *Annals of the Entomological Society of America*, 77, 651–656.

Vayssières, J.–F., De Meyer, M., Ouagoussounon, I., Sinzogan, A., Adandonon, A., Korie, S., Wargui, R., Anato, F., Houngbo, H., Didier, C., De Bon, H. and Goergen, G. (2015) Seasonal abundance of mango fruit flies (Diptera: Tephritidae) and ecological implications for their management in mango and cashew orchards in Benin (Centre and North). *Journal of Economic Entomology*, 108, 2213–2230.

Vijaysegaran, S., Walter, G.H. and Drew, R.A.I. (1997) Mouthpart structure, feeding mechanisms, and natural food sources of adult *Bactrocera* (Diptera: Tephritidae). *Annals of the Entomological Society of America*, 90, 184–201.

Vijaysegaran, S., Walter, G.H. and Drew, R.A.I. (2002) Influence of adult diet on the development of the reproductive system and mating ability of Queensland fruit fly *Bactrocera tryoni* (Froggatt) (Diptera: Tephritidae). *Journal of Tropical Agriculture and Food Science*, 30, 119–136.

Virgilio, M., Backeljau, T., Emeleme, R., Juakali, J.L. and De Meyer, M. (2011) A quantitative comparison of frugivorous tephritids (Diptera: Tephritidae) in tropical forests and rural areas of the Democratic Republic of Congo. *Bulletin of Entomological Research*, 101, 591–597.

Wang, H., Jin, L., Peng, T., Zhang, H., Chen, Q. and Hua, Y. (2014) Identification of cultivable bacteria in the intestinal tract of *Bactrocera dorsalis* from three different populations and determination of their attractive potential. *Pest Management Science*, 70, 80–87.

Weldon, C.W. and Taylor, P.W. (2011) Sexual development of wild and mass-reared male Queensland fruit flies in response to natural food sources. *Entomologia Experimentalis et Applicata*, 139, 17–24.

Wu, B., Shen, K., An, K., Huang, J. and Zhang, R. (2011) Effect of larval density and host species on preimaginal development of *Bactrocera tau* (Diptera: Tephritidae). *Journal of Economic Entomology*, 104, 1840–1850.

Yao, M., Zhang, H., Cai, P., Gu, X., Wang, D. and Ji, Q. (2017) Enhanced fitness of a *Bactrocera cucurbitae* genetic sexing strain based on the addition of gut-isolated probiotics (Enterobacter spec.) to the larval diet. *Entomologia Experimentalis et Applicata*, 162, 197–203.

Yonow, T., Zalucki, M.P., Sutherst, R.W., Dominiak, B.C., Maywald, G.F., Maelzer, D.A. and Kriticos, D.J. (2004) Modelling the population dynamics of the Queensland fruit fly, *Bactrocera (Dacus) tryoni*: A cohort-based approach incorporating the effects of weather. *Ecological Modelling*, 173, 9–30.

5 Reproductive Biology and Mating Behaviour

All dacines have separate males and females, which need to locate each other and mate in order for the female to produce fertile eggs. Males and females emerge from the pupal stage as sexually immature and need to undergo sexual maturation, a period lasting from 10 to 30 days, before mating can commence. The mechanisms by which the sexes find and recognise each other are complex, and mediated in many species by plant-derived phytochemicals, the so-called 'male lures'. Males mate with high frequency if females are available, while female re-mating is less frequent but certainly occurs. Females produce eggs throughout their lives, although daily egg production peaks at around 3 to 4 weeks after emergence.

Several important control strategies for the dacines rely on manipulating the mating system. The sterile insect technique (SIT) manipulates female access to fertile male mates; protein-bait spray technology targets both sexes need for protein during the sexual maturation period; whereas the male annihilation technique targets the male's need for male lures (see Chapter 9). Because mating and reproduction are so intimately tied to management, this chapter discusses these aspects of the flies' biology in detail. The chapter starts with a description of *Bactrocera* reproductive morphology, then describes the mating systems by which males and females find each other in the environment, and then describes individual components of the courtship process. The chapter concludes with a lengthy discussion of the male lures, one of the most unique components of dacine biology.

5.1 Reproductive Morphology

5.1.1 Males

The male reproductive system (Fig. 5.1) is a typical dipteran reproductive system with paired testes and accessory glands, a common ejaculatory duct and a

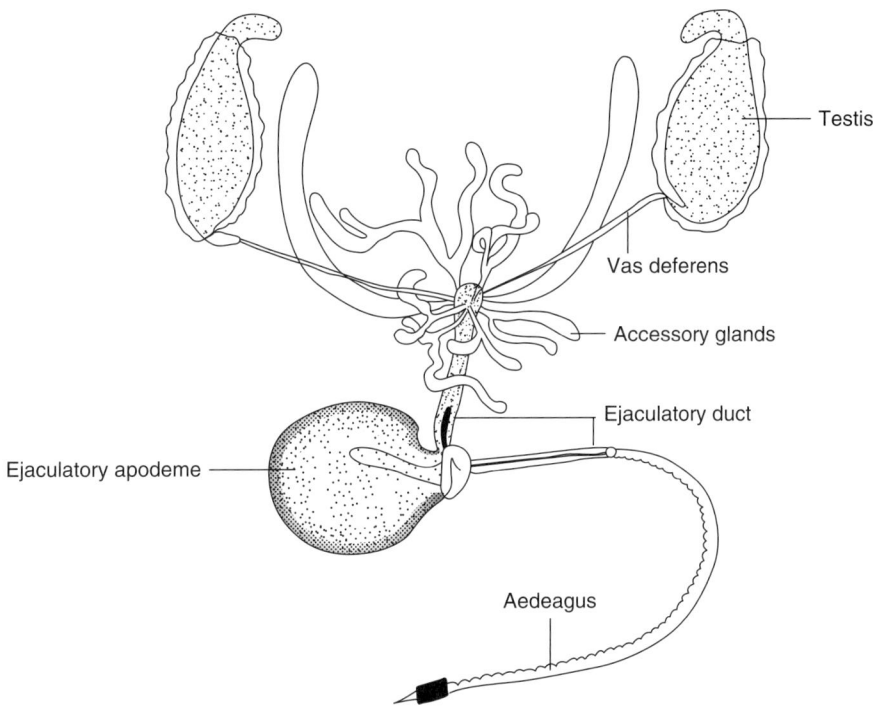

Fig. 5.1. Male reproductive system of *Bactrocera tryoni* (redrawn from Drew, 1969).

single aedeagus. The ejaculatory apodeme and other cuticular components of the male genitalia in *B. tryoni* and *B. cacuminata* have been demonstrated to grow during the sexual maturation period, continuing to expand and change in size until maturation is reached, when size and shape then remain fixed. This can potentially be used as a way of ageing immature male flies, but there is so much potential variation in maturation rate this may be highly problematic.

The length of the male aedeagus has been used as a species diagnostic marker in *Bactrocera* systematics, but it can show significant differences between populations of the same species and should be considered an unreliable character state (Fig. 5.2). That populations of *B. dorsalis* and *Z. cucurbitae* have been reported to have significantly variable aedeagus lengths is an unusual biological attribute. Stabilising selection tends to act on male genitalia to make them the same length across individuals within a species, such that male genitalia length is not correlated with changes in individual body size. An explanation of why male genitalia length is variable within a *Bactrocera* species has not been postulated.

5.1.2 Females

The mature female reproductive system is shown in Fig. 5.3. As in most insects, the female has paired ovaries, each consisting of multiple polytrophic ovarioles that produce the eggs. When mature, the ovaries fill most of the abdominal

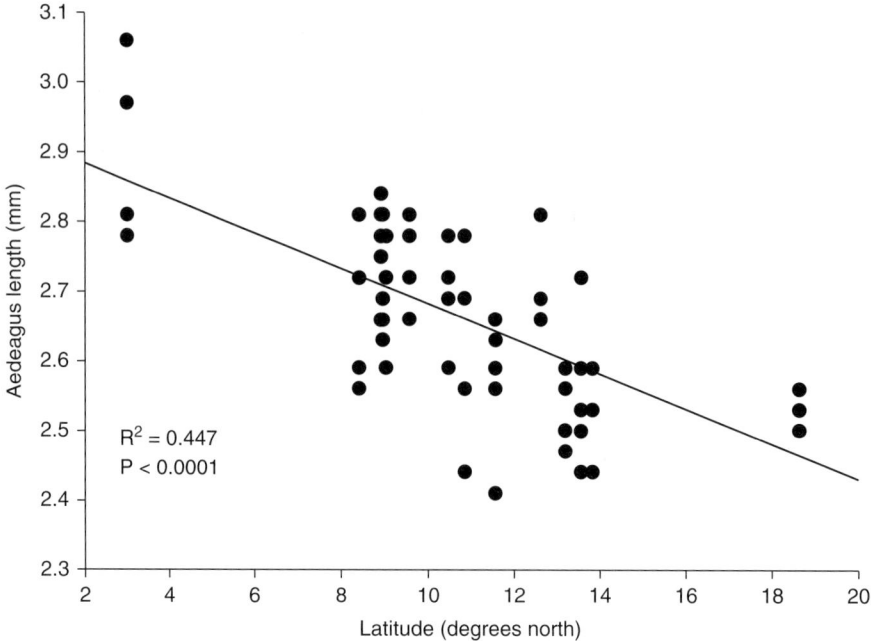

Fig. 5.2. Male aedeagal length of *Bactrocera dorsalis* illustrating a clinal reduction in length with increasing latitude (redrawn from Krosch *et al*., 2013).

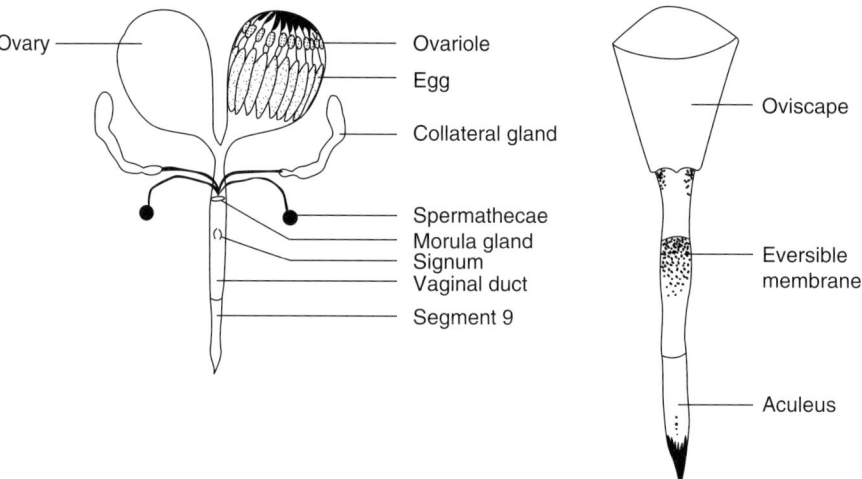

Fig. 5.3. Female reproductive system of *Bactrocera tryoni* (redrawn from Drew, 1969).

cavity. The number of ovarioles is variable across and within species and has shown to be correlated with larval host range size, with polyphagous species having more ovarioles and monophagous species fewer. In a study of 14 *Bactrocera* and *Zeugodacus* species, Fitt (1990) recorded mean ovariole number/ovary to range from 8.0 and 12.5 in the monophagous species *B. visenda* and *B. aglaiae*, respectively, to 39.1 and 37.7 in the polyphagous *B. tryoni* and *B. neohumeralis*, respectively. In the same study Fitt recorded significant difference in mean egg size across eight species, from 0.93 mm long for *B. krausii* to 1.34 mm long for *Z. cucumis*. Egg size and volume were closely correlated within a species, but egg length and body size were not correlated within or across species. Female *Bactrocera* lay more eggs when they are younger than when they are older, with maximum fecundity reached at approximately 3 weeks of age in both *B. dorsalis* and *B. tryoni*. Flies keep laying eggs until their death, even if fewer eggs are laid in old age.

Each ovary has a short lateral oviduct that joins into the common oviduct, which becomes the vagina and then the vaginal duct. The vaginal duct itself attaches to segment 9 of the ovipositor (see next paragraph). Entering the vagina are the spermathecal ducts from the paired spermatheca, which store the male sperm after mating. The spermatheca are small and black and easily observed in dissection. Crushing the spermatheca in saline will release the sperm, which can be seen under a microscope as a test for mating.

The ovipositor consists of abdominal segments 7, 8 and 9. In all three segments the tergal and sternal plates are highly modified and are tightly fused and the three segments effectively operate as single morphological units. Segment 7 is the oviscape and makes up most of what is seen of the external part of the female ovipositor when a female is not actively probing or laying eggs. Segment 8 is the eversible membrane, being long and tubular in shape and this is what provides the ovipositor with its length when extended. The eversible membrane is strengthened by two lateral, sclerotised bands, and is covered with sclerotised 'teeth' on the posterior two-thirds. The role of the teeth is unclear, but may be associated with helping maintain the position of the ovipositor when it is extended in fruit, or alternatively may play a role in keeping the ovipositor in place when held retracted within the body. The structure of the teeth is variable along the length of the eversible membrane, and also across species. The shape of the teeth as examined under scanning electron microscopy are used as taxonomic characters, but like any character state care needs to be taken to assess within and between species variation. The former *B. papayae* and *B. philippinensis* were initially separate based on teeth shape, but were later synonymised when this character was found not to vary between the taxa.

Segment 9, the terminal part of the ovipositor is the aculeus   this is the sclerotised component of the ovipositor that penetrates the fruit. Across the tephritids the shape of the aculeus is highly variable, but in the *Bactrocera* species so far studied it is dorsoventrally flattened but otherwise needle like, terminating in a highly sclerotised sharp tip. Blunting of the aculeus tip is referred to frequently in the tephritid literature and has been illustrated for *Ceratitis capitata* and some *Rhagoletis* species. In the literature, blunting is considered a negative attribute as it may limit access of tephritids to some hosts with tough peel or dense flesh. In a study on *B. tryoni*, it was found aculeus blunting did not occur in that species. Aculeus wear did occur over the life of a fly, and was detected in field populations, but for this species wear took the form of the aculeus becoming shorter and narrower (i.e. the two sides of the aculeus wore away), but the tip became no less sharp.

The gonopore, where the male aedeagus enters the female during mating, is located ventrally on segment 9. At mating the male aedeagus enters the partially extended ovipositor through the gonopore, traverses the rest of the aculeus and enters the vaginal duct, before coming to rest at the vagina where ejaculated sperm can enter the spermathecal ducts.

5.1.3   Diet and reproductive development

Reproductive development in *Bactrocera* is heavily influenced by post-teneral nutrition. Flies will die rapidly if not provided with a carbohydrate source, while lack of protein stops sexual maturation in females and males. Even limited or interrupted protein sources can delay maturation and reduce egg production. In the field it is unlikely that protein and sugar is limiting as flies actively forage for these resources, which are thought to be obtained from floral exudates, leaf surface bacteria, bird droppings and plant leachates. Nevertheless, when mass-rearing flies for SIT the prompt and adequate supply of nutritional resources to post-teneral flies can greatly affect their fitness. Provision of protein (in the form of yeast hydrolysate) and sugar to male flies of 24–48 hours old can increase their rate of sexual maturation, mating probability, copulation duration and longevity, while decreasing the re-mating propensity of females with which they mate. Nutrition thus needs to be carefully considered when manipulating flies for the SIT.

5.2   Mating Systems

Despite its importance to the SIT, surprisingly little is known about the mating behaviour of *Bactrocera* and the related genera. Within *Bactrocera*, aspects of mating behaviour have been recorded for *B. dorsalis*, *B. carambolae*, *B. minax*, *B. oleae*, *B. tryoni*, *B. neohumeralis* and *B. cacuminata*, while knowledge of the other genera are restricted to *Z. cucurbitae* and *D. longistylus*. For no single species has the full sequence of mating, from how potential partners find each other in the field to close-range courtship and mating been documented; although the work on *B. oleae* and *B. tryoni* comes close. As far as is known, the mating system of *Z. cucurbitae* appears similar to the studied *Bactrocera* species, but the mating of *D. longistylus* is quite different and more related to some *Phytalmia* species.

In many aspects of fruit fly research, for example in understanding general processes of host fruit location or a fly's response to field control strategies, the utilisation of information from other species within a genus, or even from other genera, can be done with some degree of confidence. This is because many aspects of the biology and management of the frugivorous tephritids appear to be shared across genera and species. However, this approach is fraught with danger if applied to mating behaviour. By most species definitions, the mating behaviour of a species is unique to that species. Even closely related species can have quite different mating behaviours and, even in the few dacine species studied to date, there are significant differences between them. Despite this, there is a tendency in the tephritid literature to draw general conclusions about all tephritid mating based on detailed studies of only a few species. This approach will not be followed here.

5.2.1 Introduction to mating systems

The mating system of an organism is a human-defined concept that covers behaviours involved in how males and females find each other in the landscape (mate location), through to (in some cases) the individual selection of one mate over another (sexual selection). In the context of pattern versus process, the mating system can be considered a pattern and it can be a useful way of describing generalised mating behaviour, without requiring knowledge of specific processes. In contrast, courtship and copulation (also known by some authors as the specific mate recognition system) is a process-based understanding of mating. In stark contrast to mating systems, which may be (and are) shared by many different species, the process of courtship and specific mate recognition is species specific. Using process-based knowledge to describe the mating system of a fly can be done with accuracy, but assuming specific processes occur because a particular type of mating system has been applied to a species can lead to error.

In this section I will describe the mating system terminology that has been applied to dacines, while in the following section I will detail specific courtship processes.

Resource-based mating systems

In this type of mating system, individuals of one sex (normally males) guard a resource needed by the other sex. While these resources can include sheltering sites or rare food resources, in tephritids the males are generally guarding possible oviposition sites. Males may compete with each other to guard these sites and will often only allow a female access to the oviposition site following mating with the guarding male. Males will often guard the female after mating while she is ovipositing, to stop re-mating attempts by other males.

A resource-based mating system has been described in detail for *D. longistylus*. In this species, monophagous on the woody milkweed *Calotropis procera* (Family Apocynaceae), males fight for and defend the fleshy green fruit, but not immature or over-ripe fruit. Females that come to these fruit for oviposition are treated 'cautiously' by resident males until the female extends her ovipositor to begin probing. At this point the male jumps on the female, attempts to force the female ovipositor out of the probing hole and, if successful, the male inserts his aedeagus into the female ovipositor. Mating time averages 53 minutes and for most of that time the pair (which always formed on the fruit) would move to a sheltered site under a leaf. After mating, the male would return to the fruit and attempt to guard the female until oviposition had commenced. Nevertheless, ovipositing females were mated multiple times each day; from around three to four times, but up to ten times in a day for one individual (Hendrichs and Reyes, 1987).

Bactrocera minax is another species with a resource-based mating system. As described in detail by Dong *et al.* (2014), the mating system of this citrus specialist is remarkably similar to *D. longistylus*, except of course that the host is green citrus fruit. But the behaviours of the males initially competing for a fruit, a winning male subsequently waiting quietly for a female to arrive and begin probing the fruit, and then jumping the female and forcing removal of the ovipositor from the probing hole before copulation are essentially identical.

Several *Bactrocera* species (e.g. *B. tryoni*, *B. cacuminata*, *B. dorsalis* and *B. oleae*) have also all been observed mating on fruiting host plants, but for these species there is little or no evidence of actual mating on the fruit resource, and certainly guarding of the fruit by males is not known. Therefore their mating system should not be described as resource based and they are discussed further below.

Aggregated mating systems

For all *Bactrocera* so far studied, a common pattern of their mating seems to be that it is aggregated. This is, multiple males and females come together to mate, rather than finding each other as individual pairs. This does not infer that it should be assumed to occur for all *Bactrocera* (see comments above), but it does appear to be common.

LEKS.   In the literature, one of the best-known aggregated mating systems are those known as leks. Used originally for some bird and mammal species, the term lek was applied to mating systems in which males gathered for display and to which the females came solely for the purpose of mating. The requirements to describe a mating system as a lek are that: (i) there is no evidence of male parental care; (ii) males are spatially clumped at time of mating; (iii) males do not control resources needed by the females (such as fruit or food); and (iv) females exhibit a clear choice for a particular mating partner. Other common or inferred attributes of a lek mating system are that males will fight each other to gain and defend preferred sites (=territories) within the lek; and that females come to the lek after such sites have become established. The term lek is now used broadly in the literature and other definitions have been applied, but this list of attributes are those most commonly seen in lekking species.

Japanese workers, especially H. Kuba and colleagues, have described in detail the mating system for *Z. cucurbitae*, and it appears to closely align with the theoretical expectations of a lek-based mating system. In a field cage holding a tree with fruit hanging from it, males that had been spread throughout the field cage aggregated prior to the dusk mating period on the plant, but on leaves not fruit. They engaged in male–male conflict to defend territories (single leaves). Aggregated males then 'called' (began active wing vibration) and released a pheromone to which females responded. Females did not respond to the plant in the absence of males. The role of the pheromone as a female attractant was confirmed through experimentation where calling males were visually hidden from females. Where *Z. cucurbitae* mate in the field is unclear, other than it is almost certainly not on the larval host plant.

OTHER AGGREGATED MATING SYSTEMS.   Leks are not the only form of aggregated mating system. Other well-studied aggregated mating systems include swarms, landmarking and hot spots. In swarming behaviour, seen for example in many mosquitoes and other nematoceran flies, large male aggregations occur on the wing and females enter those swarms where they are grabbed by a male for mating. The mating pair may continue flying, settle on a nearby branch or leaf, or simply fall to the ground. In some species, swarms may be so large and dense that they can be confused for clouds of smoke.

For landmarking species, males and females congregate at some obvious visual feature in the landscape. There are many different forms of landmarks that are known to be used by insects. Insects of many species will accumulate at the tops of hills (known as hill-topping) and this behaviour is particularly well documented in butterflies. In tropical forests, canopy insects may landmark around the crown of the rare emergent trees that rise above the relatively uniform canopy top. Also in forests, lower canopy species will aggregate at sun spots, places where a shaft of sun is piercing the canopy or where a tree fall has opened a canopy gap. Other landmarks may be less spectacular but are still important for aggregation, such as the tallest plant in the local environment, or for many aquatic species a small island or dead tree trunk emerging from a lake.

'Hot spotting' involves the aggregation of insects, commonly males, at sites where they are likely to encounter females. For dacines, the most obvious hot-spot location will be a fruiting host plant, where females are present for oviposition. And indeed, for many *Bactrocera* species males are commonly found on fruiting host plants during the day, even though they are not interacting with the females.

Because several frugivorous tephritids have well-documented lek mating systems, such as the *C. capitata*, some *Anastrepha* species and *Z. cucurbitae*, and because most *Bactrocera* aggregate, it has been explicitly inferred or stated that *Bactrocera* also lek. This is a common belief in the literature and experiments are often designed on the assumption that *Bactrocera* are lekking species. Experiments that test if females are more attracted to large or small caged aggregations of males (which have been done for two *Bactrocera* species) have been run under an assumption of lekking and that females will visit established male aggregations for mating. The evidence that *Bactrocera* are lekking species, however, is limited and often contradicted by actual experimental or observational data. Their mating is commonly aggregated, and several species display some characteristics of 'lek-like' behaviour, but the prevalence of true lekking is unclear. It is also clear that much confusion and apparently contradictory information exists for *Bactrocera* mating behaviour. In order to detail what is known, and also to highlight the confusion in some areas, the following section takes a species-by-species approach to describing mating in *Bactrocera*.

5.2.2   Mating systems of individual 'lekking' species

Olive fruit fly (B. oleae)
Bactrocera oleae is described as having males that aggregate in flight before settling on leaves within an olive canopy shortly before the dusk mating period.

They then show marked male–male aggression to defend their leaves, which is typical of a lekking species, and because of this aggregation and male behaviour olive fruit fly has been described in several papers as a lekking species.

However, studies have also shown that male–male aggression in *B. oleae* occurs throughout the day, peaking at late morning. So the role of male–male aggression specifically related to mating behaviour remains unclear. *Bactrocera oleae* also has a well-documented female pheromone (discussed in the pheromone section below) that acts as a male attractant over both short and long distances. Female pheromone-baited traps are used commercially to sample and trap olive fruit fly, and such traps catch almost exclusively males. A strong female pheromone attractant is not consistent with descriptions of male-dominated aggregations to which females come for mating (as is predicted for a lekking species), nor are published studies of the initial phase of olive fly courtship being a step that involves males actively seeking females, rather than females seeking males.

Oriental fruit fly (B. dorsalis) and Queensland fruit fly (B. tryoni)

Both *B. tryoni* and *B. dorsalis* are species that have been described as having lek, or lek-like mating systems. Again, however, the evidence for this is equivocal. For *B. dorsalis* there have been three studies on field mating, and each gives a different result. In one study, male aggregations in a single fruiting orange tree within a small orchard of approximately 60 trees were recorded over several successive nights, but for the entire study only one mating pair was observed. The generality of inference that the species creates male aggregations to which females come for mating is thus low. In a second study, females were observed to leave fruiting trees they had been on during the day and then make small aggregations in nearby non-host plants, with males arriving half-an-hour later for mating. In a third study, laboratory-reared flies were released into a papaya orchard, and all matings were observed on the host fruit. In both the first and third studies females were present on the plants in consistent numbers during the entire observation periods, which went from mid-afternoon to sunset, whereas male numbers increased markedly in the pre-dusk period. Females were ovipositing into the fruit on the plants and were not obstructed from that activity by males. Only the observations of the first study are consistent with traditional lekking.

For *B. tryoni*, publications on mating behaviour are similarly conflicting. In a field-cage study, male flies were shown to first swarm, and then settle on leaves in the direction of the prevailing breeze. Aggregation occurred regardless of number of flies released in the field cage; all that changed was the size of the aggregation. Males contested territories on leaves and females approached males only after the swarm had settled. Although the terms lek or lekking was not used in this study, it has since been routinely cited to support the case that *B. tryoni* is a lekking species. Apart from this study (and a personal communication cited within it), male swarming has not otherwise been observed for *B. tryoni* in the field or field cages and if it occurs it is not a common behaviour. In other papers, *Bactrocera tryoni* has been reported as mating exclusively on fruit host plants, but with no mention made of aggregations.

Ekanayake *et al.* (2017) used large field cages, containing multiple artificial trees, to explicitly examine the mating behaviour of *B. tryoni*. Rather than simply releasing flies into a uniform environment, they also modified the environment so they could experimentally test the effects (if any) of changes in plant height (to examine possible landmarking) and trees with and without fruit. At the time of mating they also recorded the abundance and behaviours of the sexes at 5-minute intervals. Their work highlighted several aspects of behaviour that help clarify previous anomalies in the literature for *B. tryoni*. Firstly, flies did aggregate at the time of mating, but there was no obvious male or female lead in the aggregation. Rather, for this species, both sexes were equally abundant (at relatively low levels) just prior to the mating period, and then larger number of both sexes (but more males) arrived at the aggregation once male calling and mating had started (Fig. 5.4). The role of fruit was unimportant in location of the aggregation site, with aggregations equally likely to occur on plants with or without fruit even though females were using the fruit for oviposition. But dramatically, plant height was important, with nearly all mating and mating aggregations occurring in tall trees over short trees. Male–male interactions were observed but were rare, and courtship (other than male calling) was not obvious with matings occurring rapidly after a female landed near a male. Rapid male mounting of females has also been reported in *B. dorsalis*.

Wild tobacco fly (B. cacuminata)

Bactrocera cacuminata is an endemic Australian non-pest species that is largely monophagous on its wild/feral host, wild tobacco (*Solanum mauritianum*). It is also a species for which there is conflict in the literature on aspects of its mating behaviour. For this fly field studies have recorded mating on the host plant, no mating on the host plant, or mating on some but not all host plants in an area; and there is conflict over whether mating is restricted to the host plant. One common factor for the two studies that reported mating on the host plant was that plants with matings were tall. In the third study where mating was not recorded, all plants were of uniform and relatively short height. Thus mating in this species may occur on the host plant, but host plants with a particular architectural attribute (i.e. tallness compared to other plants).

5.2.3 Summary of dacine-mating systems

This somewhat long section has attempted three things. First, to summarise what is known about how the sexes are brought together in *Bactrocera* and other dacines, which is surprisingly little. The second, and perhaps more important role, is to emphasise that within and between the few species studied there are differences in the observations made, so that trying to develop general predictions or expectations of *Bactrocera* mating behaviour is likely to be problematic and prone to error. Several recent reviews have made explicit statements that all tephritids are lekking species. This is patently wrong, as some have resource-defence mating systems. But even in species where there is aggregation, it does not automatically follow that lekking is the reason for aggregation, or that individual behaviours commonly associated with lekking species will occur.

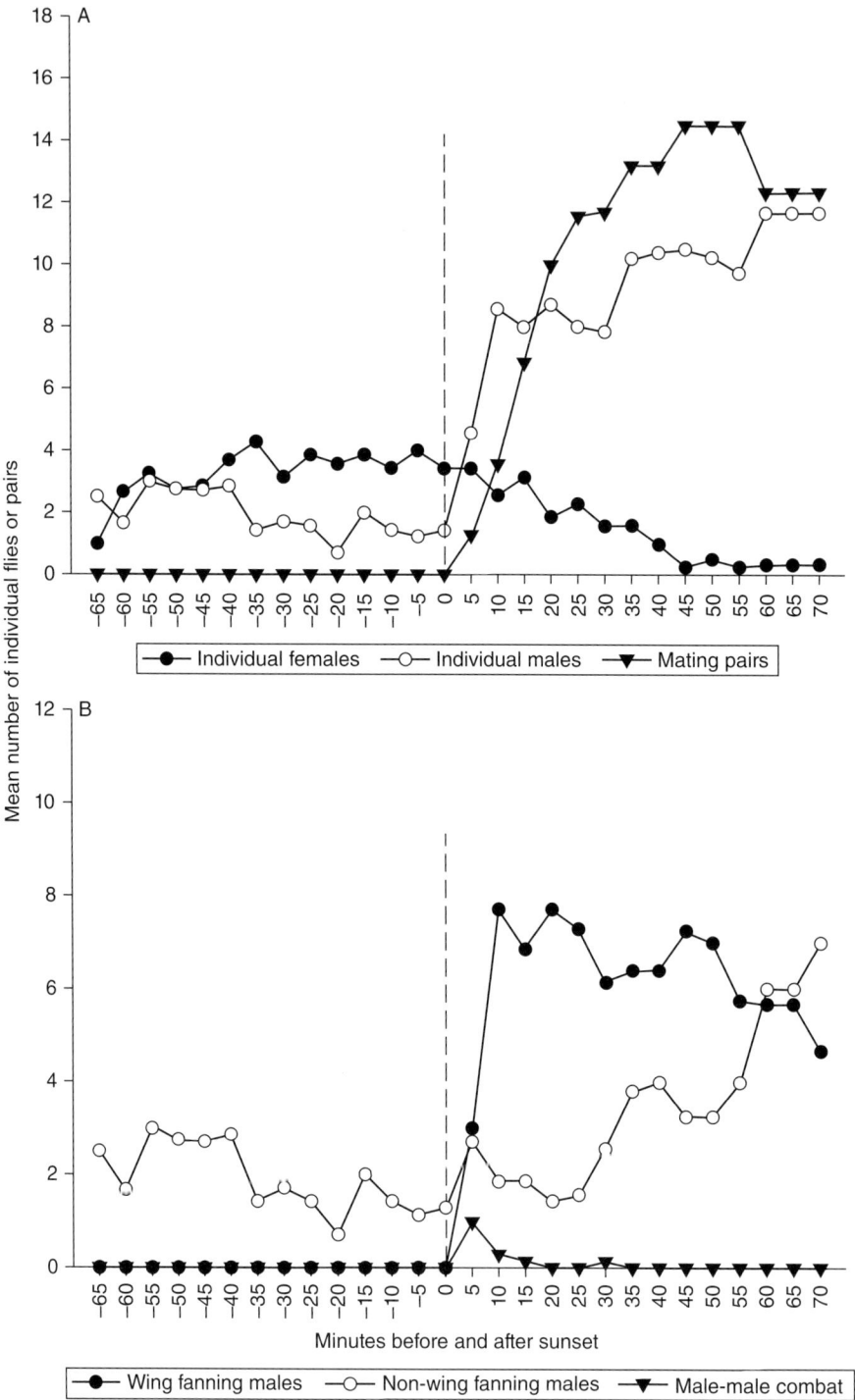

Fig. 5.4. The timing of behaviours in *Bactrocera tryoni* mating aggregations (from Ekanayake *et al.*, 2017).

5.3   Courtship

While mating systems (as described above) are largely about describing different 'patterns' of mating behaviour across species, courtship is a process-based
approach to understanding mating within a species. Courtship is a sequential
series of behaviours between males and females that, if carried to completion,
leads to the recognition and selection of a suitable mating partner. Courtship
can and does include the processes associated with how males and females
locate each other in the larger environment, but studies of courtship commonly
focus on the close distance communications between the sexes once they are
in close proximity to each other.

Courtship has been described as a signal-response chain. It is started by an
individual of one sex sending a signal to the opposite sex. The receiving individual may ignore that signal, in which case the courtship stops. Alternatively,
the receiving individual may recognise and accept the signal, and make their
own signal in return; that signal may then, in turn, be accepted or rejected
and the courtship progressed or halted. Two important points flow from this
understanding of the courtship process. First, in an evolutionary sense, courtship comes under strong 'stabilising selection'. That is, individuals that show
significant variation in some part of their courtship signalling will be selected
against as they are less likely to elicit the appropriate response from a prospec-
tive partner. For that reason the courtship process is highly conserved within a
species. The second point that flows is that to understand courtship properly,
the whole sequence of courtship behaviours need to be understood, and not
just the individual elements that make up the courtship. This is a significant
issue with *Bactrocera* mating, because while we have extensive knowledge of
several individual aspects of the courtship process, how they fit together into a
signal-response chain is still poorly understood.

5.3.1   Time of mating

The timing of mating in *Bactrocera* is perhaps one of the best-known and studied
aspects of their courtship. The majority of *Bactrocera* mate over a narrow temporal
window, often associated with dusk, and the mating is driven by critical light levels rather than an actual time. Exactly when species mate with respect to dusk varies with species. For example, *Z. cucurbitae* mates approximately 1 hour before
dusk, *B. dorsalis* commences mating approximately 45 minutes prior to dusk and
continues for 15 minutes after dusk, whereas *B. tryoni* commences exactly on
dusk and then mates over a period of approximately 45–60 minutes after dusk.

The mating time plays a key role in sexual isolation, as best demonstrated
in the *B. tryoni/B. neohumeralis* sibling pair. This species pair is closely related,
with total genetic differences less than is commonly seen between populations
of one species. With one exception, all known aspects of their courtship are
identical and, if forced to mate, they produce viable offspring across multiple
generations. However, they are sexually isolated by mating time. *Bactrocera
tryoni* courts on dusk, whereas in contrast *B. neohumeralis* mates over a period

of several hours either side of midday. Differences in mating time do not have to be as dramatic as dusk versus midday to result in isolation. Experimental lines of *Z. cucurbitae*, selected for long and short circadian rhythms, show significant isolation when their mating time differs by only 1 hour. Subtle variation in mating timing between field populations of the same species are documented. In large cage trials, two 'wildish' populations (either F1 or F2 generation) of *B. dorsalis*, one from southern Thailand and one from central Thailand, showed significant positive assortative mating in cage trials because males from the southern population initiated mating 40 minutes earlier than northern males, despite the official sunset beginning (depending on season) only 6 to 12 minutes earlier in the south than the north. Interestingly, this variation between the two populations was lost after 12 months in culture.

Two important 'clock genes', *period* (or *per*) and *cryptochrome* (or *cry*) have been sequenced for *B. dorsalis*, *Z. cucurbitae* (multiple populations), *B. tryoni*, *B. neohumeralis* and *B. oleae*. Despite their differences in mating time, for *B. tryoni* and *B. neohumeralis* no sequence differences were found between these two genes, but there were higher expression levels of the *cry* gene in *B. neohumeralis* than *B. tryoni*. Differential *cry* transcript levels were also detected between populations of *Z. cucurbitae* that had different mating times, while the gene itself was also found to have minor polymorphisms. The circadian clock mechanism of *B. oleae* has been described in detail (Bertolini *et al.* 2018) and involves not just differences in diurnal expression patterns of *per* and *cry*, but also expression of other genes and a complex neural circuitry in the brain. The clock mechanism in *B. oleae* was found to operate as for the circadian clock in *Drosophila*, which opens avenues for more fundamental research in dacine clocks.

Clocks in insects are strongly linked for different developmental and diurnal processes, and Japanese workers have made a strong case that pleiotropic changes (i.e. 'accidental' changes through gene linkage) in dacine mating time may be the result of selection acting on other life-history traits, so indirectly driving reproductive isolation. This is an area of great theoretical interest, as well as having important applied ramifications in ensuring that SIT-produced males mate at the same time as wild populations.

5.3.2 Visual cues

Across the wider Tephritidae and Tephritoidea, display of highly patterned wings by courting males is a central part of the courtship process: hence the common name of the 'picture wing' flies (or similar) for many tephritoid groups. However what role, if any, wing visual cues may play within dacine courtship is largely unknown, although whether this is because it is biologically absent or just unstudied is unclear.

The great majority of the pest dacines have clear wings with no patterns, and in mating studies of these species there is only one example of a wing display playing a role in courtship. In *B. tryoni* courtship, males and females hold their wings at right angles to their bodies and then face a potential mate.

Males that held their wings in this position for longer were more successful in courtships than males who displayed for shorter durations.

Not all dacines have clear wings and many non-pest species (plus some pests) have dramatically marked wings, and even species with predominantly clear wings may have a broad dark band along the front edge of the wing (the costal margin) and an enlarged apical spot on the tip of the wing. Because of lack of study it is unknown if these wing markings play a role in the courtship.

5.3.3 Male calling/wing vibration

While the wings may not be used for courtship visual display, they are certainly still used in the courtship process. Prior to copulation, males of most species rapidly beat their wings and make a high pitched buzzing noise. Under the right conditions, this can be clearly audible to humans. As described in *B. oleae* a single wing beat involves the wings first being held close to the abdomen and parallel to the ground. They are then raised and simultaneously rolled on their longitudinal axis, before being lowered to the start position. For *Z. cucurbitae*, the time of a single wing stroke of this sort is 0.0033 seconds. As demonstrated not just in *B. oleae*, but also in most of the better-studied pest species, this movement of the wing drags a sexually dimorphic part of the male wing over the pecten, a row of stiff bristles located on either side of the hind margin of abdominal tergite 4 of male flies (Fig. 5.5). The sound produced has been demonstrated to result from both the simple wing movement and the brushing of the wing against the pecten. Flies that have had their pecten experimentally removed still produce calls, but the call signal lacks subtle variations detected in flies with their pecten intact. As examples of the structure of calls, those for *B. oleae* and *B. tryoni* are shown in Table 5.1.

Initially postulated to be part of male–male interactions, it is now much clearer that male calls are a critical part of courtship. Calls are most commonly initiated once males are directly orientated to a female. In *B. oleae*, males that mounted a female directly without calling were unsuccessful in mating. In *Z. cucurbitae*, *B. dorsalis* and *B. oleae*, males with wings surgically removed have greatly reduced mating success, although less than 50% removal of wings in *B. dorsalis* leaving the fold at the $Cu_1 + 1A$ vein did not influence mating success. Irradiation of *B. tryoni* significantly influenced call attributes, but did not impact on mating success. However, analysis of the calls of successful and unsuccessful maters in olive fruit fly demonstrated major differences in call attributes between winners and losers.

5.3.4 Pheromones

Bactrocera have complex chemical communication during courtship and some aspects of the pheromone system are well documented. Anyone who has cultured flies knows the distinct smell of the male pheromones, and many workers can tell different species by their smell. However, despite being so well

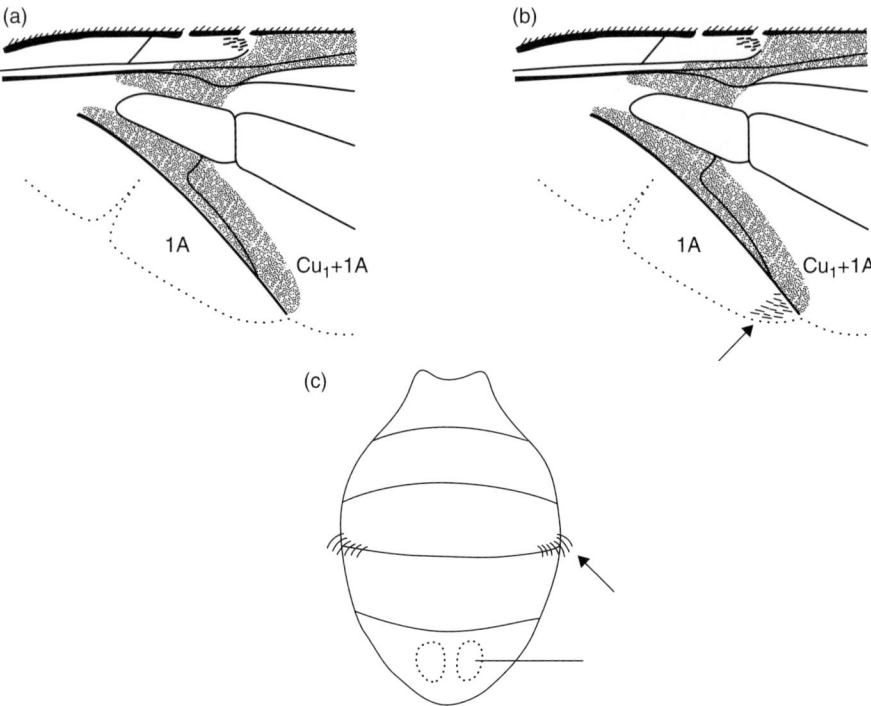

Fig. 5.5. Posterior portions of wings of (a) female and (b) male *Bactrocera tryoni* showing the sexual dimorphism (redrawn from Monro 1953). Marked with an arrow (in (b)) is the part that rubs against the pecten (arrowed in (c)) during male calling.

Table 5.1. Male call attributes of *Bactrocera oleae* and *B. tryoni*. For *B. oleae* the comparison is between successful and unsuccessful males in mate choice experiments, and for *B. tryoni* between irradiated (for the SIT) and non-irradiated males. Data from Benelli *et al.* (2012) and Mankin *et al.* (2008), respectively.

	Frequency (Hz)	Pulse duration (ms)	Inter-pulse interval (ms)
B. oleae successful male	340.7	230	360
B. oleae unsuccessful male	320.8	120	380
B. tryoni untreated	304.8	27.5	11.7
B. tryoni irradiated	303.6	29.2	16.7

known, and a large body of literature associated with *Bactrocera* pheromones, the role pheromones play in *Bactrocera* courtship is, for the most part, poorly understood.

Male pheromones
Both male and female *Bactrocera* produce pheromones although, with the exception of *B. oleae* where the female pheromone is used in trapping and so

well studied, many workers are unaware of the female pheromones. The male pheromones are produced in a modified rectal gland, consisting of a epithelium-lined secretory sac and a larger reservoir. As characteristic of insect pheromones in general, the pheromone composition varies between species, but major constituents are amides, esters, spiroacetals and long-chain fatty acids. Certain chemicals are common across species, for example, *N*-3-methylbutylacetamide is shared by eight species, while many other chemicals are unique to single species. Fletcher and Kitching (1995) remains the most authoritative review of the chemistry of dacine pheromones.

The male pheromone is distributed at the time of mating, while the male is wing fanning (as described in the previous section). The pheromone is excreted through the anus where droplets are picked up by the hind legs, which are then rubbed across the wings. Rapid wing vibration and rubbing of the wings across the pecten vaporises the pheromone that is distributed as micro-droplets. However, for a volatile agent these droplets are still relatively large and it is thought that they do not travel far. The pheromone of *Z. cucurbitae* has been described as 'smoke', so obvious can the pheromone cloud be.

Female flies may orientate positively to the male pheromone, but there is no evidence that it acts as a long-distance attractant as do, for example, moth pheromones. Numerous studies have failed to show even relatively short distance (e.g. within a field cage) orientation of females to calling males, and the classical pheromone 'test' of putting the pheromone releasing sex into a cage and recording the opposite sex coming to them does not work for dacines. One paper reported orientation of *B. dorsalis* (males and females) to male pheromones within a flight tunnel over a distance of 1.7 m, but *B. tryoni* did not show such orientation in a 1.8 m flight tunnel, although there was some upwind flight in clean air controls.

If the copiously produced male pheromones are not for long-distance female attraction, what then is their role? Basically it is unknown. The male pheromone has been described as a female 'excitation chemical', i.e. a mechanism to prepare females for mating. It may play a role in mate recognition by being a species-specific chemical 'signature'; or it may be some mechanism of 'honest signal', by which females may judge male quality. It may also serve as an aggregation signal, as conspecific males are also known to respond to male pheromones. At this stage it is simply unknown and significantly more work is needed in this area.

Female pheromones

Female pheromones are known in *Bactrocera*, although apart from *B. oleae* they are poorly studied. Contrary to the male pheromones, the female pheromone of *B. oleae* attracts males over a distance and is used commercially in monitoring traps. The pheromone is released by the female during the late afternoon/early evening when *B. oleae* mates. As for males, the pheromone is produced in a rectal gland and its major component is 1,7-dioxaspiro-[5,5]-undecane, but up to ten other components of the pheromone have been identified and four are used in the commercial olive fruit fly pheromone blend. Unusually, young males of *B. oleae* (< approximately 10 days of age) also produce 1,7-dioxaspiro-[5,5]-undecane in small amounts and these flies are courted by older males (which do not produce the chemical). The reason for this female mimicking by young olive fruit fly males is unknown.

In other *Bactrocera* species female pheromones are often reported not to occur, but this is wrong as female pheromones have been reported from *B. dorsalis*, *B. tryoni* and *Z. cucurbitae*. Their general structure is similar to those reported from males, being amides and spiroacetals. It is not clear in the literature where these chemicals are produced or when they are released during the day.

Aggregation pheromones?

One role of both the male and females pheromones may be aggregation. This has been specifically postulated as a role of the *B. oleae* pheromones, but has only been hinted at for other *Bactrocera* species. Nevertheless, *B. dorsalis* pheromones attract males as well as females, while behavioural evidence (see Fig. 5.4) strongly suggests a rapid aggregation of both males and females for mating. This is another area of dacine chemical ecology that urgently needs more research, especially given the potential applied importance.

Cuticular hydrocarbons

As the name suggests, cuticular hydrocarbons (CHCs) are long-chain hydrocarbons produced by epidermal secretory cells of the cuticle. Their primary function in insects is water preservation, but in several orders of insects (e.g. Hymenoptera, Blattodea, Diptera) CHCs play an important chemical communication role as, like volatile pheromones, their exact chemical composition is species specific. In *Drosophila*, CHCs have been shown to be involved in sexual selection, with females being able to assess male quality based on subtle variation in a male's CHC profile. CHCs have been well researched in some tephritid groups, such as *Anastrepha* and *Ceratitis* for chemotaxonomic purposes, and during these studies the CHC profiles within a species have been shown to differ significantly between males and females and immature and mature flies, suggesting that the CHCs may play a role in sexual communication in tephritids. Species-specific CHCs have been documented for a small number of *Bactrocera* species, but no attempt has been made to test if CHCs are involved as part of the courtship process, although they do differ within a species between the sexes. Given the close contact between males and females following mounting, the physical ability of individuals to assess another's CHC profile exists and might, potentially, be used as part of courtship.

5.3.5 Post-mounting behaviour

In many insects where courtship and mounting has been studied in fine detail, the courtship commonly continues after mounting and before introgression, and then during introgression. Male 'pacification' of females can reduce female rejection before copulation, or during copulation may increase the length of mating allowing more sperm to be transferred. In other tephritids, with males mounted above females, such behaviours can include 'kissing' (touching of the labella), males rubbing their legs over the female, males tapping with their labellum on the female thorax and wing movements.

Such behaviours are occasionally noted in the dacine-mating literature, but no effort seems to have been made to formally quantify this for any species. One frequently observed behaviour of *Bactrocera* is that the female will often reject a male after mounting, with some authors noting 50% or more of mountings failing to lead to copulations. The reason for this rejection is unknown, but is presumably related to a male having failed to 'send' an appropriate signal in the signal-response chain. Significantly more needs to be done in post-mounting behaviour.

Courtship summary

As stated at the start of this section, some aspects of mating and courtship are well understood, yet others are largely black holes. Male 'calling' is well documented and has been demonstrated to play a critical part in successful courtship and mating by a male fly. Male volatile pheromones have been chemically identified for many species, but what role(s) they play in the courtship process is far from clear. The role of the female pheromone in *B. oleae* seems reasonably well established, but what role female pheromones play in other dacines is, as yet, entirely unknown. Finally, the high level of female post-mounting rejection of males remains entirely unstudied.

5.4 Sexual Selection

Sexual selection was first considered by Charles Darwin to explain ornate secondary sexual features (e.g. the peacock's tail or the red comb on roosters) and examines why some individuals within a species may have higher mating success than other individuals. Because of the greater physiological investment made by females into eggs than is made by males to produce sperm, sexual selection generally acts through females choosing high-quality males with which to mate. Evidence for sexual selection is some consistent pattern of certain individuals (normally males) getting more matings than other individuals and is relatively easy to confirm or deny. However, the mechanisms that underpin sexual selection can be harder to identify. For example, sexual selection may involve direct female choice of males based on some 'honest signal' produced by the male (such as larger body size, a better call or a larger nuptial gift); or alternatively it may involve male–male competition for access to females. It can be easily seen that such features can be confounded: a bigger male may be chosen by a female as likely to be genetically fitter; but similarly, a bigger male may also be competitively stronger than other males to get greater access to females. This can be illustrated in true leks. Males may compete with each other to gain a superior physical position within the lek (often the top of the lek), and then females may show a choice preference for approaching the top of the lek over the bottom. In such a situation it is almost impossible to determine if sexual selection is through male–male competition, female choice, or a combination of the two.

With the exception of studies related to the male lures (see Section 5.7), there is a relative lack of information on sexual selection in the dacines, but has been confirmed to occur in *B. dorsalis*, *B. oleae*, *B. tryoni* and *Z. cucurbitae*. For *B. dorsalis*, preferred males are more successful in male–male aggressive interactions and are more likely to attract females and then proceed successfully from attraction through to mating. In *B. oleae*, *B. tryoni* and *Z. cucurbitae*, larger males have greater mating success over smaller males, attributed in *B. oleae* to a different male call, with successfully mating large males (in comparison to small males and unsuccessful large males) having higher frequency calls with shorter inter-pulse durations.

5.5 Mating Frequency

Once mature, males have the capacity to mate with high frequency if females are available, reportedly every day in some species (e.g. *B. dorsalis*, *B. oleae*), or at least once every 3 days in others (*B. cacuminata*). However, female re-mating is more complex. In older literature, females are often reported as being monandrous (i.e. only mating once), with perhaps a second mating only if they lived for 4 weeks or longer. This is now known to be incorrect. Within the dacines, female polyandry (i.e. multiple mating) has been reported from *Z. cucurbitae*, *Z. cucumis*, *B. tryoni*, *B. dorsalis*, *B. carambolae* and *B. cacuminata*. Re-mating in these species can occur several times over the life of an individual female, and females that have re-mated lay more eggs than females that have mated only once.

The reasons for polyandry are uncertain and some studies have related the frequency of female re-mating to being a species, or even a population-specific characteristic. For example, under identical holding reconditions, *B. dorsalis* females had a significantly greater number of re-matings than did *B. carambolae* females. Seventy-six per cent of the experimental *B. dorsalis* females re-mated, with 17% re-mating within 1 to 3 days of the first mating, whereas only 10% of the *B. carambolae* females re-mated, with a minimum of 8 days between the first and second matings.

Independent of species level differences, most studies consider re-mating as being linked to factors affecting the female 'refractory period', or that period of time after mating when the propensity to re-mate is suppressed by male attributes passed at the time of mating. Some studies show the refractory period of *Bactrocera* females to be 50 days or more, but other studies have demonstrated no refactory period following mating. For example, in one study 15% of *B. tryoni* females re-mated the following day after mating with poor quality, mass-reared factory males. While re-mating inhibition has been most studied in *Bactrocera* with respect to the quality of mass-produced males, multiple matings do occur in wild populations. A paternity study using microsatellites on the offspring of 22 wild caught, gravid *B. cacuminata* found that females had re-mated on average 1.72 times.

Male attributes affecting the refractory period include the diet the males had been exposed to prior to mating (protein fed males induce a longer refractory period), the duration of the first mating (longer mating may lead to a longer refractory period), the quality and quantity of sperm in the spermatheca (more sperm the longer the refractory period), inhibitory effects caused by male accessory gland material, or a combination of these factors. Most of these factors can be shrunk down to two key ones – amount of sperm and male accessory gland material. Separating the effects of these two is difficult as they are biologically linked, but where attempts have been made to do so (which has been done in both B. tryoni and B. dorsalis) the inhibitory effects of the male accessory gland material have been confirmed, while the sperm deposited or held in the spermatheca is unimportant.

5.6   Methoprene and Mating Behaviour

Methoprene (1-methylethyl (E,E)-11-methoxy-3,7,11-trimethyl-2,4-dodecadienoate) is a juvenile hormone (JH) analogue that is produced commercially for use as an insecticide. It is non-toxic to vertebrates. Juvenile hormones regulate growth and moulting in insects, as well as several other insect attributes including pheromone production. Juvenile hormone titre is high in larvae immediately following a moult, where it promotes rapid growth. However, the JH titre decreases as a larva grows and it is low to absent at the time of moulting between instars, or at the time of pupal emergence to adult. Methoprene is used as an insecticide against several insects through its ability to interrupt normal growth and development, and because of its low-to-absent mammalian toxicity. It would probably be more widely used as an insecticide, except that it is costly to synthesise and is a relatively unstable chemical.

Developed for use in the SIT, exposure of immature adult flies to methoprene increases the rate of sexual maturation and male mating advantage, but only when done in conjunction with offering protein in the diet. Topical application of methoprene plus protein has been found to increase the rate of sexual maturity and mating competitiveness in male Z. cucurbitae, without decreasing male longevity. Mating with a methoprene-treated male did not, however, alter female Z. cucurbitae's propensity to re-mate. Similar results of increasing development rate have been demonstrated in B. tryoni, although it was noted that attempts to treat flies while pupae increased levels of failed pupal emergence or adult deformities. The effect of methoprene appears synergistic with protein in the diet. Early male access to protein and sugar increases maturation rate and subsequent mating success, while protein plus methoprene enhances these effects. Methoprene on its own, however, produces limited or no development enhancement, and certainly less than by offering protein on its own.

It should be noted that one study of methoprene on the development rate of B. dorsalis failed to detect any methoprene treatment effect (even though protein was supplied). The authors did not know the reason for that and could only speculate that the lack of significant effects may have been due to the fly population used, treatment protocols or dosage rates. This is an important outcome to illustrate that the methoprene effects should not automatically be assumed to work, but need to be validated for each new system.

5.7 Male Lures

One of the best-known behavioural traits of male *Bactrocera* is their response to the so-called male lures, the best known of which are methyl eugenol (ME) and raspberry ketone (RK) (or its synthetic analogue cue-lure [CL]). Sexually maturing and mature male flies will fly upwind to locate these chemicals, and having found it will feed upon that source. Both ME- and RK/CL-responsive flies show a diurnal foraging pattern to these lures, foraging for the lures strongly in the morning and then with declining or ceased foraging activity from late morning into the remainder of the day. That foraging for lure is a repeated, diurnal activity suggests that the lures are a resource required by male flies no less than other more obvious resources, such as food, shelter and mates.

So strong is the response of male *Bactrocera* to these chemicals, that when used as a lure-and-kill device they can be used to drive local populations to extinction, particularly when used in combination with protein baits to target females. Some *Dacus* and numerous *Zeugodacus* species also respond to the male lures, but the trait is not as common in these two genera and for *Zeugodacus* it seems largely restricted to RK/CL responsiveness. While it is estimated that only 60% of *Bactrocera* species respond to ME or RK/CL, it is possible (probable?) that all *Bactrocera* species (and *Dacus* and *Zeugodacus*?) have a male lure response, it is just that the exact chemical attractant is not known. Evidence for this statement comes from recent trials with a range of related chemicals that are attracting many species traditionally considered as 'non-lure responders'. This section describes in detail the biology of *Bactrocera* male lure response.

5.7.1 Discovery, chemistry and general fly response to lures

The initial discovery of the response of *Bactrocera* males to chemical lures is attributed to F.M. Howlett, who in 1912 identified that male *B. dorsalis* and *B. zonata* responded to citronella oil, and 3 years later published the identification of ME as the attractive element. The attractiveness of CL/RK was not identified until nearly 50 years later, with the near simultaneous discovery of the attractiveness of cue-lure to *B. cucurbitae* in Hawaii and to *B. tryoni* in Australia (where it was known as Willison's lure); the name cue-lure is an abbreviation of the term cucurbitae-lure.

Both methyl eugenol [4-allyl-1,2-dimethoxybenzene] and raspberry ketone [4-(p-hydroxyphenyl) butan-2-one] are secondary plant chemicals and are chemically related, with ME a phenylpropanoid and RK a phenylbutanoid. Cue-lure [4-(4-acetoxyphenyl)-2-butanone] is rare in nature (until recently it was not considered to occur naturally at all), but is an analogue of raspberry ketone (the hydroxyl equivalent of cue-lure), which does occur in nature. Methyl eugenol is a common secondary plant chemical, having been documented in the essential oils from 450 plant species across 80 plant families. While occurring in only trace amounts in some of these plants, in others it can constitute up to 90% of the essential oil volume. Raspberry ketone, the essence of raspberries, is less common in nature but is still recorded from plants in eight angiosperm orders.

The bulk of *Bactrocera* literature classifies flies as either ME responsive, RK/CL responsive, or non-lure responsive. This is because it has traditionally been considered that males of a species only ever respond to one of these attractants and not the other – or they simply do not respond to lures at all. Based on this belief, lure response has been used as a diagnostic character to separate groups of related flies, and it is always given as part of a species description if known. While nearly all species do respond to only one of these two lures, changed understanding in *Bactrocera* lure response suggests that the system is far from being so black and white.

The first evidence that this might be the case stems from differential lure response. Under controlled feeding conditions, the ME-responsive and close-sibling species *B. dorsalis* and *B. carambolae* have different dose-dependent response rates to ME. This suggests that even within the one lure, not all species react to them the same. This is a lab result that accurately reflects what field workers have often determined based on experience, that some *Bactrocera* species are 'easier to trap' than others.

The second line of evidence that lure response is not one thing or other, comes from work on new lures. A small number of *Bactrocera* species have been previously recorded in the literature as responding to both ME and RK/CL, and while such records have been dismissed as lure contamination or misidentifications, there is now no reason to consider them as such. The *Bactrocera* male attractant zingerone [4-(4-hydroxy-3-methoxyphenyl)-2-butanone], first identified from the flowers of the *Bactrocera*-pollinated orchid *Bulbophyllum patens*, is known to attract ME-responsive species, RK/CL-responsive species, and some traditionally non-lure responsive species. Similarly, other tests have shown that yet more of the phenylbutanoid/propanoid-related chemicals, including isoeugenol, methyl-isoeugenol and dihydroeugenol, attract *Bactrocera* and *Dacus* species that had previously been considered non-lure responsive, or attracted more individuals of traditionally ME- or RK/CL-responsive species than did traps containing those lures.

Both lines of evidence (differential response by different species to a single lure and response by one species to multiple lures) suggest that different fly species are probably evolutionary attuned to one or a small group of closely related phenylpropanoids/butanoids, but 'loose receptors' allow *Bactrocera* species to respond at least partially to ME, RK/CL or zingerone. Ongoing research is currently finding more attractive lures for major pest species (i.e. other RK analogues that are significantly more attractive to *B. tryoni* and *Z. cucurbitae*), as well as finding novel lures for 'non-lure responsive' species.

5.7.2 Functional role of lures in mating

While ME and RK/CL have been used for fruit fly monitoring and control for at least the last 50 years, why flies sought the lures and what functional role they served for the flies (if any) is a much more recent field of study, initially developing in the mid-1990s and which is still maturing as new information becomes available.

It is now agreed that the lures play a role in the mating system of *Bactrocera* species, but how they do this varies with fly species and lure. Only a small number of species have been studied with respect to lures and mating behaviour (*B. dorsalis*, *B. carambolae*, *B. cacuminata*: ME responsive; *B. cucurbitae* and *B. tryoni*: RK/CL and weakly zingerone responsive; and *B. jarvisi*: zingerone responsive) and even in this small mix of species and lures there are both commonalities and differences. The most common pattern seen is that for all the above species except for *B. cacuminata*, males fed on lures have a mating advantage over lure-unfed males – that is lure-fed males get more females. Other than this commonality, many other things vary as described below.

- *Lure ingested and broken down:* With Oriental fruit fly feeding on ME, the individually best studied system, imbibed ME is converted into two main products, 2-allyl-4,5-dimethoxyphenol (DMP) and (E)-coniferyl alcohol (CF). Conversion is near immediate after ingestion and occurs in the insect's crop. The DMP and CF are transported via the haemolymph system to the rectal gland (i.e. the pheromone gland of *Bactrocera*) where they are released along with an endogenous pheromone compound at time of mating. CF, particularly, is highly attractive to females and the assumed mechanism for enhanced male mating success is because of the enhanced male pheromone. This pattern also occurs in the sibling species *B. carambolae*, except that only CF is produced. *Difference 1:* The same lure can be broken down in different ways in different fly species.

- *Lure ingested and not broken down:* In the RK/CL-responsive *B. cucurbitae* and *B. tryoni*, ingested lures are not broken down. In *B. cucurbitae*, ingested RK and zingerone are accumulated and stored in the rectal glad in an unaltered state whereas in *B. tryoni*, ingested CL had the minor conversion to RK but was otherwise sequestered unaltered, as was zingerone. *Difference 2:* ME- and RK/CL-responsive species seem to handle ingested lures differently.

- *Lure increases pheromone attractiveness – or not:* In *B. dorsalis* and *B. carambolae* (fed on ME), *B. cucurbitae* (fed on CL and zingerone) and *B. tryoni* (fed on CL), male pheromones of calling males (in some cases the extracted pheromone was tested, in other cases a calling male, in some cases both) had increased female attraction, supportive of the 'enhanced male pheromone' hypothesis. However, *B. tryoni* fed zingerone did not have a significantly enhanced male pheromone, even though zingerone-fed males had increased mating success. *Difference 3:* Lure feeding does not automatically lead to a 'sexier' pheromone.

- *Lure feeding increases male mating success – or not:* In *B. cacuminata* ME feeding did not increase male mating success, but there was direct evidence from well replicated trials that an ME source was used as a mate rendezvous signal. This has been postulated, but never demonstrated for other *Bactrocera* species. *Difference 4:* Lures can be associated with the mating system of *Bactrocera* species in ways other than linked to the male pheromone.

- *Mating with a lure-fed male alters the female – or not:* In studies of *B. dorsalis* and *Z. cucurbitae* mating with a ME- or CL-fed male, respectively, did not alter female longevity or fecundity. In contrast, female *B. tryoni* mating with either a CL- or zingerone-fed male laid more eggs over her life time, laid them more quickly, and died sooner. Further, her sons from lure-fed fathers could themselves find lure faster. *Difference 5:* Females may or may not be directly affected by mating with a lure-fed male.

- *ME and CL/RK effects last for different times:* Several studies have demonstrated major difference in the duration of lure effects. For ME feeding, male mating competitive advantages can last for up to 30 days. In stark contrast, RF/CL-feeding effects last for about 3 days. *Difference 6:* Different lures have different underlying physiological effects.

- *Flies response to lures when sexually mature or immature:* For nearly all species studied, adult male response to lures only occurs when the males are sexually mature. However, for *B. jarvisi* responding to zingerone, males began responding before sexual maturity. *Difference 7:* Different species can respond to lures at different states of sexual maturity.

5.7.3 Metabolic effect of lures

Recent work with *B. tryoni* males fed on zingerone and CL identified that the lures are metabolism and energy enhancers. Outside of fruit fly systems, RK and zingerone are both known to have nearly identical metabolic effects to caffeine consumption, including the depression of appetite and the enhancement of short-term energy by increasing baseline metabolic rates. So well-known is this effect that entry of 'raspberry ketone diet' into an internet search engine returns several pages of 'fad' diet regimes where raspberry ketone tablets are sold as magic diet pills.

When a lure metabolism effect was tested on *B. tryoni*, it was indeed seen that both RK and zingerone increased male locomotor activity and caused increased weight loss even in the presence of food. More tellingly, a comparative transcriptome analysis of zingerone-fed males compared to lure-unfed males demonstrated massive upregulation of transcriptomic activity following lure feeding, including upregulation of transcripts associated with some 344 different metabolic pathways, multiple biological processes and specific pathways associated with mating, including inter-male aggression, courtship and pheromone release. This enhanced activity may be expected to provide lure-fed males a mating advantage through their being able to better physically compete in male/male competition and in more active display and courtship activities. Subsequently, CL and zingerone fed to sexually immature *B. tryoni* has been shown to increase their sexual maturation rate.

It is not yet known if ME produces the same metabolic enhancement as do RK/CL and zingerone, but there is indirect evidence in the literature that it might. Across several papers there are explicit statements about ME-fed males wing fanning more frequently, beginning courtship earlier and having greater female mounting success (in *Bactrocera* males are often physically rejected by the female at the mounting stage). While recorded, how the lure may have influenced these observations has never been discussed. Metabolic enhancement through lure action now seems a possible answer.

As for all other parts of the male lure story, metabolism modification is not a one-answer-fits-all solution for understanding lures. Trials of the type done for *B. tryoni* have been repeated in *Z. cucurbitae* and no metabolic effect was observed. This reinforces the point that research on lure effects is increasingly showing that they need to be investigated at a species-by-species level. The research on metabolism enhancement should also not be seen as a conflicting hypothesis to the well-documented pheromone enhancement research. Rather, as has been shown for *B. tryoni*, both may play a role. *Bactrocera tryoni* males fed CL have a sexier pheromone and are 'hyped up' and so have a mating advantage over unfed males. *Bactrocera tryoni* males fed zingerone do not have a sexier pheromone, but are hyped and so still have a mating advantage over unfed males, although less of an advantage than CL-fed males.

5.7.4 Lure effects on predation

It has been proved conclusively that ME-fed *B. dorsalis* are toxic when fed exclusively to vertebrate predators (Asian house geckos). Conditioned geckos will avoid not only lure-fed males, but also female flies in the presence of such males. Physically disturbed males also secrete a rectal gland secretion that can deter geckos. The combined male mating advantage and predator protection offered by ME feeding has been suggested as the reason for the evolution of male lure response. However, the weakness of this argument is that no predation data has been presented to support an argument that vertebrate predators are a major mortality source for adult *Bactrocera*. For the day-mating Mediterranean fruit fly it has been suggested that dragonflies may be important predators of mating pairs. In contrast, most *Bactrocera* mate at dusk and almost certainly in environments well removed from the normal habitat of geckos. While not questioning the results that ME can deter geckos, the ecological and evolutionary importance of such an effect does need more data and consideration.

5.7.5 Evolution of lure response

Evolutionary theories are impossible to confirm and can change greatly as new information gained provides new insights. This small section should thus be read critically and with an assumption it may well change.

A great deal of the early focus on the evolution of *Bactrocera* lure response focused on the possibility that response to the phenylpropanoids/butanoids were evolutionary 'hangovers' of early host-location mechanisms. This assumed that primitive fruit flies bred in decaying organic matter and the phenylpropanoids/butanoids were produced by fermentation of that organic matter and hence were an attraction signal. However, current systematics suggests that invertebrate parasitism may be the ancestral state for fruit flies, with both the family Pyrgotidae (the evolutionary closest family to the Tephritidae) and the

Tachiniscinae (the most basal sub-family of the Tephritidae) being parasitoids. Mapping of lure response (ME or RK/CL) against molecular phylogenies of the Dacini also does not show an obvious signal that lure response is evolved from some common ancestral state, although such interpretation is confounded by placing lures into a binary structure (i.e. ME or RK/CL) that should probably no longer be seen as biologically meaningful as more lures are discovered.

Given this, the currently most plausible evolutionary argument for male lure response stems from its role in mating behaviour. The *B. tryoni* system, particularly, provides almost a perfect model for what is known in evolutionary biology as 'sexy-son run-away sexual selection'.

Sexy-son run-away sexual selection is the theoretical model used to explain the peacock's tail and the antlers of the Irish elk, which grew so large that they may have led to the elk eventually becoming extinct. In this model, males have some characteristic that makes them more attractive to females. Females preferentially choose such males, and so the preferred male characteristic spreads and becomes reinforced in the population. With each successive generation selection for the male trait becomes stronger and the trait can become more extreme. In the Irish elk it led to huge antlers, disproportionally larger than the males could sustain; in *Bactrocera* it may have led to a lure response so strong that by linking it to lure-and-kill control we can cause a population to drive itself to extinction.

The theory for *B. tryoni* runs as such (but it can be modified slightly for any *Bactrocera* where there is a male mating advantage following lure feeding). A male finds a natural source of RK or zingerone, for example in the petals of orchids. That male becomes more likely to be successful in mating (i.e. it gains an evolutionary fitness advantage) because it is metabolically 'hyped up' and/ or produces a sexier pheromone. Females that mate with the lure-fed males produce more offspring (demonstrated in *B. tryoni* but no other species to date) and so she gains an evolutionary fitness advantage. The sons of these matings are themselves more likely to find lures because they inherited the skill from their fathers (or in the case of *B. tryoni* because there is a direct transgenerational epigenetic effect), thus becoming 'sexier sons'. Simultaneously, natural selection is likely to act on those daughters who can detect the new 'modified' pheromone now produced by the 'sexier sons'. In each generation selection will act on the cycle, eventually leading to the male lure response acting on the entire species. Well at least that is my current best guess, and as for most things in evolutionary biology hard to prove or disprove.

5.8 Further Reading and References Cited

Benelli, G., Canale, A., Bonsignori, G., Ragni, G., Stefanini, C. and Raspi, A. (2012) Male wing vibration in the mating behavior of the Olive fruit fly *Bactrocera oleae* (Rossi) (Diptera: Tephritidae). *Journal of Insect Behaviour*, 25, 590–603.

Benelli, G., Daane, K.M., Canale, A., Niu, C.Y., Messing, R.H. and Vargas, R.I. (2014a) Sexual communication and related behaviours in Tephritidae: current knowledge and potential applications for integrated pest management. *Journal of Pest Science*, 87, 385–405.

Benelli, G., Giunti, G., Canale, A. and Messing, R.H. (2014b) Lek dynamics and cues evoking mating behavior in tephritid flies infesting soft fruits: implications for behavior-based control tools. *Applied Entomology and Zoology*, 49, 363–373.

Benelli, G., Donati, E., Romano, D., Ragni, G., Bonsignori, G., Stefanini, C. and Canale, A. (2016) Is bigger better? Male body size affects wing-borne courtship signals and mating success in the olive fruit fly, *Bactrocera oleae* (Diptera: Tephritidae). *Insect Science*, 23, 869–880.

Bertolini, E., Kistenpfennig, C., Menegazzi, P., Keller, A., Koukidou, M. and Helfrich-Förster, C. (2018) The characterization of the circadian clock in the olive fly *Bactrocera oleae* (Diptera: Tephrtidae) reveals a *Drosophila*-like organization. *Scientific Reports*, 8, 816.

Collins, S.R., Reynolds, O.L. and Taylor, P.W. (2014) Combined effects of dietary yeast supplementation and methoprene treatment on sexual maturation of Queensland fruit fly. *Journal of Insect Physiology*, 61, 51–57.

Dong, Y., Wan, L., Pereira, R., Desneux, N. and Niu, C. (2014) Feeding and mating behaviour of Chinese citrus fly *Bactrocera minax* (Diptera, Tephritidae) in the field. *Journal of Pest Science*, 87, 647–657.

Drew, R.A.I. (1969) Morphology of the reproductive system of *Strumeta tryoni* (Froggatt) (Diptera: Trypetidae) with a method of distinguishing sexually mature adult males. *Journal of the Australian Entomological* Society, 8, 21–32.

Ekanayake, E.W.M.T.D., Peek, T., Jayasundara, J.M.S.H., Clarke, A.R. and Schutze, M.K. (2017) The mating system of the true fruit fly *Bactrocera tryoni* and its sister-species, *Bactrocera neohumeralis*. *Insect Science*, 24, 478–490.

Ekanayake, E.W.M.T.D., Clarke, A.R. and Schutze, M.K. (2019) Close-distance courtship of laboratory reared *Bactrocera tryoni* (Diptera: Tephritidae). *Austral Entomology*, in press.

Fitt, G.P. (1990) Comparative fecundity, clutch size, ovariole number and egg size of *Dacus tryoni* and *D. jarvisi*, and their relationship to body size. *Entomologia Experimentalis et Applicata*, 55, 11–21.

Fletcher, B.S. (1968) Storage and release of a sex pheromone by the Queensland fruit fly, *Dacus tryoni* (Diptera: Trypetidae). *Nature*, 219, 631–632.

Fletcher, B.S. (1969) The structure and function of the sex pheromone gland of the male Queensland fruit fly, *Dacus tryoni*. *Journal of Insect Physiology*, 15, 1309–1322.

Fletcher, M.T. and Kitching, W. (1995) Chemistry of fruit flies. *Chemical Reviews*, 95, 789–628.

Fuchikawa T., Sanada S., Nishio R., Matsumoto A., Matsuyama T., Yamagishi M., Tomioka K., Tanimura T. and Miyatake T. (2010) The clock gene cryptochrome of *Bactrocera cucurbitae* (Diptera: Tephritidae) in strains with different mating times. *Heredity*, 104, 387–392.

Hendrichs, J. and Reyes, J. (1987) Reproductive behaviour and post-mating female guarding in the monophagous multivoltine *Dacus longistylus* (Wied.) (Diptera: Tephritidae) in southern Egypt. In: Economopoulos A.P. (ed.), *Fruit Flies: Proceeding of the Second International Symposium, 16-21 September 1986. Colymbari, Crete*. Elsevier Science, New York, NY, pp. 303–313.

Krosch, M.N., Schutze, M.K., Armstrong, K.F., Boontop, Y., Boykin, L.M., Chapman, T.A., Englezou, A., Cameron, S.L. and Clarke, A.R. (2013) Piecing together an integrative taxonomic puzzle: microsatellite, wing shape and aedeagus length analysis of *Bactrocera dorsalis s.l.* (Diptera: Tephritidae) find no evidence of multiple lineages in a proposed contact zone along the Thai/Malay Peninsula. *Systematic Entomology*, 38, 2–13.

Kumaran, N.K., Prentis, P., Mangalam, K.P., Schutze, M.K. and Clarke, A.R. (2014) Sexual selection in true fruit flies (Diptera: Tephritidae): transcriptome and experimental evidences for phytochemicals increasing male competitive ability. *Molecular Ecology*, 23, 4645–4657.

Mankin, R.W., Lemon, M., Harmer, A.M.T., Evans, C.S. and Taylor, P.W. (2008) Time-pattern and frequency analyses of sounds produced by irradiated and untreated male *Bactrocera tryoni* (Diptera: Tephritidae) during mating behaviour. *Annals of the Entomological Society of America*, 101, 664–674.

Monro, J. (1953) Stridulation in the Queensland fruit fly *Dacus* (*Strumeta*) *tryoni* Frogg. *The Australian Journal of Science*, 16, 60–62.

Pike, N. and Meats, A. (2002) Potential for mating between *Bactrocera tryoni* (Froggatt) and *Bactrocera neohumeralis* (Hardy) (Diptera: Tephritidae). *Australian Journal of Entomology*, 41, 70–74.

Radhakrishnan, P. and Taylor, P.W. (2007) Seminal fluids mediate sexual inhibition and short copula duration in mated female Queensland fruit flies. *Journal of Insect Physiology*, 53, 741–745.

Raghu, S. (2004) Functional significance of phytochemical lures to dacine fruit flies (Diptera: Tephritidae): An ecological and evolutionary synthesis. *Bulletin of Entomological Research*, 94, 385–399.

Royer, J.E. (2015) Responses of fruit flies (Tephritidae: Dacinae) to novel male attractants in north Queensland, Australia, and improved lures for some pest species. *Austral Entomology*, 54, 411–426.

Shelly, T. (2010) Effects of methyl eugenol and raspberry ketone/cue lure on the sexual behavior of *Bactrocera* species (Diptera: Tephritidae). *Applied Entomology and Zoology*, 45, 349–361.

Song, S.D., Drew, R.A.I. and Hughes, J.M. (2007) Multiple paternity in a natural population of a wild tobacco fly, *Bactrocera cacuminata* (Diptera: Tephritidae), assessed by microsatellite DNA markers. *Molecular Ecology*, 16, 2353–2361.

Tan, K.H. and Nishida, R. (2012) Methyl eugenol: Its occurrence, distribution, and role in nature, especially in relation to insect behavior and pollination. *Journal of Insect Science*, 12, 56.

Taylor, P.W., Pérez-Staples, D., Weldon, C.W., Collins, S.R., Fanson, B.G., Yap, S. and Smallridge, C. (2013) Post-teneral nutrition as an influence on reproductive development, sexual performance and longevity of Queensland fruit flies. *Journal of Applied Entomology*, 137 (s1), 113–125.

ul Haq, I., Cáceres, C., Hendrichs, J., Teal, P.E.A., Stauffer, C. and Robinson, A.S. (2010) Methoprene modulates the effect of diet on male melon fly, *Bactrocera cucurbitae*, performance at mating aggregations. *Entomologia Experimentalis et Applicata*, 136, 21–30.

Wee, S.L., Tan, K.H. and Nishida, R. (2007) Pharmacophagy of methyl eugenol by males enhances sexual selection of *Bactrocera carambolae*. *Journal of Chemical Ecology*, 33, 1272–1282.

Wee, S.L., Abdul Munir, M.Z. and Hee, A.K.W. (2018) Attraction and consumption of methyl eugenol by male *Bactrocera umbrosa* Fabricius (Diptera: Tephritidae) promotes conspecific sexual communication and mating performance. *Bulletin of Entomological Research*, 108, 116–124.

6 Host Use

The common name of the *Bactrocera*, 'the tropical fruit flies', says a great deal about the importance of the larval host fruit to the biology of these insects. Fruit is only a single resource among several (e.g. mates, protein, carbohydrates, resting sites) needed by dacines, and this resource is only needed by two life stages: the gravid female and the larvae. Nevertheless, without this resource the flies would not exist and so it can be rightly regarded as a primary resource. Therefore, studying both the patterns of dacine host use, and the processes that drive those patterns, is fundamental to understanding the biology of these flies. Further, as it is fruit damage that makes *Bactrocera* and other fruit flies pests, knowledge of how they locate and utilise the fruit is fundamental to many control techniques.

This chapter first looks at the patterns of larval host use by *Bactrocera* and other dacines, and then at the processes that underpin those patterns. The section on patterns examines host specialism versus generalism, and the relationship between female preference and offspring performance. After examining pattern, the chapter looks at the processes behind these patterns, starting first with host location. Host location in *Bactrocera* is similar to most herbivorous insects and uses both visual and olfactory cues, the importance of which likely vary with changing proximity to the fruit. Having found a particular fruit piece, the 'decision' to use that fruit for oviposition depends on intrinsic factors pertinent to the fruit fly such as physiological state and experience, as well as extrinsic factors such as fruit peel properties or ripeness stage. The process component of the chapter finishes by returning to pattern, with a section focused on the polyphagous dacines and how to interpret host usage in such flies.

Having discussed the patterns and processes of host use, the chapter concludes with sections relevant to application of this information: host use as a tool for market access; and plant resistance breeding.

6.1 Patterns of Host Use

6.1.1 Monophagy, oligophagy and polyphagy

The simplest descriptor of host use for any herbivore, and used particularly for insects, are the terms *monophagy*, *oligophagy* and *polyphagy*. Monophagous insects exclusively attack one host plant species only, oligophagous insects attack a small range of closely related plant species, whereas polyphagous species attack multiple plant species across two or more plant families. Across insect herbivores, monophagy is the most common host-usage pattern, with approximately 80% of herbivore species being host specialists. Depending on definition, approximately 15–18% of the remainder are oligophagous and the remaining few per cent polyphagous. Also seen increasingly in the literature are the terms generalism and specialism, where generalism = polyphagy and specialism = monophagy + oligophagy.

Bactrocera is a lineage that runs contrary to these percentages. If polyphagy is defined as the ability to feed on plants belonging to two or more plant families, then approximately 40% of *Bactrocera* are polyphagous, approximately 24% are oligophagous and 36% monophagous. Even species that are commonly thought of as host specialists, such as the olive fruit fly (*B. oleae*), banana fruit fly (*B. musae*), and breadfruit fly (*B. umbrosa*) are all actually oligophagous and feed on a small number of closely related species within the genera *Olea*, *Musa* and *Artocarpus*, respectively. In rainforest-restricted species, the percentage of true specialists may be higher, but the difficulty of sampling rare, non-pest, host specialists within a rainforest may mean that they are under-represented in published host lists. In an intensive fruit-rearing study in Papua New Guinea rainforests, 66% of *Bactrocera* sampled were collected from more than one host species (similar to other calculated figures), but most of these (88%) were restricted to different plant species within a genus. Thus in an intensively studied natural system, oligophagy was most common, followed by monophagy, followed by polyphagy.

As equally unusual as high levels of oligophagy and polyphagy is the relative abundance of 'extreme polyphags' within *Bactrocera*. This term has been defined in the literature as the ability to feed across more than 20 plant families and it is considered extraordinarily rare in insects – restricted to as few as 150 species globally (i.e. 0.003% of the approximately five million estimated insect species). However, extreme polyphagy is quite common in *Bactrocera* with 11 species having published host lists of this size or greater: *B. carambolae*, *B. correcta*, *B. curvipennis*, *B. dorsalis*, *B. fascialis*, *B. frauenfeldi*, *B. krausii*, *B. neohumeralis*, *B. passiflorae*, *B. tryoni* and *B. xanthodes*. There are also several other species that have host ranges nearly as large, e.g. *B. kirki* (19 families) and *B. melanotus* (18 families), and from a management and evolutionary perspective they need to be considered in the same way.

Why so many generalist Bactrocera?
In an evolutionary sense, monophagy is commonly assumed to be an evolved characteristic, with polyphagy expected (and often observed) in more basal lineages. Evolution is expected to favour host specialisation for several reasons.

These include specialisation of the herbivore in order to overcome plant chemical or physical defences. Plant–herbivore interactions are commonly considered in the context of an evolutionary 'arms race'. Herbivore feeding is considered to negatively impact on plant reproductive fitness and so plants should evolve mechanisms to deter herbivores. Herbivores, in turn, should evolve ways to overcome these defences, and so on. The outcome is increasing specialisation of a herbivore on a particular plant. The second reason for a herbivore to specialise is to refine host location mechanisms, so as to locate a host plant more effectively. Specialisation is also thought of as a way of reducing feeding competition with generalist herbivores (who may be less able to feed on defended plants), and as a way of escaping from generalist natural enemies who may not forage on particular plants or in certain habitats. As most herbivores are specialists, then it is assumed these theoretical predictions of the way specialisation occurs are correct – and indeed there is much empirical evidence to support the theory.

But if the same logic is applied to *Bactrocera*, then it may be seen that the drivers for specialisation are not strong. While it has not been intensively studied, available evidence suggests that *Bactrocera* do not negatively impact on the evolutionary fitness of their hosts as few species actively predate on seed. Rather, the presence of maggots in fruit may attract vertebrate seed dispersers, or make fruit break down faster and so enhance germination. There is thus no evolutionary driver for a plant to evolve a defence against fruit flies. As ripe fruit (the stage *Bactrocera* preferentially attack) are also evolved primarily as attractants for vertebrate seed dispersers, plants have evolutionary constraints against defending fruit and nearly all fruit are non-toxic at the ripe stage (although many are toxic when immature). Given all this, there is little evolutionary 'need' for *Bactrocera* to host specialise in order to overcome plant defence.

The need to specialise in order to develop good host location mechanisms may also not be critical. Within a rainforest, many tree species are rare and so mass flower only once every 2 to 3 years as a way of maximising cross–pollination. The mass flowering leads to 'mast fruiting', with hugely abundant fruit loads. The fruit itself, for some 70% of old world rainforest species, is designed for vertebrate dispersal, and so is colourful, often rich in odours and commonly hung on the ends of branches: effectively becoming a large signal saying, 'here I am, come and eat me'.

This then leads to the need to specialise to overcome generalist competitors and generalist natural enemies. Among insects, frugivory is relatively uncommon as a generalist feeding strategy. Why this may be so is unclear, in many ways they are an ideal resource, but nevertheless generalist feeding on fruits is not that common. In their rainforest environment *Bactrocera* are rare and specialism to avoid competition with other *Bactrocera* species is probably unnecessary. Generalist invertebrate natural enemies are also not a problem for *Bactrocera*, as the larvae are concealed within fruit. Larvae do suffer, however, from specialist parasitoids who use host plant cues to find larvae. Evolving to use a new fruit may therefore be a mechanism to move into 'enemy-free space' and so avoid these parasitoids.

In summary then, a disproportionately large number of generalist host users among *Bactrocera* may be because: (i) a co-evolutionary 'arms race' between *Bactrocera* and their host plants is unlikely, hence there is no driver for increasing specialisation within an arms-race model; (ii) many rainforest fruits are 'advertised' to attract vertebrate frugivores and are 'easy targets' for host location, so lessening the need to specialise to increase host location ability; (iii) competition or mortality from generalist invertebrate frugivores and natural enemies is also minimal, reducing the need for specialisation to avoid them; and (iv) the ability to move into enemy-free space from specialist parasitoids when using a novel host may be a driver to increase host range. This is a complex hypothesis that needs to be tested, but the individual components are already demonstrated.

6.1.2 Preference/performance relationships

When discussing host use patterns in fruit flies and other insect herbivores, it is common to break it down into two parts: (i) parental female oviposition preference; and (ii) offspring performance. A dominant theory in insect herbivore ecology, often referred to as *'mother knows best'*, predicts a positive relationship between the two. The mother-knows-best theory proposes that in species where the female oviposition choice strongly influences where her offspring will feed, then there should be a close and positive correlation between the hosts that the females most prefer to lay in, and the suitability of those hosts for their offspring development; the link is presumed, logically, to be driven by natural selection. Fruit flies, where the ovipositing mother plays a 100% role in determining where her offspring will subsequently feed, would be predicted to have a strong preference/performance relationship.

Despite the theoretical predictions, direct tests of the mother-knows-best theory in *Bactrocera* have commonly failed to detect positive preference/performance relationships. Some studies find such a relationship, others do not, and some find both. For example in a study of *B. dorsalis* host use of two mango varieties, each at three ripening stages, a positive preference/performance relationship was found for the mango variety Namdorkmai, but not for the variety Oakrong. In the latter variety the females showed a strong oviposition preference for fully-ripe fruit, but larvae did best in fruit just on colour change (Fig. 6.1). After studying multiple *Bactrocera* species across multiple host fruit, one author concluded that host use patterns in *Bactrocera* are best understood by studying female preference, rather than offspring performance, as the larvae could normally survive across multiple hosts regardless of whether placed there by females or not. The females, however, were selective in the hosts they used.

6.2 Host Location and Oviposition Behaviour

Location of a host fruit can be considered a sequential series of steps that involve a combination of visual and chemical cues. The two sensory modalities of vision and smell work together for finding a host fruit, although the relative importance of one to the other, at different steps of the host location and utilisation process,

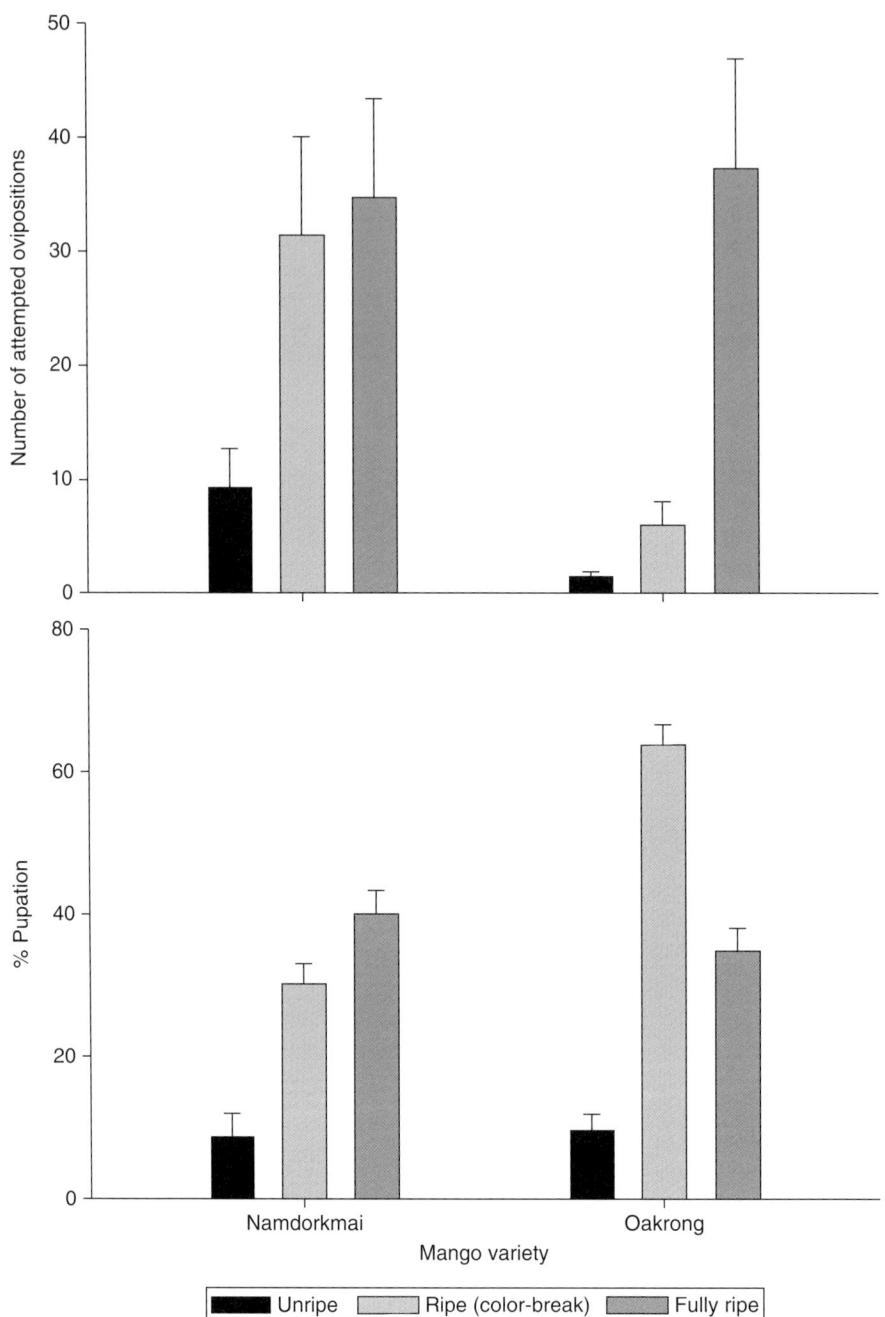

Fig. 6.1. *Bactrocera dorsalis* host preference and performance in two varieties of mango, at three stages of ripeness. Note the match in adult preference (oviposition) and offspring performance (% pupation) in variety Namdorkmai, but the mismatch in adult preference and offspring performance in variety Oakrong. In Oakrong mangoes the females show significant preference for full-ripe mangoes, but offspring do best in ripe (i.e. colour break) mangoes. Redrawn from Rattanapun *et al.* 2009

is unknown. Once a fly is on a fruit piece then olfaction remains important, although contact chemoreceptors (i.e. taste sensors) located on different locations on the body also play a role. Rather than divide this section into olfactory and chemical cues, the section describes the key behaviours and sensory mechanisms associated with the different steps of host location and utilisation.

6.2.1 Host tree location

It is unknown how *Bactrocera* first locate a fruiting host plant from a distance. It is possible that they initially search for silhouettes of plants on the horizon, as there is behavioural evidence that they will orientate to complex black and white patterns. It may also be that initial foraging is random, as many mark-release-recapture studies record largely non-directional dispersal from a release site. A capacity to learn about host fruit from exposure to them (see Section 6.3.2) may also increase a fly's capacity to orientate to a host plant from a distance; or alternatively to keep the fly in the vicinity of a fruiting host plant while it remains in fruit.

Olfaction

Olfactory cues, both the odours of host fruit and possibly host tree foliage are known to be important in host plant location, and it is presumed they play a key orientation role once a fly is within olfactory range. What the 'olfactory range' of a plant is remains unknown. Presumably it will be highly variable based on the amount of odour being produced by the plant (itself dependent on the amount of fruit, size of fruit, odours produced by the fruit), and wind speed and direction. For the polyphagous *B. tryoni*, stage of fruit ripeness also plays a major role in how chemically attractive fruit are, with flies responding strongly and positively to very ripe guava fruit, but significantly less so to under-ripe or green fruit (Fig. 6.2).

There is an extensive literature associated with identifying fruit volatile odours that are associated with host fruit location, largely because of the interest in developing fruit-based odour attractants. The standard research protocol involves screening the odour of fruit through a coupled gas chromatograph/mass spectrometer (GC-MS) to isolate and identify individual fruit odours. Identified chemicals are then blown across an electroantennogram (EAG) to determine if the chemicals produce a neural response in the fly (i.e. are they biologically active). These two steps can be combined using a linked GC-EAG, with subsequent MS done only on the biologically active chemicals. Blends of the most common chemicals are then behaviourally tested against flies for their attractiveness. Chemical classes that are known to be biologically active to *Bactrocera* include alcohols (e.g. ethanol), aromatic esters (e.g. isoamyl acetate, hexyl acetate), phenylbutanoids (eugenol, methyl eugenol) and sesquiterpenes (e.g. (E)-β-farnesene, (Z,E)-α-farnesene).

Increasingly, research is showing that the behavioural response is to a chemical blend, not to individual chemical(s), even though individual chemicals elicit electroantennogram responses. In the design of fruit-odour-based

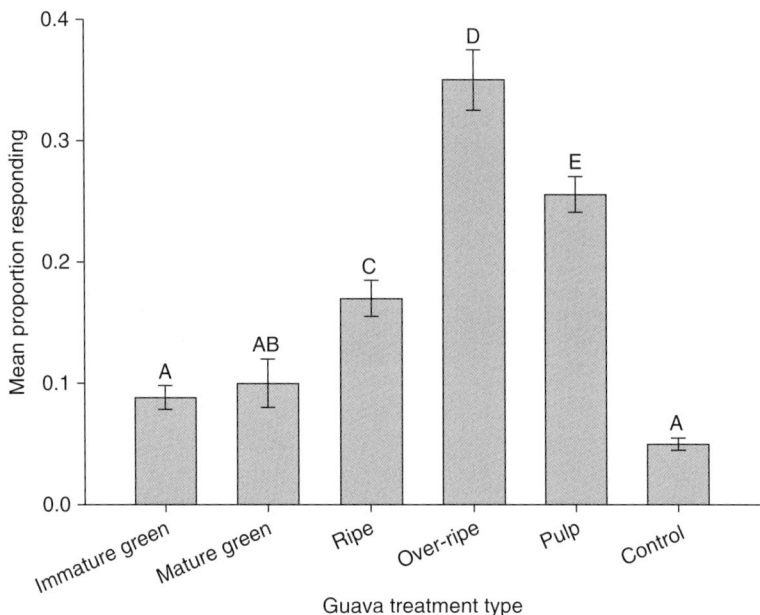

Fig. 6.2. Response of *Bactrocera tryoni* females to the volatiles of four ripening classes of guava. (redrawn from Cunningham *et al.*, 2016).

lures for female *Bactrocera,* it has become clear that getting the blend right (in both the individual chemicals used and their correct ratios to each other) is vital to getting a working lure. For the dacines, a commercial fruit-odour-based lure has so far only been developed for *Z. cucurbitae,* based on chemicals derived from cucumber. Fruit-based odour lures for the polyphagous *Bactrocera* have so far not been shown to work well in the field, and it is hoped that new advances in both the theory and practice of odour blends, as well as including volatiles from fruit fly associated bacteria and fungi, may increase this success rate.

Visual
In addition to olfactory response, *Bactrocera* also have strong visual orientation ability. Flies respond strongly to colour, with blue, yellow and red the preferred colours depending on trial and species. Yellow appears most attractive when painted on a panel, but dark colours (blue, red and black) are most attractive when used on spheres. This is almost certainly related to host location in their endemic South-east Asian rainforest, where 70% of bird-dispersed fleshy fruits (the primary hosts of *Bactrocera*) are black or red in colour. However, in a study of *B. tryoni,* the size of painted spheres was as or more important than their colour, with larger spheres preferred over a cluster of smaller spheres, reinforcing the importance of visual cues other than colour alone in *Bactrocera* host location. Colour itself is also only part of the orientation mechanism, as the same colour (blue) with elevated reflectance was more attractive to *B. tryoni* than blue without elevated reflectance. This was speculated to be a link with the waxy 'blooms' that often cover rainforest fruits.

Knowledge of visual orientation in fruit flies has been applied to practical ends. The 'Ladd' trap is a commercially manufactured 'sticky trap' that, while initially developed based on host location research for *Rhagoletis pomonella*, has also been shown to be an effective trap for both *B. dorsalis* and *B. tryoni*. A Ladd trap consists of a yellow panel with a red sphere in the centre, all of which is covered with a non-drying glue (Fig. 6.3). The assumed mechanism of the Ladd trap, which makes it attractive to fruit flies, is that the yellow panel is a 'super-foliage' mimic (this has been widely demonstrated in other insect systems), while the central red sphere mimics a piece of fruit. While the red sphere was originally designed as an apple mimic for the apple maggot fly, research on *B. tryoni* showed changing the colour mimic to blue did not increase attractancy. That the red sphere and yellow panel are considered to have different attractant properties is supported by trap data. For *B. tryoni*, the yellow panel caught both males and females, but slightly more males; the catch on the red sphere was strongly sex biased towards females. This trap, originally developed based on research by the late Ron Prokopy and colleagues in Massachusetts, is an excellent example of how a standard tool (i.e. a yellow sticky trap) can be improved based on biological knowledge of the target organism.

Fig. 6.3. Ladd trap in the field (photo credit Dr Mark Schutze).

6.2.2   Host fruit location within a tree

Bactrocera do not appear to fly directly to a fruit from outside a plant canopy. Rather, like other frugivorous tephritids (e.g. *Rhagoletis*, *Anastrepha*), they land on canopy foliage and then enter a searching mode that is based on upwards movement by walking along branches and/or short 'hopping' flights between neighbouring branches. In this way they gradually move upwards within a canopy until they detect a fruit, at which point they move directly towards it. The threshold distance at which a *Bactrocera* is likely to respond to a fruit has not been determined, but for *R. pomonella* it is approximately 80 cm. In experimental studies with *B. tryoni*, failure to detect a fruit within a canopy within 15 minutes caused the fly to leave the canopy, but the presence of even a few fruit on a tree will greatly extend the time spent within the canopy. The longer time spent by flies in canopies with fruit eventually results in flies being aggregated in trees with fruit.

The mechanisms by which the fruit is recognised at short distance are not known, but most likely involves both visual and olfactory cues. Fruit surrounded by foliage is more likely to be detected by flies than fruit clear of foliage, suggesting that a fruit/foliage contrast effect is important (a biological basis of the Ladd trap design, where red spheres on a yellow panel are more attractive than red spheres on their own). Olive fruit fly shows selectivity of individual fruit within a tree based on the fruit's size, shape and a combination of the fruit's hue and reflectance. Within a canopy, female *B. tryoni* respond differently to visual fruit mimics versus real fruit, suggesting that vision alone is insufficient to find a fruit within a canopy. Carbon dioxide produced from fruit wounds has been identified as a short-range attractant and oviposition cue for *B. tryoni*, and so this chemical may play a role in within-canopy foraging.

6.2.3   Oviposition behaviour

Pre-oviposition
Arriving on a fruit piece, a female fly will explore nearly all the surface of the fruit by walking back and forward across it. This is a relatively quick behaviour that is commonly reported across different fly species. Flies show a distinct preference for fruit that has existing mechanical damage, including oviposition puncture wounds caused by other flies, so long as the previous eggs have not hatched (see Sections 6.4.5 and 6.4.6). The cues likely used to detect such wounds are CO_2, detected by CO_2 receptors on the antenna, and fructose, detected by tarsal and labellar gustatory receptors.

Flies have been shown to exhibit preferences for laying on the shady side of trees, or individual fruit that are shaded versus those that are not. Flies also prefer to lay on the leeward side of fruit (i.e. out of the wind), and the combination of preference for shade and wind shelter can lead to a non-random distribution of eggs in an orchard, especially if there is a common prevailing wind direction. *Bactrocera dorsalis* prefers the top (i.e. stalk end) of fruit over the bottom of fruit, and this may be because, in climacteric ripening fruit, the top of the fruit will always be slightly riper than the bottom. In laboratory experiments it is common behaviour for flies to show strong preference for laying on the edge of the abscission scar on fruit (i.e. the location where the fruit and stalk connect), and it is presumed that this is because the peel is naturally thin in this area. So strong is this response that many experimental papers refer to the abscission area as being experimentally covered with wax or similar, so as to get more realistic oviposition behaviour from the flies.

Oviposition
Having located a likely oviposition site on a fruit, Pritchard (1969, p. 295) describes in detail the oviposition behaviour of an individual fly.

> With ovipositor extended, the female applies the oral region of her head to the fruit surface. The head is then lifted and the female walks forward, dragging her ovipositor up to the area tested by the head. As she walks, her body is arched so that the ovipositor comes to form an angle of 60-70° with the fruit surface at the point of penetration. ... the point at which the ovipositor is eventually pushed is the same point tested by the head. ...Having penetrated the surface, or entered the existing hole, the fly moves backwards slightly, bringing the ovipositor to an angle of 90° with the fruit surface. During the initial part of the move forward the fly walks normally, following the typical tripod procedure, but in the last stage arching of the body is the main cause of forward movement of the ovipositor, accompanied by movement of the forelegs, while the middle and particularly the hind legs are extended to accommodate.
>
> The significance of the contact between the head and the fruit surface is not clear. The head is lowered far enough for the antennae to touch the fruit surface, for the antennae are noticeably bent back as the fly moves forward. ... The behaviour of the ovipositing female suggests that the probing site is found by chemical and perhaps tactile stimulation of sensilla on the head.

In addition to sensilla on the head and feet, the ovipositor is innervated with mechanoreceptors, hygroreceptors and gustatory receptors. What role these receptors play immediately prior to and during fruit penetration is unknown.

Following insertion of the ovipositor, the female aculeus makes a small cavity within the fruit into which the eggs are laid. At its maximum, the depth of the cavity is limited by the physical length of the female ovipositor, and if this allows egg placement directly into the fruit flesh then eggs will be placed there. However, eggs will be laid into the peel itself if the peel is too thick to be penetrated. Depth of the egg cavity is also inversely correlated with increasing fruit peel and/or flesh density. In a detailed study of *B. tryoni* oviposition across five *Citrus* species/varieties, depth of oviposition was negatively correlated with increasing cell density in the peel layer. In Eureka lemon, particularly, the density of the peel meant that eggs were placed just beneath the outermost layer of peel cells.

Clutch size varies dramatically within and across *Bactrocera* species, with some reports of as few as one egg in a clutch (for *B. oleae*), to in excess of 100 eggs in a clutch (for *B. dorsalis*). 'Normal' range for a polyphagous *Bactrocera* is probably between 2 to 25 eggs per clutch. Factors causing clutch size variation within a species include: the existing female egg load, more eggs are laid if egg load is high; the availability of fruit, fewer eggs per clutch but more clutches are laid as fruit availability increases; and fruit/peel density, clutches become larger the longer it takes a female to bore into fruit.

Post-oviposition

Unlike many frugivorous tephritid genera, marking of the fruit with an oviposition deterring pheromone (ODP) is not known in *Bactrocera*. ODPs are applied from a special gland associated with the ovipositor, and with species that show such behaviour the ODP deters other females from ovipositing: this is not known in nearly all *Bactrocera*. The only exception to host marking recorded from *Bactrocera* involves the olive fruit fly. In *B. oleae*, the female after oviposition uses her labellum to spread olive juice escaping from the oviposition wound around the wound site and the fruit. This deters other females and limits olive fruit infestation to only one larva per fruit.

After oviposition females will spend a significant amount of time cleaning the ovipositor with the hind legs. Females will rest on the fruit, or on leaves near the fruit, between oviposition bouts. The exception to this appears to be in cucurbit-infesting species such as *Z. cucurbitae*, where resting occurs on vegetation away from the vining host.

6.2.4   Other adult behaviour on fruit

Research has shown that the presence of an individual fly on a fruit piece does not influence the likelihood of subsequent females alighting on that fruit, and so during the initial fruit-exploration phase female flies may encounter other females already on the fruit. Resident females may defend the fruit, even if already ovipositing, with flies elevating themselves on their tarsi, holding wings out at right angles to the body, and waving the forelegs. The fly may move from side-to-side during this display. Failure to remove an intruder can escalate to head butting, in which either the resident or intruder may become the victor (Fig. 6.4). Aggressive behaviour of this type is commonly observed in dacines and other tephritids, and may take place on the underside of leaves, as well as on fruit.

Aggressive behaviour on fruit is not invariable, and in some cases multiple females may simultaneously use a single fruit piece. Where a second female arrives on a fruit piece where a female is already ovipositing, and aggression does not occur, the time taken for the second female to start ovipositing is shorter than if she were alone on the fruit (referred to as facilitation of oviposition). Why some interactions lead to competition, while others lead to facilitation, has not been researched.

6.3   Intrinsic Factors Influencing Host Use

Factors internal or intrinsic to a fly (age, experience, etc.) will modify its host use patterns, but this area is poorly researched in *Bactrocera*, with only a small

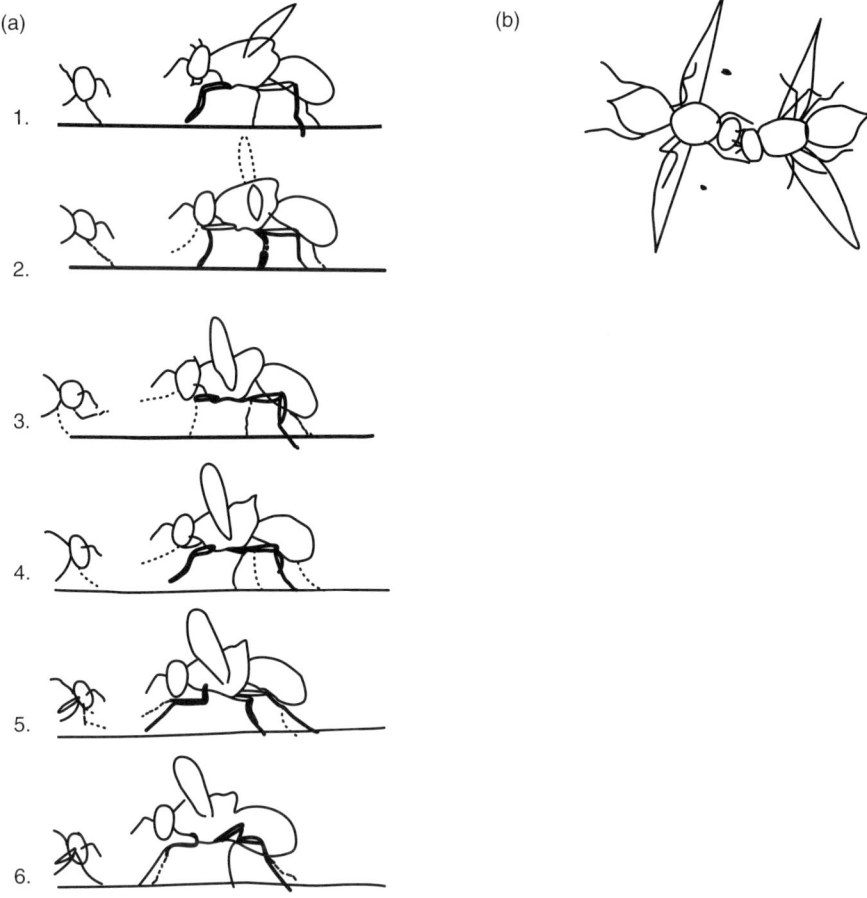

Fig. 6.4. The stereotyped, aggressive interactions between female *Bactrocera tryoni* when on the same fruit. The behaviour involves simultaneous body orientation, wing orientation and leg movements. Part (a) is the 'tip-toe' stance and involves stereotyped wing and leg movements, part (b) is an escalated response involving headbutting (redrawn from Pritchard, 1969). Reproduced from Pritchard (1969) with permission from CSIRO Publishing.

number of papers in the field. All such papers deal with polyphagous species, so it is unclear how applicable the findings are to host specialists.

6.3.1 Age

In tephritids, there is a general pattern that young flies lay more eggs than old flies. In *B. dorsalis*, younger females carry out more oviposition bouts than older females if fruit is not limiting, thus laying more clutches of smaller average size. If fruit is limiting, younger flies simply lay more eggs into the fruit that is available, meaning that average clutch sizes are larger.

6.3.2   Experience and learning

A positive influence on host use based on prior experience has been demon-strated for both *B. dorsalis* and *B. tryoni*. In an experiment using kumquats and green apples, a greater percentage of female *B. dorsalis* responded to kumquat if they had prior exposure to that fruit for a period of 3 to 5 days. The learning effect lasted for 3 to 5 days, even when the flies were exposed to an alternative host (apple) in the intervening period. Similar results have been found for *B. tryoni*, although the memory period for the first host differed depending on the second host offered. Thus a memory for pear (after exposure for 3 days) was retained for up to 4 days when tomato was offered in the intervening period, but only retained for 2 days if grapes were offered in the intervening period. Experience to any host also increases host location ability in comparison to an oviposition-ally naïve fly.

6.3.3   Egg load

In polyphagous *Bactrocera* species, high egg load will lead to decreased selec-tivity of host, that is, hosts of less preferred species/variety/ripeness will be used than if the egg load is small. If hosts are available but limited, high egg load will also lead to larger clutches being laid.

6.4   Extrinsic Factors Influencing Host Use

A great deal of work has been done on the external (i.e. extrinsic) factors that influence differential host use by *Bactrocera*. These factors include attributes derived directly from the fruit (e.g. its ripening stage or chemical properties) and external factors that have modified a fruit, such as mechanical damage or prior fruit fly oviposition.

6.4.1   Host availability

Host availability can impact on several attributes of host use. If all available fruit are of the same species, then altering host availability can change the number of oviposition bouts undertaken by a fly, and the number of eggs laid per fly. Specifically, if more fruits are available, flies will lay more, smaller-sized clutches, than if fewer fruits are available. This results in eggs being more widely distributed across host fruits and presumably decreases the possibility of larval competition between siblings.

An absence of host fruits appears to impact on generalist versus specialist *Bactrocera* species differently. Specialist species in an absence of preferred hosts are unlikely to change their host preferences, and in the absence of a preferred host, maturation of eggs carried by females is slowed or stopped. In contrast, host deprivation of preferred hosts leads to generalist species accumulating eggs in their ovaries and an acceptance of previously less or non-preferred hosts. At its extreme, and seen frequently in rearing cages, polyphagous species such as *B. tryoni* and *B. dorsalis* will simply dump eggs.

6.4.2 Host physical attributes

The physical attributes of a host fruit, especially peel properties, are one of the major determinants of host use by ovipositing *Bactrocera* females. A number of tropical fruits (e.g. rambutan and mangosteen) have flesh that is perfectly suitable for fruit fly larvae, but they are considered non-hosts because the hard peel cannot be penetrated by the ovipositing fly. Green mango is another suitable (although poor) larval host, but green fruit is essentially a non-host because of the tough peel. Peel characteristics are not just restricted to toughness. In tomatoes, *B. tryoni* resistance has been linked to peel elasticity. In this case, like pushing a blunt pencil into a balloon, the peel is elastic enough to bend before the female ovipositor without being punctured.

6.4.3 Host chemical attributes

In citrus, varietal resistance to *Bactrocera* is correlated with two peel attributes: the number of oil glands and the density of cells in the zest (i.e. flavedo) layer. If the cells in the flavedo layer are densely packed then the ovipositor cannot penetrate all the way into the oil gland-free pith (i.e. albedo). The larvae hatching in the flavedo layer are then exposed to the citrus oils, which are highly toxic to the young larvae. Valencia lemon, which has both dense cell packing and a large number of oil glands, is an extraordinarily bad host for *B. tryoni* because of these two attributes, which result in almost 100% larval death. This is despite the fact that the fruit will regularly get oviposited into in the field. Contrary to statements found in many early studies, there is generally not a relationship between sugar content (i.e. brix) and *Bactrocera* larval survival. Survival of larvae is no better or worse in the flesh of acidic fruit (such as lemons) than it is in sweet fruit.

Several fruits, such as mango and papaya, have toxic sap in the peel layer when the fruit is still green, and this offers another form of combined chemical/physical defence. High sap load in immature fruit may also offer a straight mechanical defence, as eggs can be pushed out of the fruit by leaking sap (Fig. 6.5). Because the fruit of many plants are designed to attract vertebrate seed dispersers, toxic components of fruit (such as sap) are often 'turned off' at maturity, making fruit that was toxic and not a susceptible host when immature, non-toxic and susceptible when ripe.

6.4.4 Ripening stage

Bactrocera preferentially lay into mature fruit at colour break or slightly later. This has been demonstrated formally in several fruits, including banana, guava,

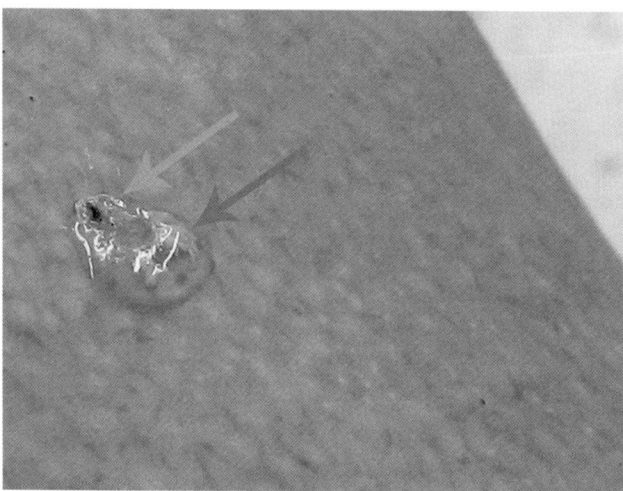

Fig. 6.5. An egg batch of *Bactrocera dorsalis* (bottom arrow) physically pushed out of an oviposition puncture (top arrow) in green mango by exuding sap (photo credit Dr Wigunda Rattanapun).

avocado and papaya, but it is also a simple observation made by anyone who has worked with these flies in the field. Not only do flies prefer mature fruit, but both *B. dorsalis* and *B. tryoni* have distinct behaviours of preferentially laying into the stem end of fruit, which for acclimatic (or gradually ripening) fruit is always the ripest part of the fruit.

A mistake is made, however, in assuming that *Bactrocera* only lay in mature fruit. *Bactrocera dorsalis* and *B. tryoni* (probably among others) will actively lay, or attempt to lay, in immature fruit, and larvae can develop through on green fruit. In the laboratory, *B. dorsalis* egg laying into immature (hard green) fruit has been demonstrated to produce viable offspring in mango, although at a much reduced rate of survival than in mature fruit. In the field in Thailand, *B. dorsalis* has been recorded from 20–30% of immature green guava, although the yield of pupae was <1–4/kg of fruit (in contrast, for ripe guava infestation was 85–97% and pupal yield 13–20/kg of fruit). The initial detection of an incursive (and since eradicated) *B. dorsalis* population in far north Queensland in the mid-1990s was through a farmer reporting fruit fly maggots in green papaya, an unusual behaviour but one that *B. dorsalis* is known to exhibit. Also in Australia, nectarines are regularly laid into as immature fruit by *B. tryoni*, which is evidenced by a unique wound response: sap comes out and then dries in a hair-like filament. At the extreme end of laying into immature fruit, *B. minax* will begin laying into tiny juvenile citrus once the fruit is equal to or greater than 11 mm in diameter. This specific size is the minimum size the fruit must be for the fly to gain an adequate purchase for egg laying. Several of the cucurbit specialist *Zeugodacus* preferentially lay into the juvenile fruit located at the base of female cucurbit flowers, and *Z. cucurbitae* has even been recorded from young cucurbit stems and seedlings.

While laying into immature fruit is sometimes recorded in the lab, and may occasionally be observed in the field, it must be recognised that this behaviour is rare in the field. Several crops have international market access based on picking at the hard green stage, and such market access is only gained following the confirmation from thousands of individual fruit samples that they do not host fruit fly. Such large, negative records are a more accurate reflection of how nearly all *Bactrocera* prefer to lay in mature fruit, than are the few (and often lab derived) positive records of immature fruit use.

6.4.5 Prior use

Bactrocera have been found to avoid reusing fruit that has been previously oviposited into by other *Bactrocera* females, of their own or another species. This effect is not immediate and flies will use an existing oviposition wound if the previously laid eggs have not yet hatched. This means that multiple oviposition bouts may be made into an oviposition wound over a period of approximately 48 hours from the initial egg lay to hatch. However, once larvae begin feeding repeat oviposition normally stops.

Behavioural work with *B. tryoni* showed that re-use of existing oviposition holes was associated with a marked decrease in 'host handling time'. That is, a female arriving at a recently stung piece of fruit spent much less time exploring the fruit surface before beginning to oviposit, and finished ovipositing sooner, than females laying into sound fruit. However, once the eggs hatch and the larvae commence feeding, new females are strongly deterred from using the fruit. The chemical acetoin, a compound produced by rot bacteria and captured from the head-space of *Bactrocera*-infested fruit, has been identified as having strong deterrent effects on ovipositing *B. tryoni*. The deterrence effect, while marked in laboratory studies, may break down in the field. In large fruit, or where fruit availability is limiting and fly populations are high, it is not uncommon to get mixed-aged larval cohorts in the one fruit piece.

6.4.6 Physical damage

Most frugivorous tephritids will preferentially oviposit into sites where the integrity of the fruit peel has been broken or lessened, for example, in puncture wounds caused by heteropteran feeding, where rubbing by a branch has broken or thinned peel, where fruit rot or disease has weakened the peel, or at sites where birds, bats or other vertebrates have fed. In a detailed study of *B. tryoni*, 70% of eggs were laid into existing fruit wounds when the fruit was hard, but in softer fruit where the fruit surface had protuberances, only 20% of eggs were laid into existing wounds.

The reason for using existing wounds is unclear. One study concluded the variable use of holes in hard and soft fruit was simply related to the mechanical ability of the fly to work the ovipositor into the fruit. In hard, smooth-surfaced fruit the fly could not get sufficient purchase on the fruit peel to insert the ovipositor and so preferentially used existing wounds. Alternatively, in soft fruit with surface protuberances to lodge against, then insertion of the ovipositor was easier and so existing wounds were used less often. In other tephritid literature the preference for existing wounds is attributed to fly behaviours designed to lessen wear on the ovipositor aculeus (i.e. to prevent bluntening), but in the only study to explicitly test this (an unpublished PhD study of *B. tryoni*) ovipositor wear seemed unimportant, but reduced host handling time was important.

6.5   Understanding Host Use in Polyphagous *Bactrocera*

While species such as *B. dorsalis*, *B. tryoni* and *Z. cucurbitae* may have large host ranges, this does not mean all available hosts are used equally. It is important to understand that even the most polyphagous species will use certain hosts more frequently than others.

6.5.1   Host ranking

In laboratory trials and field sampling, both *B. dorsalis* and *B. tryoni* show distinct ranking preferences for some fruits over others. Indeed, in every laboratory experiment the author and his students have ever run with *B. tryoni*, using different combinations of fruit, a female preference ranking is always found. Table 6.1 illustrates this point with field data. Both *B. dorsalis* and *B. correcta* are highly polyphagous flies, and if they used all available host plants equally then it would be expected that in a sample of fruit from a given location or region, the percentage of fruit fly reared from a given host plant would be proportional to the amount of that fruit type collected. The table shows this is not the case for either

Table 6.1. Examples of host use by two polyphagous *Bactrocera* species in Bangkok, Thailand (extracted from Table 1 in Clarke *et al.*, 2001). For each fruit species, the first number is the number of flies collected from that host as a percentage of the total number of each fly species collected. The (second number) is the number of infested fruit pieces of that fruit type as a percentage of all fruit collected infested by that fly species.

		B. dorsalis	*B. correcta*
Total number of flies reared		24 833	19 233
Total number of fruit sampled		53 352	51 957
Anacardiaceae	*Mangifera indica*	4.2 (2.90)	
Capparaceae	*Maerua siamensis*		2.4 (0.47)
Combretaceae	*Terminalia catappa*	62 (16.90)	31.3 (17.02)
Elaeocarpaceae	*Muntingia calabura*		3.2 (8.19)
Fabaceae	*Parkia speciosa*	3.0 (0.18)	
Myrtaceae	*Psidium guajava*	8.2 (3.81)	23.6 (3.84)
	Syzygium samarangense	7.1 (5.68)	18.4 (5.72)
Rhamnaceae	*Ziziphus jujuba*	4.6 (34.45)	14.7 (34.72)

of these species. For example, 62% of *B. dorsalis* and 31% of *B. correcta* came from only one host, *Terminalia catappa*, which constituted approximately 17% of infested fruit sampled for both fly species: thus this plant species is disproportionally preferred by both flies. At the other end of host use, *Ziziphus jujube* yielded on 4.6% of the *B. dorsalis* sampled, despite being 34% of the fruit collected: this fruit is lowly ranked.

6.5.2 No-choice host use

An important counterpart to this observation of host ranking, however, is what happens when the flies have no choice in egg laying. In no-choice laboratory trials, *B. tryoni* will use all offered hosts equally, even though it will rank the same hosts if they are offered simultaneously. Thus the polyphagous *Bactrocera* seem to have an extraordinary capacity to use non-preferred hosts if they have no choice (which can lead to highly unusual oviposition records in both the laboratory and field), but when preferred hosts are common they will use those and ignore the lesser hosts. A good example of this is *B. dorsalis* in coffee. When *B. dorsalis* invaded north Queensland in the 1990s, it badly damaged the small local coffee industry by infesting the ripe coffee berries. When a few years later the fly started moving into the coffee-rich highlands of Papua New Guinea, it was expected to have similar devastating results. Yet it never did and despite large coffee berry collections, only a few flies were ever reared from Highlands coffee. Why the difference? In north Queensland the coffee was grown in an isolated, dry sub-coastal site with few, if any, other suitable fruits around. In contrast, the tropical highlands of Papua New Guinea are rich with alternative host fruits. This is a field example of where the lack of preferred hosts meant coffee was attacked in Queensland, but the presence of alternative hosts in Papua New Guinea means coffee is ignored as a host.

6.5.3 Quantifying host records

A final point about understanding polyphagy is recognising what the published host records mean. Nearly all published host lists for *Bactrocera* spp., for which there are several comprehensive regional collations available, lack critical information. The most informative lists give the number of records made for a particular fly species/fruit species combination, and this can be valuable for polyphagous fruit fly species for providing a 'feel' of what may be regularly used hosts versus rarely used hosts. But these are the best lists, others provide no such information or use qualitative terms such as 'major host' or 'occasional host'. Importantly, no list provides information on negative rearing records, which are as important as positive records. For example, if *Bactrocera* Species 1 has published host records from Plant A (50 observations) and Plant B (5 observations), then the inference drawn from that data is that the fly uses Plant A ten times more often than Plant B (or at least a lot more often than Plant B). But if the same records said Plant A (50 positive observations from 500 fruit samples) and Plant B (5 observations from 8 fruit samples), then the inference that will be drawn is likely to be different despite the original data being the same.

Published host lists also do not have other critical information. Cultivar or variety of fruit is one example. Thin-skinned avocado varieties, particularly if over-ripe, can be heavily damaged by *Bactrocera*, but thick-skinned varieties can be conditional non-hosts because of the mechanical protection the skin provides: this information is lost from a host list that simply records *Persea americana*. Host lists, particularly if using older records, may also not mention if the records were derived from field collections or laboratory tests, which provides different information. The status of the piece of fruit providing a record, for example, its ripeness stage or presence or absence of fruit damage, is also not provided (despite such data often being recorded by the initial field collectors). This is important as, for example, lychee is not a *Bactrocera* host, but can potentially become one if the peel integrity is broken through disease or mechanical damage. If all that is recorded in a published list is *Litchi chinensis* then an erroneous host record may be being made.

The purpose of this section is not to denigrate the available host lists for *Bactrocera*. *Bactrocera* workers are extraordinarily fortunate by the large host lists that are available for the genus and related groups. Such lists are commonly absent for other pest groups and have only been developed for *Bactrocera* based on a huge amount of dedicated fruit collection, fruit rearing, fly identification and databasing. Rather, the purpose of this section is simply to highlight that the lists should not be used naïvely (especially by regulators) and that for understanding the polyphagous species they provide a starting point for further refinement, rather than a definitive never-to-be-questioned list.

6.6 Applications of Host Plant Usage

6.6.1 Host status and market access

Non-host status
If a fruit is not a host of fruit fly then it might be thought that this should be relatively easy to determine and then used as a basis for negotiating market access. However, based on a biological understanding of host use by *Bactrocera*, it can be quickly understood that determining host status or non-host status is not so easily assessed. For polyphagous flies, particularly, a lack of preferred hosts and high egg load may result in those flies utilising hosts that under most circumstances they would never use. Similarly, fruits may have peel that is normally impervious to the fly, but mechanical damage may break the skin and make the fruit susceptible. High fly numbers in a laboratory cage, with no other host choice, may also make flies use a host they would never use in the field. All such situations, and several more, may make flies utilise a diverse range of host fruit that may or may not be 'normal hosts'.

The biological problems of assessing host versus non-host status for tephritid flies have been developed at length in a review paper by Aluja and Mangan (2008), whereas a new international phytosanitary standard 'Determination of host status of fruit to fruit flies (Tephritidae)' (Secretariat of the International Plant Protection Convention, 2016) provides guidelines for developing data sets upon which to base market access negotiations based on non-host status. Both documents use the term 'natural host' to describe a plant species or cultivar for which unequivocal *field* data shows that the plant is used by the fly and can support fly growth from egg to adult. The term 'conditional host' is used in ISPM 37 to describe a host that, when exposed to a fly in semi-field conditions (i.e. whole plants, fruit on plant, field cage or other large cage allowing normal fly movement) can support oviposition and offspring development to viable adults. Fruits that do not produce viable adults under such conditions should now be regarded as non-hosts under international regulations. Small cage laboratory tests that report host use are now no longer sufficient evidence to claim the fruit is a host without further testing done under semi-field conditions. Laboratory tests are sufficient, however, to claim non-host status as it is assumed that fruit flies in a small cage are a 'worse case' scenario for trying to determine non-host status (i.e. polyphagous flies in a small cage will attempt to lay into anything). Further discussion of ISPM 37 is presented in Section 11.3.5, and the operational difficulties and complexities of determining host status are further discussed in Section 10.3.

Conditional non-host status (early harvest)

Bactrocera are widely recognised as preferentially attacking fruit only when fully mature, commonly at colour change or later. This is not true for all fruit, but it is for many. If the fruit is suitable to be picked before it becomes susceptible to fruit fly attack, then this obviously becomes a valuable control strategy. With appropriate field validation, harvesting at 'mature green' can be used as market access protocol based on what is often referred to as 'conditional non-host status'. Banana, papaya, mango and avocado are just some examples of fruits that are traded (in at least some countries) based on conditional non-host status to *Bactrocera* when picked at the hard green stage. Quarantine risk may arise if individual fruit ripen sooner that most of the crop (e.g. an individual banana finger within a bunch), or if peel integrity is lost through plant disease or physical damage allowing fruit flies to lay into fruit they would otherwise not be able to penetrate. Such fruit must be culled by pickers or in the packing shed. Armstrong (2001) provides an example of market access gained for mature green bananas. ISPM 37 does not use the term 'conditional non-host', rather it states to cover this situation:

> If no infestation is found after field surveillance by fruit sampling, and no further information indicates that the fruit has the potential to become infested, taking into consideration the conditions in which the commodity is known to be traded, such as physiological condition, cultivar and stage of maturity, the plant may be categorised as a non-host (ISPM 37, General Requirement Item C3).

6.6.2 Resistance breeding

Little research has been put into developing varietal resistance or resistance breeding as a management tool for *Bactrocera*. This is quite surprising given the pest status of *Bactrocera* species, how much variation is known in crop varietal susceptibility to fruit fly, and how much traditional plant breeding and newer biotechnology approaches have been used to manage other plant pests and diseases.

Indian research against melon fly has found across the cucurbit species musk melon (*Cucumis melo*), watermelon (*Citrullus lanatus*), ridge melon (*Luffa acutangula*) and bitter melon (*Momordica charantia*) that significant varietal resistance occurs in each against *Z. cucurbitae*. For each of these cucurbits there are varieties that are resistant to the fly (0% infestation), and varieties that are highly susceptible (80–100% infestation). Resistance has been correlated with both antixenotic traits (i.e. those that reduce or stop oviposition) and antibiotic traits (i.e. those that kill or weaken eggs and larvae). Antixenotic traits were found to include increased fruit hairiness, peel toughness and peel thickness, whereas antibiotic traits included increased levels of phenolics and tannins. In *Bactrocera* species these attributes have also been linked to differential host susceptibility, along with other traits such as peel flexibility and peel oils (in citrus).

Why haven't traditional and modern plant breeding processes been applied more for fruit fly control? The reason is unclear, but the answer received when this question was once asked of a professional plant breeder (who worked with subtropical fruits and routinely incorporated disease resistance into his selection programme) may give insights. His answer was 'if I make it taste bad to a fruit fly, it will taste bad to a human'. This answer suggests that entomologists have not done a good job of communicating with their plant breeding colleagues about resistance mechanisms, or an over-reliance on old literature that often reported (incorrectly) that fruit susceptibility to fruit fly was positively correlated with fruit sweetness (i.e. brix level), and hence to make a fruit resistant it had to be sour. Several studies have demonstrated that fruit brix level is rarely correlated with *Bactrocera* susceptibility, although brix may be auto-correlated with other fruit traits that are important for resistance. For example, during ripening, fruit susceptibility levels increase, as does brix level. But at the same time peel toughness and phenolic levels decrease (i.e. changes in brix, peel toughness, phenolic concentrations are biologically linked during ripening and so are statistically 'auto-correlated'). When each variable is experimentally assessed independently of the others it is the decrease in peel toughness and phenolic concentration, and not the increasing brix, that makes the fruit more susceptible to fruit fly attack.

As the factors that make a fruit resistant to fruit fly become better known, and there is increasing awareness that such factors may have little or no impact on fruit eating quality, there should be nothing to stop fruit fly resistance becoming an important part of traditional or biotechnology-driven selection programmes. With the exception of the Indian work referred to above, this is not being actively pursued by *Bactrocera* workers. However, it is a major thrust of far-sighted Mexican work to control *Anastrepha* fruit flies and I consider it a key future area for *Bactrocera* workers.

6.7 Further Reading and References Cited

Allwood, A.J., Chinajariyawong, A., Drew, R.A.I., Hamacek, E.L., Hancock, D.L., Hengsawad, C., Jipanin, J.C., Jirasurat, M., Kong Krong, C., Kritsaneepaiboon, S., Leong, C.T.S. and Vijaysegaran, S. (1999) Host plant records for fruit flies (Diptera: Tephritidae) in South East Asia. *Raffles Bulletin of Zoology*, Suppl. No. 7, 1–92.

Aluja, M. and Mangan, R.L. (2008) Fruit fly (Diptera: Tephritidae) host status determination: critical conceptual, methodological, and regulatory considerations. *Annual Review of Entomology*, 53, 473–502.

Armstrong, J.W. (2001) Quarantine security of bananas at harvest maturity against Mediterranean and Oriental fruit flies (Diptera: Tephritidae) in Hawaii. *Journal of Economic Entomology*, 94, 302–314.

Balagawi, S., Vijaysegaran, S., Drew, R.A.I. and Raghu, S. (2005) Influence of fruit traits on oviposition preference and offspring performance of *Bactrocera tryoni* (Froggatt) (Diptera: Tephritidae) on three tomato *(Lycopersicon lycopersicum)* cultivars. *Australian Journal of Entomology*, 44, 97–103.

Balagawi, S., Drew, R.A.I. and Clarke, A.R. (2013) Simultaneous tests of the preference-performance and phylogenetic conservatism hypotheses: Is either theory useful? *Arthropod-Plant Interactions*, 7, 299–313.

Clarke, A.R. (2017) Why so many polyphagous fruit flies (Diptera: Tephritidae): a further contribution to the 'generalism' debate. *Biological Journal of the Linnaean Society*, 120, 245–257.

Clarke, A.R., Allwood, A., Chinajariyawong, A., Drew, R.A.I., Hengsawad, C., Jirasurat, M., Kong Krong, C., Kritsaneepaiboon, S. and Vijaysegaran, S. (2001) Seasonal abundance and host use patterns of seven *Bactrocera* Macquart species (Diptera: Tephritidae) in Thailand and Peninsular Malaysia. *Raffles Bulletin of Zoology*, 49, 207–220.

Cugala, D., Ekesi, S., Ambasse, D., Adamu, R.S. and Mohamed, S.A. (2014) Assessment of ripening stages of Cavendish dwarf bananas as host or non-host to *Bactrocera invadens*. *Journal of Applied Entomology*, 138, 449–457.

Cunningham, J.P., Carlsson, M.A., Tommaso, F.V., Dekker, T. and Clarke, A.R. (2016) Do fruit ripening volatiles enable resource specialism in polyphagous fruit flies? *Journal of Chemical Ecology*, 42, 931–940.

Dalby-Ball, G. and Meats, A. (2000) Effects of fruit abundance within a tree canopy on the behaviour of wild and cultured Queensland fruit flies, *Bactrocera tryoni* (Froggatt) (Diptera: Tephritidae). *Australian Journal of Entomology*, 39, 201–207.

Drew, R.A.I., Prokopy, R.J. and Romig, M.C. (2003) Attraction of fruit flies of the genus *Bactrocera* to colored mimics of host fruit. *Entomologia Experimentalis et Applicata*, 107, 39–45.

Fitt, G.P. (1984) Oviposition behaviour of two tephritid fruit flies, *Dacus tryoni* and *Dacus jarvisi*, as influenced by the presence of larvae in the host fruit. *Oecologia* 62, 37–46.

Fitt, G.P. (1986a) The influence of a shortage of hosts on the specificity of oviposition behaviour in species of *Dacus* (Diptera, Tephritidae). *Physiological Entomology*, 11, 133–143.

Fitt, G.P. (1986b) The roles of adult and larval specialisations in limiting the occurrence of five species of *Dacus* (Diptera: Tephritidae) in cultivated fruits. *Oecologia*, 69, 101–109.

Goergen, G., Vayssières, J.-F., Gnanvossou, D. and Tindo, M. (2011) *Bactrocera invadens* (Diptera: Tephritidae), a new invasive fruit fly pest for the Afrotropical Region: host plant range and distribution in West and Central Africa. *Environmental Entomology*, 40, 844–854.

Gogi, M.D., Ashfaq, M., Arif, M.J., Sarfraz, R.M. and Nawab, N.N. (2010) Investigating phenotypic structures and allelochemical compounds of the fruits of *Momordica charantia* L. genotypes as sources of resistance against *Bactrocera cucurbitae* (Coquillett) (Diptera: Tephritidae). *Crop Protection*, 29, 884–890.

Haldhar, S.M., Choudhary, B.R., Bhargava, R. and Meena, S.R. (2015) Antixenotic and allel-ochemical resistance traits of watermelon against *Bactrocera cucurbitae* in a hot arid region of India. *Florida Entomologist*, 98, 827–834.

Hancock, D.L., Hamacek, E.L., Lloyd, A.C. and Elson-Harris, M.M. (2000) *The Distribution and Host Plants of Fruit Flies (Diptera: Tephritidae) in Australia*. DPI Publications, Brisbane, Australia.

Leblanc, L., Tora Vueti, E., Drew, R.A.I. and Allwood, A.J. (2012) Host plant records for fruit flies (Diptera: Tephritidae: Dacini) in the Pacific Islands. *Proceedings of the Hawaiian Entomological Society*, 44, 11–53.

Migani, V., Ekesi, S. and Hoffmeister, T.S. (2014) Physiology vs. environment: what drives oviposition decisions in mango fruit flies (*Bactrocera invadens* and *Ceratitis cosyra*)? *Journal of Applied Entomology*, 138, 395–402.

Pritchard, G. (1969) The ecology of a natural population of Queensland fruit fly, *Dacus tryoni II*. The distribution of eggs and its relation to behaviour. *Australian Journal of Zoology*, 17, 293–311. https://doi.org/10.1071/ZO9690293

Prokopy, R.J., Green, T.A. and Vargas, R.I. (1990) *Dacus dorsalis* flies can learn to find and accept host fruit. *Journal of Insect Behaviour*, 3, 663–672.

Rattanapun, W., Amornsak, W. and Clarke, A.R. (2009) *Bactrocera dorsalis* preference for and performance on two mango varieties at three stages of ripeness. *Entomologia Experimentalis et Applicata*, 131, 243–253.

Rwomushana, I., Ekesi, S., Gordon, I. and Ogol, C.K.P.O. (2008) Host plants and host plant preference studies for *Bactrocera invadens* (Diptera: Tephritidae) in Kenya, a new invasive fruit fly species in Africa. *Annals of the Entomological Society of America*, 101, 331–340.

Schutze, K.M., Cribb, C., Cunningham, J.P., Newman, J., Peek, T. and Clarke, A.R. (2016) 'Ladd traps' as a visual trap for male and female Queensland Fruit Fly, *Bactrocera tryoni* (Froggatt) (Diptera: Tephritidae). *Austral Entomology*, 55, 324–329.

Secretariat of the International Plant Protection Convention. (2016). ISPM 37: Determination of host status of fruit to fruit flies (Tephritidae). International Plant Protection Convention and FAO, Rome, Italy, 18 pp.

Shivashankar, S., Sumathi, M., Krishnakumar, N.K. and Rao, V.K. (2015) Role of phenolic acids and enzymes of phenylpropanoid pathway in resistance of chayote fruit (*Sechium edule*) against infestation by melon fly, *Bactrocera cucurbitae*. *Annals of Applied Biology*, 166, 420–433.

Siderhurst, M.S. and Jang, E.B. (2006) Female-biased attraction of oriental fruit fly, *Bactrocera dorsalis* (Hendel), to a blend of host fruit volatiles from *Terminalia catappa* L. *Journal of Chemical Ecology*, 32, 2513–2524.

Siderhurst, M.S. and Jang, E.B. (2010) Cucumber volatile blend attractive to female melon fly, *Bactrocera cucurbitae* (Coquillett). *Journal of Chemical Ecology*, 36, 699–708.

Wang, M., Cribb, B., Clarke, A.R. and Hanan, J. (2016) A generic individual-based model as a novel tool for investigating insect-plant interactions: a case study of the behavioural ecology of frugivorous Tephritidae. *PLOS ONE*, 11: e0151777.

Xu, L., Zhou, C., Xiao, Z., Zhang, P., Tang, Y. and Xu Y. (2012) Insect oviposition plasticity in response to host availability: the case of the tephritid fruit fly *Bactrocera dorsalis*. *Ecological Entomology*, 37, 446–452.

7 *Bactrocera* as Invasive Organisms

Dacini species are quarantine pests of global concern. Successful invasions by *Bactrocera* species over the last 20 years include, but are not restricted to, incursions in continental North America (several species, multiple occasions, eradicated and established [*B. oleae*]), South America (*B. carambolae*, established), Australia (*B. dorsalis* [twice], eradicated), New Zealand (*B. tryoni* multiple times, eradicated), China (*B. correcta*, established) and Africa (several species, established). One paper has reported Chinese quarantine figures of 19 000 interceptions of fruit fly-infested fruit over a 10-year period, of which 97.7% were *Bactrocera* species. Given this track record, it is not surprising that tens of millions of dollars are spent annually around the globe on border quarantine surveillance, early detection and emergency response for exotic *Bactrocera* species. An excellent review of tephritids as invasive organisms and the problems they cause is provided by Papadopoulos (2014).

Movement by *Bactrocera* species that are of quarantine importance is not just restricted to large-scale movement between continents. It also involves movement between neighbouring countries that share land borders, and movement within a country if the pest is regulated in one area and not in another. Such movement need not be human assisted, that is, it can be natural dispersal and so confounds quarantine efforts. A further confounding aspect of *Bactrocera* quarantine and invasions, is that *Bactrocera* species seem to be highly aggressive invaders, often rapidly displacing tephritid species that already exist in an area.

This chapter covers our understanding of *Bactrocera* species as invasive organisms, with a particular focus on their movement capacity, and then the related quarantine aspects of surveillance, early detection and response.

7.1 Biological Invasions

Biological invasions are broadly defined as the movement of organisms into areas where they did not previously occur, while invasion biology is the field of research that documents and analyses such movement. Commonly, but not exclusively, invasion biology focuses on organisms that have the demonstrated or potential capacity to cause human health, environmental or economic damage in their new location, and which have been moved with direct or indirect human assistance. Natural, rapid movement of pest organisms (e.g. on storm fronts) is also covered under invasion biology. The slow, gradual range expansion of native, non-pest animals into new environments is not considered within invasion biology, as different ecological and biological drivers operate.

Invasive organisms are considered major threats to human, environmental and agricultural health. Ever increasing levels and speed of human movement and trade increases the chances of pest organisms being carried, and if carried reaching a destination alive. Similarly, within agriculture and silviculture, the globalisation of major crop and tree species means that even host-specific pest organisms have increased likelihood of finding suitable hosts within a new environment. As an example of the damage that can be caused by invasive organisms, the massive losses of indigenous and first nation peoples around the world to novel diseases spread during the era of European colonisation are a stark and painful reminder. Global warming is increasing the threat posed by invasive organisms, particularly from the tropics and sub-tropics, as formally temperate areas are now becoming climatically suitable for tropical insect vectors and the diseases they carry.

While varying slightly between authors, there is a general commonality in how the invasion process is understood to occur, and in all cases it is considered a sequential or step-wise process with distinct phases (Fig. 7.1). The first step is the movement (i.e. immigration) of the organism from its former range to the new range. The second stage is growth and reproduction of at least one individual, followed by the third stage of population growth to a minimum viable population. Some authors link these two phases as 'establishment'. Also treated under establishment, or sometimes as a separate component, is local adaptation. Plants, especially, often show long lag phases (of 50 to 100 years) in an invasion process where the incursive population remains at low levels

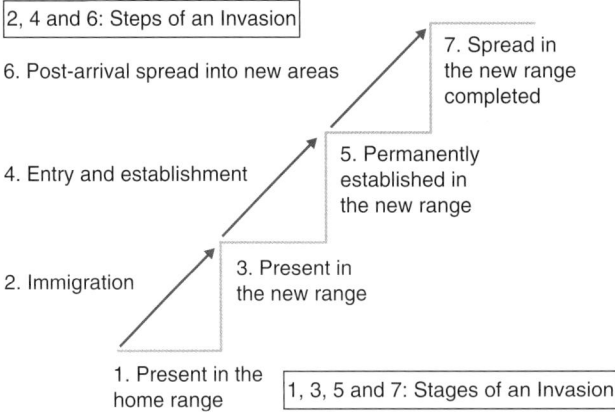

Fig. 7.1. A diagrammatic representation of the biological invasion process.

until eventually exploding. This has been explained through a local adaptation process. The final stage of the invasion process is considered the spread of the now established population in its new environment.

The invasion process can be illustrated by the well-documented invasion of Africa by *B. dorsalis*. The immigration step occurred sometime before 2003 into an unknown location in East Africa. The incursive flies were initially postulated to be from Sri Lanka, but later evidence has suggested other parts of the larger Indian sub-continent. Wherever they came from, the immigrating flies had to cover a large distance (i.e. the Indian Ocean) and their movement was almost certainly human assisted, most likely through the carriage of infested fruit. An initial population successfully established and built up, such that a large and already locally dispersed population was already present when first identified by Lux and colleagues in Kenya in 2003. Since then the now permanent population has continued to expand its range, the third step of the invasion process, through all of sub-Saharan Africa except for parts of South Africa. The range expansion following initial establishment has been both dramatic and fast. It now only requires the fly to continue moving into central and southern South Africa, and across Mediterranean North Africa, and the invasion of an entire continent by one fruit fly species will be complete.

7.1.1 Human-assisted movement: The first step of the invasion process

Quarantine targets the first step of the invasion process: immigration from an outside source to the new locality. Stopping *Bactrocera* immigration means that an invasion has never started. Unfortunately, the evidence is that *Bactrocera* are good travellers and even the most stringent border security can be breached by natural and human-assisted movement.

It is assumed that most movement of pest *Bactrocera* species is through human carriage of infested fruit. Such carriage can either be in commercial fruit shipments or by private travellers. Amongst fruit fly workers, it is considered that the risk of fruit fly movement via regulated, commercial commodity trade is low due to the strict phytosanitary requirements that need to be met in order to gain and maintain international market access. There is no such thing as nil risk in fresh commodity trade, but so long as there is no failure in the application of the phytosanitary treatments required then it is assumed that risk through commercial commodity movement is minimised. In an analysis of New Zealand data, Framptom and Nalder (2009) confirmed this general assumption, finding that of 104 'new-to-New Zealand' insect and mite records made over the period 1990–2007, only 12 were intercepted on fresh produce. They concluded that commercially imported fresh produce did not constitute a major pathway for exotic pests into New Zealand. However, non-regulated commercial movement within and between adjoining countries may be an extremely important form of fruit fly movement. The rapid spread of *B. dorsalis* throughout sub-Saharan Africa in a 10-year period is almost certainly due to large-scale non-regulated fruit movement within countries and, even if illegal, across borders. Risk of fruit fly movement in commercial fruit may also depend heavily on source country. An analysis of US interceptions reported 30% of exotic plant pests detected were associated with commercial shipments, but this figure varied dramatically by country and region of origin.

In contrast to authorised, commercial fruit movement, there is significant data to illustrate that large amounts of fruit infested with fruit fly are moved by private travellers. In the 16 years from 1984 to 2000, US quarantine made 2616 interceptions of tephritid-infested fruit from suitcases of travellers arriving into only seven of the US's international airports. The same study acknowledged 'only a tiny fraction of incoming passengers are subjected to inspections', and that actual carriage is likely to be substantially larger than the data reported.

The importance of traveller carriage in the movement of fruit flies can be illustrated not just by international carriage, but equally by domestic fruit movement. In a series of papers, Dominiak and colleagues analysed data from road blocks put in place to restrict the movement of potentially fruit fly-infested fruit from areas of Australia where Queensland fruit fly was then endemic (central and northern New South Wales [NSW], Queensland) to the then regulated, pest-free areas in southern NSW and Victoria. These analyses showed significant movement of fruit. For example, in one small pilot study that halted 1455 vehicles at one site over a period of less than 2 months, 5.3% of vehicles were carrying fruit into the regulated area, despite signage and public awareness campaigns prohibiting such. The interceptions amounted to 51.3 kg of fruit, including 18.5 kg of apples, 8.9 kg of oranges and 6.9 kg of tomatoes. While no fruit fly infestations were detected in this particular sample, this was not the case in larger data sets analysed. Analysing a data set running from 1976–1983, Dominiak and colleagues recorded 388 detections of infested fruit; approximately 16% of all intercepted fruit. The amount of fruit movement also varied significantly between studies, but the 5.3% reported above was low as other studies reported up to 20% fruit carriage into restricted areas. These high amounts of fruit movement are despite the fact that Australia, as a nation, has a high degree of quarantine awareness and the institutional and legal capacity to carry out surveys and implement spot fines. In countries where this is not the case, fruit movement and the risk of spreading flies is even greater.

Papua New Guinea (PNG) is a developing nation of the South Pacific. While it has both endemic fruit fly pests (e.g. *B. frauenfeldi* and *B. musae*), as well as exotic flies (e.g. *B. dorsalis* and *Z. cucurbitae*), not all are spread entirely across the country. PNG has a number of island provinces (e.g. East and West New Britain, Manus) from which some of these pests are absent and for which internal quarantine could play a great role in restricting their spread. Nevertheless, in a survey of 1904 domestic airline passengers, it was found that 39% of passengers were carrying fresh commodities. On certain airline routes, particularly those leaving the highly productive 'highlands' provinces, the percentage of carriage increased to 70–80%. Passengers carrying commodities were likely to be carrying two or more types, and nearly all exotic and native fruit and vegetables grown in PNG were identified as being carried. Further, nearly all fruit and vegetables came from backyard gardens or local markets, where infestation rates could be 20% or higher. For one particular example for which risk analysis was done, it was estimated that every second flight into a province then free of banana fly was likely to have a passenger carrying infested bananas. Multiple passengers of every flight were carrying bananas. That province is no longer *B. musae* free.

In PNG, fruit and vegetables are carried as cultural gifts, and such gifts are also customary in many parts of the Pacific, Asia and Africa. Nevertheless, they carry huge risk and may play a large part in the observed rapid spread of exotic fruit flies when they arrive in countries.

7.1.2   Population establishment: The second step of the invasion process

Initial establishment

One researcher has coined the phrase for tephritids 'good travellers, poor colonisers', and this sums up well what appears to be two important aspects of *Bactrocera* invasion biology. As eggs or larvae in fruit, *Bactrocera* can travel within a protected micro-environment, and the frequent formal and informal movement of fresh produce means that there is high likelihood of flies being moved. Thankfully, every such movement of infested fruit does not lead to establishment, and there is good evidence that most potential *Bactrocera* invasions fail at the establishment phase.

If an immigration of *Bactrocera* begins as larvae in fruit, then those larvae have to complete development, successfully pupate, emerge and then develop through to sexual maturity and locate a mate, before there is any possibility that the population can establish. It is this sequence of steps, and something known as the Allee effect, that makes initial establishment of *Bactrocera* populations rare. The Allee effect, a population genetics term, says that the genetic fitness of an individual within a population increases with increasing population size. The Allee effect describes the outcome of one or more factors that act on populations, and will vary between species, or even between populations of the same species in different places or at different times. At low population levels, populations can be threatened because most of the population is past breeding age, or a negative mutation is common, or simple chance events (e.g. a wild fire) destroy many individuals. All such things will impact more on a small population than a large population – thus the Allee effect is the bane of conservation biologists.

The relevance of the Allee effect to *Bactrocera* relates to the chances of a small incursive population reaching permanent establishment. If it is assumed that a single infested piece of fruit is brought into a country, then the number of maggots within that fruit may be as few as one or two, maybe as many as ten, but rarely more. By the time sexual maturity is reached, the cohort may have already died out, or at most there will be only a few individuals left. For reproduction to occur then there has to be at least one female and one male, and they have to be able to locate each other. The chances of this happening is low and so it can be easily seen that the fitness (i.e. likelihood to pass on genes to the next generation) for any individual within that population will be zero, or close to zero. The Allee threshold is a theoretical population size, below which Allee effects will eventually drive a population to extinction, but above which the population will survive and grow. In ecological literature, the Allee threshold is also often referred to as the minimum viable population.

Operationally, the Allee effect is often seen in intensive fruit fly monitoring programmes to demonstrate area freedom. A single fly, or maybe two flies, might appear in a surveillance trap, but then no further flies are ever caught and it is assumed that the population has simply died out because a viable population could not establish. For this reason, many surveillance programmes need multiple flies to be captured before a full response is triggered, as it is presumed below the threshold catch the population will go to extinction naturally.

It is clear that the Allee effect is driven by the number of individuals in a population, and for this reason 'propagule pressure' is widely regarded as the single most important factor influencing invasion success. Propagule pressure is a fancy phrase to simply say how many invasive organisms are there, and/or how often do these invasive organisms appear. For pretty much any biological system, high propagule pressure will eventually result in successful invasion, assuming the new environment is suitable. Low propagule pressure does not guarantee an invasion never happens, but it will lower the likelihood. Border quarantine and risk reduction treatments (Chapter 10) act to reduce propagule pressure.

Early population growth

Assuming that an incursive population has got through the first generation, operational and theoretical evidence suggests that there is a window of only a few generations before an invasive *Bactrocera* population is large and well above minimum viable size. This causes problems, because experience shows that except under the most intensive surveillance systems, populations can be well established before detection occurs. For example, *B. dorsalis* was widely distributed with locally large populations in southern Kenya when first detected in 2003, but a general fruit fly survey of Kenya and neighbouring east African countries in 2000 did not detect the species. In an invasion of East New Britain Province (PNG) by *B. musae*, the first detection of the fly (through infested fruit) was in September 1999, by mid-2000 the fly was widely established through a densely populated coastal strip, and by December 2000 the fly was spread through nearly the entire Gazelle Peninsula and considered beyond local capacity to eradicate (Fig. 7.2). While even well-established populations of invasive *Bactrocera* have been eradicated (see Section 7.2.2), it is recognised that the earlier an invasive population is detected, the greater the chance of successful eradication and the lower the eradication cost.

7.1.3 Population spread after establishment: The final step of the invasion process

Invasion spread models

Once a viable population has established, the final step of the invasion process is the spread of individuals into new areas of the recently invaded environment. There are classically considered two ways that this can occur, and both are

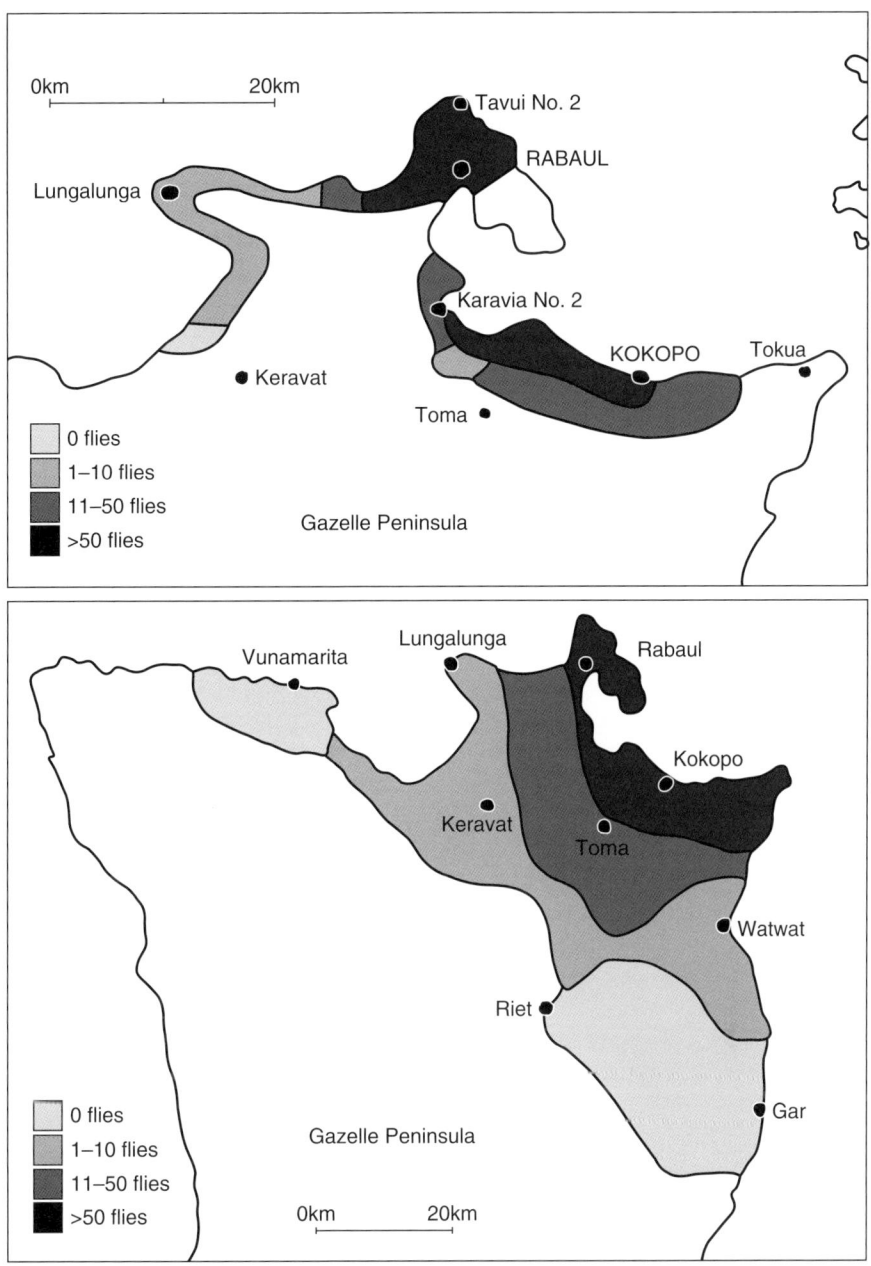

Fig. 7.2. The spread of banana fruit fly, *Bactrocera musae*, on the Gazelle Peninsula, East New Britain Province, Papua New Guinea. First detected through an infested banana in September 1999, a survey in mid-2000 found the fly widely spread throughout, but still restricted to, the densely settled coastal strip (top). By December 2000, the fly had dispersed throughout large areas of the peninsula, well away from the coast (bottom) (redrawn from Mararuai *et al.*, 2003).

relevant in *Bactrocera*. The first is known as the neighbourhood diffusion model, and this describes natural outward spread from a point, with each new invaded area being close to the last invaded area (Fig. 7.3a). For invasive *Bactrocera*, this model would describe a spread pattern where movement is through the natural dispersion of individual flies. The second model is known as the hierarchical diffusion model, where there are big jumps between infested areas, and the gaps in between are subsequently filled by natural dispersal (Fig. 7.3b). Such a spread pattern can occur with *Bactrocera* in a new invaded region if internal quarantine is not put in place. Infested fruit may be purchased from a market, and then transported long distances by truck, subsequently introducing the pest to a whole new district. The ability of *B. dorsalis* to spread over the whole of Africa in a decade is almost certainly due to a combination of natural and human-assisted dispersal.

Natural Bactrocera *movement*

While placed within the context of invasion biology, this section is equally relevant to understanding movement for landscape ecology (see Section 4.5) and area-wide management (see Section 9.11.2) .

Adult *Bactrocera* are mobile, but there is great discrepancy in the literature about how far an individual fly will move. Older literature infers that flies will move tens of kilometres easily in their lifetimes, but the consensus of more recent literature suggests that life-time movement of a few hundreds of metres (200–500 m, maximum 1 km) is more likely. One mark–release–recapture study identified a single marked fly 90 km from the initial release site and this was used for a long period as evidence for large distance dispersal, but a recent critique suggests that fly may have been moved in the car of the person clearing the traps. Most *Bactrocera* do not spend a great amount of their time flying. One field cage study reported that *B. tryoni* spends less than 1% of its time flying, the remainder of their time is spent walking (68%), grooming (14%) or simply being inactive (18%). This seems typical of several reports across different *Bactrocera* species, which spend most of their time within canopies, where

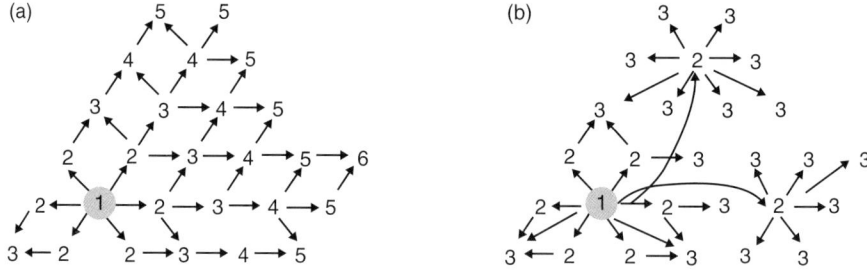

Fig. 7.3. Schematic representation of the (a) neighbourhood diffusion and (b) hierarchical diffusion models of population spread after an invasion. In neighbourhood diffusion an organism spreads sequentially from the nearest source, while in hierarchical diffusion there are larger jumps followed by diffusion. Because of human assisted carriage, many invasive *Bactrocera* may follow a hierarchical diffusion model (after Hengeveld, 1989)

movement is through walking or by short, hopping flights between branches. One consistency of much *Bactrocera* and *Zeugodacus* data, however, is that there is a great deal of within-population variation in propensity to move: this is reported in both field and laboratory studies. Thus while the great majority of individuals within a population may not move more than a few hundred metres, a small number of individuals may move several to even tens of kilometres.

There is no evidence in the *Bactrocera*, or indeed wider tephritid literature, that fruit flies are migratory in the sense of migratory birds or butterflies, or that dispersion is directional in any way. Dispersive movement by *Bactrocera* appears to be often gradual and linked to local foraging movements for resources, such as fruit hosts or resting sites.

Why flies may disperse from a site is unclear. A dispersal flight by newly emerged flies (i.e. the post-teneral dispersal flight) is often referred to in the literature, but the scientific justification for this claim is limited. B.S. Fletcher, in studies of *B. tryoni*, claimed that the post-teneral dispersal flight was 'obligatory', but he based this not on active flight interceptions, but on an absence of recaptured teneral flies in his experimental orchard. His explicit assumption was they had left his experimental orchard and therefore dispersed. However, in later laboratory studies of tethered *B. tryoni* on flight mills, as well as untethered flies, no evidence was found to support an innate post-teneral dispersal flight. Post-teneral flies do have a strong need to find sugar and protein for maturation, and there may well be local movement to find these resources if they are not in the immediate environment, but why immature flies would immediately undertake active dispersal flights is unclear as well as physiologically unlikely (they lack energy reserves). Nearly all studies that have been used to support the concept of post-teneral dispersal cannot separate dispersal by immature (post-teneral) versus mature flies.

There is direct evidence from the field and indirect laboratory evidence that mature flies will leave a resource patch once fruit are unavailable, as well as migrating into a patch as fruit (even immature fruit) becomes available. Drying of sites also leads to dispersal. However, over what distance *Bactrocera* will travel to find a new fruiting resource is unknown, and is a critical unanswered question for many areas of applied research, especially area-wide management. In contrast to dispersal, there is direct field data that show flies are relatively stationary within a habitat so long as fruit, moisture and roosting sites are available. In a Hawaiian study of *Z. cucurbitae*, 95% of released flies did not move more than 500 m in 4 weeks in a mixed agroecosystem with abundant larval hosts and roosting sites.

Wind and weather play an unclear role in assisting the movement of *Bactrocera*. An excellent analysis of captures of *B. dorsalis* in the Japanese prefecture of Okinawa from 1986–2012 showed that weather fronts, typhoons, high-pressure ridges and tropical depressions were highly correlated with *B. dorsalis* trap captures in the otherwise *B. dorsalis*-free prefecture. Backward trajectory analysis could place many of these captures as likely originating from *B. dorsalis* endemic areas of the Philippines, Taiwan and southern China. This infers that weather systems can be responsible for moving *Bactrocera* over large distances. At local levels the role of wind in *Bactrocera* movement is less clear. Different studies have reported positive correlations, negative correlations, or no correlations between mark–release recaptures and wind direction. One study of B. tryoni found no correlation of trap catch and wind direction for wind speeds <2 km/h, but a positive correlation for wind speeds >4 km/h. The effect of wind on local movement may be heavily influenced by other local variables, including habitat quality and fly physiological condition.

7.1.4   Population genetic structuring: Signatures of movement and invasion

While invasions or border detections are clear indications that *Bactrocera* species can move large distances, either through human carriage or natural movement, how frequently populations move at large scales is another issue. The field of population genetics can answer these questions, and it is extremely well developed for the pest dacines.

Population genetics uses molecular tools (e.g. microsatellite analysis, COI haplotype diversity analysis) to determine the relationship between populations within a species at scales ranging from tens of kilometres (landscape genetics) to thousands of kilometres (biogeographic analysis). Such tests can estimate the amount of gene flow (i.e. movement of reproducing adults) between populations, and estimate the degree of relatedness or difference between populations. The number of such studies for the major dacine pests is large, as similar studies are often duplicated in different regions or countries for the one species, are repeated across species, and then repeated again when new analytical tools become available.

Across these many studies some key themes appear. First, panmixia is common in many species across large geographic areas. Panmixia is the genetic term for 'all the same' and infers that the sampled populations are continually interbreeding. For example, within Australia, *B. tryoni* is considered a single panmictic population for the whole of the state of Queensland and far-northern NSW. Such data does not imply that a fly from northern Queensland will physically mate with a fly from southern Queensland (separated as they are by a distance of some 2000 km), but rather that through the continuous exchange of genes between *B. tryoni* populations across Queensland, the genetics of the species has become homogenised in that large geographical region. Panmixia across large geographic areas is routinely used as evidence for large amounts of long-distance, human-assisted fruit fly movement, but this is inference drawn from the data rather than a demonstrated fact. Quite low levels of cross-mating are required between populations to maintain panmixia, and another interpretation of data (nearly always ignored) is that long-distance dispersal by individual flies may happen more commonly than is detected in mark–release–recapture studies.

Closely related to the panmixia issue is that of isolation by distance. A significant isolation-by-distance signal identifies that populations geographically near to each other are genetically more similar than populations further away. A panmictic population will have only a weak or no isolation-by-distance signal in its genetic structure. In *Bactrocera*, positive isolation-by-distance signatures can be dependent on the scale of sampling. Thus *B. dorsalis* did not demonstrate a positive isolation-by-distance signature when sampling was restricted to Thailand (rather the population was panmictic), but a significant (albeit weak) signal was found when populations were sampled from both Thailand and Peninsula Malaysia.

There are exceptions to large-scale panmixia in the published data. For example, in the relatively small Indian Ocean island of Réunion, there are three distinct genetic clusters of *Z. cucurbitae*, with some evidence for spatial structuring of these clusters in different parts of the island (which have different environments). Melon fly is invasive on this island (first record 1972) and it is unclear if the population structuring observed is a historical signature of multiple invasions that may disappear over time, or a true spatial structuring that may strengthen over time driven by climate. In another example of invasive flies on islands, in this case *B. dorsalis* in the Hawaiian island chain, there is no evidence of any population structuring.

While panmixia is common in endemic populations of *Bactrocera*, which makes genetic analysis often uninformative, analysis of genetic population structuring can still be useful in understanding fruit fly invasions. At large geographic scales (e.g. across continents) sampled populations can often be genetically discriminated from each other. This is because the isolation-by-distance effect has become so large that populations effectively become isolated from each other, and so develop individual microsatellite or COI haplotype signatures. If the genetic fingerprint of an invasive population can be matched against suspected source population(s), then it can be a useful way of inferring invasion source. In California, approximately 40% of sampled invasive *B. dorsalis* were matched against the Hawaiian *B. dorsalis* genotype, which is quite unique. This gives strong evidence of one or more incursions of *B. dorsalis* from Hawaii to California. Where the remaining 60% of specimens came from is currently unknown, but even given this lack of information it still provides evidence that California has had a minimum of two invasions (at least one from Hawaii and at least one from elsewhere).

Invasions can also lead to strong genetic signals because of bottleneck effects. If an invasion stems from only a small incursive population (literally a few individuals), then by chance much of the genetic variation in the initial source populations will be lost. This leads to a new population with a highly restricted genetic diversity, which will be characterised by a domination of only one or two haplotypes or by microsatellites that have minimal variation. Genetic bottlenecking is seen commonly not only in biological invasions, but also in rebounding populations that had been previously reduced by hunting or similar harvesting (e.g. many whale populations). The unique and simplified genetic signature of *B. dorsalis* in Hawaii is a classical genetic signature of invasion. The interesting things about bottlenecks is that their signature can last over long periods (hundreds of years), and long after the initial small population has fully expanded in its new range

Bottlenecking effects can have both good and bad points in trying to work with invasive populations. As a plus, multiple invasions may lead to multiple sets of invasion fingerprints. For example in *B. tryoni*, while the northern endemic population in Queensland is panmictic, the exact opposite occurs in south-eastern Australia (southern NSW and Victoria). In this part of its range, where it is a recent invasive, repeated incursion events have led to many populations having their own microsatellite or COI haplotype fingerprints, and these can be used to help track the origin of southern populations. At much larger temporal and spatial scales, a series of recent and historical invasions and associated changes in gene frequency has allowed identification that *B. oleae* is most likely originally endemic to southern and eastern Africa, invading the Mediterranean in pre-history, and then invading California in recent decades from the Mediterranean. The down side of bottlenecking in invasions is that the genetic signature of the invasive population may become such a genetic 'subset' of the original parental population, that trying to genetically match populations can be difficult. The same olive fly study, for example, sampled flies from Pakistan, but they had such a reduced and unique genetic fingerprint that they could not be accurately assigned to a likely source region.

So, in summary, what can genetic approaches tell us about *Bactrocera* movement? Firstly, that often over large geographic areas there is commonly minimal genetic variation between populations of a species. This infers either regular human-assisted movement, or a greater degree of natural long-distance movement than anticipated from mark–release–recapture experiments. At continental scales (for the few widely distributed *Bactrocera* and *Zeugodacus* pest species), genetic population structuring can occur and these genetic 'signatures' can be used to help identify source populations, if known. Through bottlenecking, invasive populations can often develop unique genetic structuring and these can be used to map pathways of both recent and historical invasions.

7.1.5 Competition between species

An important aspect of the invasion biology of *Bactrocera* is their competitive interactions with existing fruit fly species already in a region, be those species native or previously introduced. As reviewed by Pierre-Francois Duyck and colleagues, and then demonstrated again during the invasion of Africa by *B. dorsalis*, the polyphagous *Bactrocera* (especially *B. dorsalis*), can markedly change the abundance and physical location of existing tephritid species. There appears to be a distinct hierarchy in this relationship, with certain *Bactrocera* species always competitively superior over other *Bactrocera* species, and then the *Bactrocera* all seem competitively superior over other tephritid genera, such as *Ceratitis* (Fig. 7.4). Certain life-history traits, notably large body size, long life span and low mobility (all *Bactrocera* attributes), may be correlated with this enhanced competitive ability.

Contrary to full hierarchical competition theory, however, fruit fly competitive interactions do not appear to drive species to extinction, but cause less competitive species to be physically displaced to other environments, or host plants, where the dominant species did not do as well (i.e. niche displacement). For *B. dorsalis* this means it has displaced other tephritids to higher, cooler altitude areas, drier areas, or onto hosts such as solanums or cucurbits that are poorly preferred. Work in the Union of Comoros, a nation consisting of three

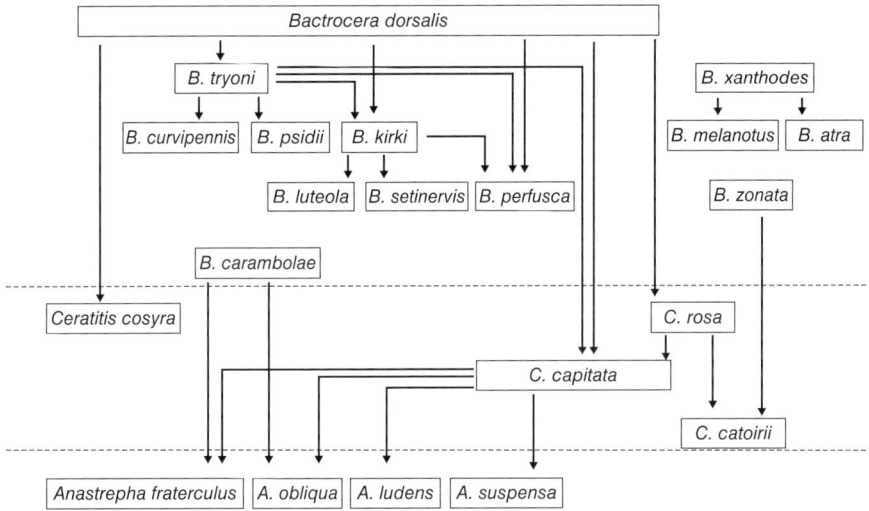

Fig. 7.4. Postulated competitive hierarchy of *Bactrocera*, *Ceratitis* and *Anastrepha* species developed by Pierre-Francois Duyck and colleagues from a review of literature. Since this work the competitive advantage of *B. dorsalis* over *Ceratitis* species has been confirmed repeatedly and dramatically in Africa and Indian Ocean islands (from Duyck *et al.*, 2004)

large islands in the western Indian Ocean, is suggestive that invasive *B. dorsalis* may have driven some rare native fruit fly species to extinction, but the nature of the sampling undertaken cannot confirm this.

As recognised by Duyck and colleagues in their 2004 review, much of the information they reviewed was weak on hard data, and so the argument for competitive displacement often had to be inferred, and the potential competitive mechanism unknown. The invasion of Africa by *B. dorsalis* has since provided the missing hard data. Displacement of native pest fruit flies by *B. dorsalis* in Africa has been identified by a series of independent researchers. The most dramatic data is that published by Ekesi and colleagues, who fortuitously had a long-term fruit fly project running before, during and after the *B. dorsalis* invasion. This data set, running over 8 years (Fig. 7.5), clearly shows the displacement of *Ceratitis rosa* by *B. dorsalis* in Kenyan mangoes. The competitive advantage of *B. dorsalis* over native flies has been experimentally demonstrated to be due to a scramble competition advantage of *B. dorsalis* larvae in fruit, seasonally earlier use of fruit by *B. dorsalis* females, and interference competition between *B. dorsalis* females and *C. rosa* females on fruit.

7.2 Quarantine and Response

7.2.1 Border quarantine and risk

Because fruit flies can move so easily, detection at the border, or early post-border, is critical for maintaining their long-term exclusion. It thus follows that strict quarantine, involving inspection at points of entry and intensive surveillance

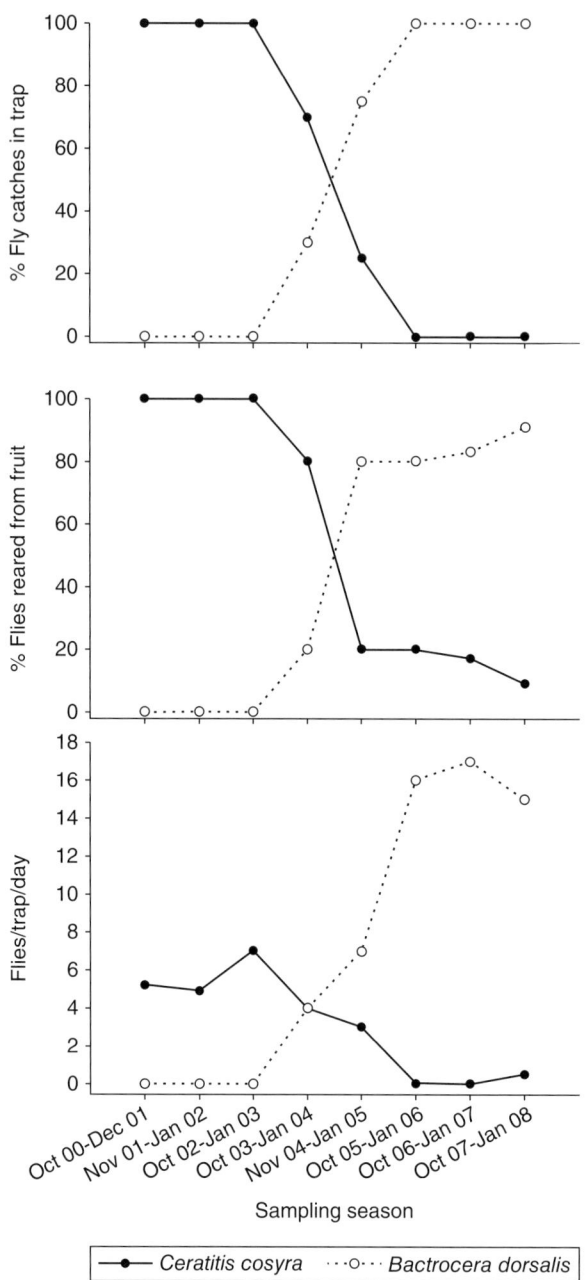

Fig. 7.5. The replacement of *Ceratitis cosyra* by *B. dorsalis* in Kenyan mangoes between the 2000/01 and 2007/08 mango seasons. This data set, a rare 'before-during-after' the invasion process is supplemented by detailed laboratory studies that show female *B. dorsalis* as more aggressive on fruit, utilise fruit earlier in the growing season, and lay large egg clutches (redrawn from Ekesi *et al.*, 2009).

around potential points of entry, offers the greatest opportunity to keep a country or region free of exotic fruit flies. However, quarantine is expensive and the investment made into a quarantine programme will depend on the perceived biosecurity risk of the exotic organism(s), the probability of detection and successful action, and the funding capacity of a government.

Risk

The biosecurity risk posed by Dacini fruit flies varies from extreme to nil, where risk = likelihood × impact. Likelihood in this case refers to the likelihood of entry, and that is dependent on the movement pathway from a source point to the end point. As an example, *Dacus dorjii* is a non-pest dacine known only from mid-altitude areas of Bhutan. The risk it poses to most countries is nil: possible pathways of the fly from Himalayan Bhutan to other regions are effectively absent, and then any impact if it arrived and established in a new area are again minimal because, even in Bhutan, it has no known pest status. In contrast, *B. dorsalis* is a risk of major importance to nearly all areas of the world where it is not already established. For Australia, as one example, possible pathways are numerous. In far-northern Australia entry of the fly from neighbouring Indonesia and PNG can be through adult flight and/or informal human carriage. Through all major ports, the fly many enter through passenger baggage or via fresh commodity shipment. At lesser northern ports of call, fishing boats and private vessels could bring the fly from South-east Asia and PNG. If entering and establishing in Australia, international experience shows the fly could rapidly increase its population, spread widely, and it will attack nearly all commercial fruits and fleshy vegetables, probably displacing the endemic fruit flies already present. The risk of this species to Australia is therefore extreme, as both likelihood and impact are high. For any individual country, risk prioritisation should be a primary part of prioritising fruit fly quarantine activities.

Probability of detection

Likelihood of detection for Dacini flies, or the fruit they may be carried in, are high compared to many plant pests. At airports, or other points of entry, visual luggage inspection, x-ray examination, and sniffer dogs can all be used to detect and intercept fruit that may be carrying flies. The effectiveness of such action is dependent on the number of passengers searched. Post-border, the males of many dacine species respond strongly to the male lures, and as new lures are being discovered for previously 'non-lure responsive' flies, the ability to run effective monitoring arrays for a large percentage of the pest dacines is high, so long as a country has the infrastructural and financial capacity. There is a large literature on optimising fruit fly trapping arrays. The designs of trapping arrays vary depending on the purpose of the arrays (e.g. early detection versus demonstration of regulatory pest freedom versus eradication), the lures being used, and any regulatory requirements. Because this literature has been extensively covered elsewhere (Shelly *et al.*, 2014), the reader is referred to that source for detailed information on tephritid surveillance and trapping.

Successful action

If detected early, the potential for eradication of a *Bactrocera* incursion is high (see next section). For this reason maintaining quarantine surveillance programmes is well justified against flies such as *B. dorsalis*, which have high impact if they establish.

7.2.2 Emergency response

Advance planning

Biological invasions are, by their nature, unexpected events. The 'law' of such things is that the first detection occurs late on a Friday afternoon, at the start of a public holiday period, or when some key scientist or administrator has just begun a period of extended leave. Avoiding a crisis is about having a response plan in place, including: having an identified chain of command and process for decision making; an identified reporting chain; an identified line of emergency funding; and an operational action plan that involves how delimiting surveys (trapping and fruit dissection) will be undertaken, emergency diagnostics, the size of quarantine zones, how emergency staff are recruited and who they report to, eradication treatment protocols, and ideally stock piles of material needed (e.g. lure, protein, etc.). See Manrakhan *et al.* (2012) in the reading list as an example.

While this may seem self-obvious, experience shows that it is not always done, even in countries or states that have the technical and financial capacity to undertake and afford such planning. The New Zealand government has extraordinarily well-developed protocols and action plans for response and eradication (if required) following detection of a fruit fly in their national trapping grid. As a result New Zealand has, to the start of 2018, a 100% success rate in eradicating incursions of exotic fruit flies. In contrast, Australia has well-developed generic response protocols following detection of exotic plant pests, but specific post-border response plans to key threats such as *B. dorsalis* or *Z. tau* are currently lacking; this puts the country at increased risk if an invasion occurs.

First response

In event of a post-border detection, either through trapping or infested fruit, the normal first response following positive identification of an exotic fly is to intensify sampling in the immediate area, and if required carry out a larger delimiting survey. Most countries with a permanent trapping grid follow strict protocols of the increased sampling effort required following an initial, single fly detection, with the response dependent on perceived national risk and, often, prior negotiated agreements with trading partners. High-intensity trapping (grid based, with traps every 50–100 m) is normally carried out at the point of the initial detection to determine if other flies are in the immediate vicinity. Fruit collecting and dissection to assess larval infestation, and to see if females are spotted sitting on fruit, is normally also undertaken at this fine scale. Involving the public in visually searching for, and ideally catching, females at this stage can greatly increase coverage. Movement rules should, ideally, be immediately enforced to stop the possible carriage of infested fruit out of the emergency zone. The trapping intensity, protocols for fruit dissection, processes for ongoing identification, and processes for public awareness and implementation of movement rules should all, ideally, be known in advance as part of advance planning documents.

Larger delimiting surveys may be carried out in parallel, or immediately following intensification of local trapping, again depending on perceived risk or prior planning. Commonly this is done by locating male traps along transects, or in grids, at spacings of multiple kilometres. The purpose of this is to quickly get an idea of how large the infested area might be. While, in an ideal world, an incursive population is detected early and the area involved is small, this is unfortunately often not the case and many *Bactrocera* incursions have not been detected until populations are large and well dispersed. For example, the delimiting survey for *B. dorsalis* in Africa, undertaken by Lux and colleagues within days of first detection, showed the fly to be already widely established across the whole of southern Kenya and almost certainly established in neighbouring Tanzania and Uganda.

Action: walk away, contain or eradicate
Action following first response is dependent on the results of the delimiting survey, assessment of likely impact and resources available. For pests of likely high impact, such as *B. dorsalis* or *B. tryoni*, eradication is a highly desirable outcome if at all possible, as future costs associated with permanently living with the pest are huge. An analysis by Suckling and colleagues using a data-base of prior eradication programmes showed that historical success of eradication programmes against *Bactrocera* species was high, at 88% (*n* = 108 programmes). However, cost increased logistically against the infested area being treated. Failing to act promptly, and failing to detect early, also increased the likelihood of an attempted eradication programme failing, or simply not being started. Most eradication programmes used a combination of tools, these being male annihilation technique, protein bait spray, quarantine and movement control, fruit removal and destruction, and SIT. A combination of approaches is thought to increase the chances of successful eradication by making better use of the Allee threshold – the theoretical population size below which a population will go to extinction due to demographic and chance environmental factors. A combination of control tools will put greater external pressure on a population, and so reduce the Allee threshold population size more than any single control tool will do on its own. The determination of successful eradication varies from country to country, but is often declared when no flies are detected for the time taken to complete from three seven generations.

In event that eradication is not considered operationally or financially viable, an invasive fruit fly can still be managed as a regulated pest. In such a situation, if the fly is shown through delimiting surveys to have a constrained geographical distribution, it may be declared by the state as a regulated pest. In this case formal mechanisms are put in place (road block inspections, fruit movement restriction orders, etc.), along with trapping grids and a management buffer zone, to ensure the fly remains restricted to the particular declared area. Under international agreements, the remainder of the country should then still be considered free of the pest.

In the final scenario, that an invasive pest fly is too wide spread to try to eradicate or limit when first detected, all that can be done is to walk away from any emergency or regulatory treatment, and deal with the new fly as an established pest which needs to be dealt with as part of routine pest management practices. Lux and colleagues reached this decision about *B. dorsalis* in East Africa after their first delimiting surveys.

7.2.3   The Californian situation and wider implications

One significant issue around quarantine detection, eradication and regulatory control continues to plague the tephritid world, and it involves *Bactrocera*, *Zeugodacus* and *Dacus* species, as well as *Ceratitis* and *Anastrepha*. The US state of California has a horticultural industry that, with a value of US$25 billion, is greater than the entire horticultural production of most countries. The state gains significant market-access advantage by being recognised as free of all pest fruit flies except for the exotic *B. oleae*. Exotic fruit flies of up to five genera are routinely detected in California through trapping grids and (less commonly) infested fruit, and when this occurs official eradication programmes are triggered.

For California, acrimonious debate continues between two groups, each of which contains leading members of the international fruit fly community. One camp argues that each new outbreak represents a new incursion, which is then successfully eradicated. The second camp argues that many fruit fly species are permanently established in California, and the eradication programmes simply push the target flies to below detectable limits, until they next build up and are 'rediscovered'. The primary argument from the first camp is that tephritids do not exist at sub-detectable levels for multiple years in a row; rather new incursions build up quickly to damaging levels. For this reason each new detection should be considered a new incursion. The primary argument from the second camp is that the number and frequency of repeat detections are too many to explain based solely on new incursions, particularly when other areas of the US and Europe have similar levels of passenger arrivals, but do not have the same number of outbreaks. References for further reading are provided in the reading list.

It is not my intent here to promote one side or other of the debate, but the Californian issue does raise numerous points pertinent when considering the Dacini as quarantine pests.

First, the flies are massive pests. Introduction and permanent establishment of just one pest species can result in altered food security in developing countries, or for food-secure countries such as the US they can still impact on individual well-being through disruption, or even loss, of rural-based industries. Personalities and politics have long ago entered the Californian debate because of the huge market-access situation involved. Has this helped resolve the problem? No, it has not, but it is to be expected and is commonly observed in any emergency plant-pest incursion situations where grower livelihoods are at stake. When it comes to fruit fly incursions, it needs to be remembered that whatever response occurs, politics will be involved.

Second, while the dacines have much in common biologically, each species is still unique. Clumping a dozen or so fruit fly species into a single argument, as has been done by both sides of the Californian debate, simply confuses the specific issues and does not result in better quarantine or post-border management. This situation is far from unique to the Californian debate. The Australian Government similarly treats all exotic fruit flies as a combined risk, without strategising and prioritising the high risk from the low risk. So while it is always tempting to treat 'fruit flies' as a single group, it is highly dangerous to do so.

Third, the issue of sub-detectable populations. All trapping programmes reach a point where they cannot detect rare individuals. A new generation of models are showing this might happen more often than expected for fruit fly populations. In any quarantine programme, do not assume that no flies in your traps mean that they are entirely absent: they may be there but rare.

7.3   Further Reading and References Cited

Barr, N.B., Ledezma, L.A., Leblanc, L., San Jose, M., Rubinoff, D., Geib, S.M., Fujita, B., Bartels, D.W., Garza, D., Hauser, P.K.M. and Gaimari, S. (2014) Genetic diversity of *Bactrocera dorsalis* (Diptera: Tephritidae) on the Hawaiian Islands: implications for an introduction pathway into California. *Journal of Economic Entomology*, 107, 1946–1958.

Boontop, Y., Schutze, M.K., Clarke, A.R., Cameron, S.L. and Krosch, M.N. (2017) Signatures of invasion: using an integrative approach to reveal the spread of melon fly, *Zeugodacus cucurbitae* (Diptera: Tephritidae), across Southeast Asia and the West Pacific. *Biological Invasions*, 19, 1597–1619.

Carey, J.R., Papadopoulos, N. and Plant, R. (2017) The 30-year debate on a multi-billion-dollar threat: Tephritid fruit fly establishment in California. *American Entomologist*, 63, 100–113.

Chapman, M.G. (1982) Experimental analysis of the pattern of tethered flight in the Queensland fruit fly, *Dacus tryoni*. *Physiological Entomology*, 7, 143–150.

Chapman, M.G. (1983) Experimental analysis of the propensity to flight of the Queensland fruit fly, *Dacus tryoni*. *Entomologia Experimentalis et Applicata*, 33, 163–170.

De Meyer, M. and Ekesi, S. (2016) Exotic invasive fruit flies (Diptera: Tephritidae): In and out of Africa. In: Ekesi, S., Mohamed, S.A. and De Meyer, M. (eds.) *Fruit Fly Research and Development in Africa. Towards a sustainable management strategy to improve horticulture.* Springer, Basel, Switzerland, pp. 127–150.

Dominiak, B.C., Campbell, M., Cameron, G. and Nicol, H. (2000) Review of vehicle inspection historical data as a tool to monitor the entry of hosts of Queensland fruit fly *Bactrocera tryoni* (Froggatt) (Diptera: Tephritidae) into a fruit fly free area. *Australian Journal of Experimental Agriculture*, 40, 763–771.

Duyck, P.-F., David, P. and Quilici, S. (2004) A review of relationships between interspecific competition and invasions in fruit flies (Diptera: Tephritidae). *Ecological Entomology*, 29, 511–520.

Duyck, P.-F., David, P., Junod, G., Brunel, C., Dupont, R. and Quilici, S. (2006) Importance of competition mechanisms in successive invasions by polyphagous tephritids in La Réunion. *Ecology*, 87, 1770–1780.

Ekesi, S., Billah, M.K., Peterson, W., Nderitu, W., Lux, S.A. and Rwomushana I. (2009) Evidence for competitive displacement of *Ceratitis cosyra* by the invasive fruit fly *Bactrocera invadens* (Diptera: Tephritidae) on mango and mechanisms contributing to the displacement. *Journal of Economic Entomology*, 102, 981–991.

Ekesi, S., De Meyer, M., Mohamed, S.A., Virgilio, M. and Borgemeister, C. (2016) Taxonomy, ecology, and management of native and exotic fruit fly species in Africa. *Annual Review of Entomology*, 61, 219–238.

El-Sayed, A.M., Suckling, D.M., Byers, J.A., Jang, E.B. and Wearing, C.H. (2009) Potential of 'lure and kill' for long-term pest management and eradication of invasive species. *Journal of Economic Entomology*, 102, 815–835.

Fletcher, B.S. (1973) The ecology of a natural population of the Queensland fruit fly, *Dacus tryoni IV*. The immigration and emigration of adults. *Australian Journal of Zoology*, 21, 541–565.

Fletcher, B.S. (1974) The ecology of a natural population of the Queensland fruit fly, *Dacus tryoni V*. The dispersal of adults. *Australian Journal of Zoology*, 22, 189–202.

Fletcher, B.S. (1989) Movements of tephritid fruit flies. In: Robinson, A.S. and Hooper, G. (eds.) *Fruit flies. Their Biology, Natural Enemies and Control*. Elsevier, Amsterdam, The Netherlands, pp. 209–219.

Frampton, E.R. and Nalder, K. (2009) A novel analysis of the risk of fresh produce imports. *New Zealand Plant Protection*, 62, 114–123.

Hengeveld, R. (1989) *Dynamics of Biological Invasions*. Chapman & Hall, London, UK.

Khamis, F.M. and Malacrida, A.R. (2016) Role of microsatellite markers in molecular population genetics of fruit flies with emphasis on the *Bactrocera dorsalis* invasion of Africa. In: Ekesi, S., Mohamed, S.A. and De Meyer, M. (eds.) *Fruit Fly Research and Development in Africa. Towards a sustainable management strategy to improve horticulture*. Springer, Basel, Switzerland, pp. 53–70.

Khamis, F.M., Karam, N., Ekesi, S., De Myer, M., Bonomi, A., Gomulski, L.M., Scolari, F., Gabrieli, P., Siciliano, P., Masiga, D., Kenya, E.U., Gasperi, G., Malacrida, A.R. and Guglielmino C.R. (2009) Uncovering the tracks of a recent and rapid invasion: the case of the fruit fly pest *Bactrocera invadens* (Diptera: Tephritidae) in Africa. *Molecular Ecology*, 18, 4798–4810.

Liebhold, A.M., Work, T.T., McCullough, D.G. and Cavey, J.F. (2006) Airline baggage as a pathway for alien insect species invading the United States. *American Entomologist*, 52, 48–54.

Lux, S.A., Copeland, R.S., White, I.M., Manrakhan, A. and Billah, M.K. (2003) A new invasive fruit fly species from the *Bactrocera dorsalis* (Hendel) group detected in East Africa. *Insect Science and Application*, 23, 355–361.

Manrakhan, A., Venter, J.-H. and Hattingh, V. (2012) Action plan for the control of the African Invader fruit fly, *Bactrocera invadens* Drew, Tsuruta and White. Department of Agriculture, Forestry and Fisheries, Pretoria, South Africa.

Mararuai, A., Allwood, A., Balagawi, S., Dori, F., Kalamen, M., Leblanc, L., Putulan, D., Sar, S., Schuhbeck, A., Tenakanai, D. and Clarke, A. (2003) Introduction and distribution of *Bactrocera musae* (Tryon) (Diptera: Tephritidae) in East New Britain, Papua New Guinea. *PNG Journal of Agriculture, Forestry and Fisheries*, 45, 59–65.

McCullough, D.G., Work, T.T., Cavey, J.F., Liebhold, A.M. and Marshal, D. (2006) Interceptions of nonindigenous plant pests at US ports of entry and border crossings over a 17-year period. *Biological Invasions*, 8, 611–630.

McInnis, D.O., Hendrichs, J., Shelly, T., Barr, N., Hoffman, K., Rodriguez, R., Lance, D.R., Bloem, K., Suckling, D.M., Enkerlin, W., Gomes, P. and Tan, K.H. (2017) Can polyphagous invasive tephritid pest populations escape detection for years under favorable climatic and host conditions? *American Entomologist*, 63, 89–99.

Meats, A., Fay, H.A.C. and Drew, R.A.I. (2008) Distribution and eradication of an exotic tephritid fruit fly in Australia: relevance of invasion theory. *Journal of Applied Entomology*, 132,
406–411.

Nardi, F., Carapelli, A., Dallai, R., Roderick, G.K. and Frati, F. (2005) Population structure and
colonization history of the olive fly, *Bactrocera oleae*. *Molecular Ecology*, 14, 2729–2738.

Otuka, A., Nagayoshi, K., Sanada-Morimura, S., Matsumura, M., Haraguchi, D. and Kakazu, R.
(2016) Estimation of possible sources for wind-borne re-invasion of *Bactrocera dorsalis*
complex (Diptera: Tephritidae) into islands of Okinawa Prefecture, southwestern Japan.
Applied Entomology and Zoology, 51, 21–35.

Papadopoulos, N.T. (2014) Fruit fly invasion: Historical, biological, economic aspects and
management. In: Shelly, T., Epsky, N., Jang, E.B., Reyes-Flores, J. and Vargas, R. (eds.)
Trapping and the Detection, Control, and Regulation of Tephritid Fruit Flies. Springer,
Dordrecht, The Netherlands, pp. 219–252.

Papadopoulos, N.T., Plant, R.E. and Carey, J.R. (2013) From trickle to flood: the large-scale,
cryptic invasion of California by tropical fruit flies. *Proceedings of the Royal Society (B)*,
280, 20131466.

Peck, S.L., McQuate, G.T., Vargas, R.I., Seager, D.C., Revis, H.C., Jang, E.B. and McInnis, D.O.
(2005) Movement of sterile male *Bactrocera cucurbitae* (Diptera: Tephritidae) in a Hawaiian
agroecosystem. *Journal of Economic Entomology*, 98, 1539–1550.

Putulan, D., Sar, S., Drew, R.A.I., Raghu, S. and Clarke, A.R. (2004) Fruit and vegetable
movement on domestic flights in Papua New Guinea and the risk of spreading fruit flies
(Diptera: Tephritidae). *International Journal of Pest Management*, 50, 17–22.

Shelly, T., Epsky, N., Jang, E.B., Reyes-Flores, J. and Vargas, R. (eds.) (2014) *Trapping
and the Detection, Control, and Regulation of Tephritid Fruit Flies*. Springer, Dordrecht,
The Netherlands.

Suckling, D.M., Tobin, P.C., McCullough, D.G. and Herms, D.A. (2012) Combining tactics
to exploit Allee effects for eradication of alien insect populations. *Journal of Economic
Entomology*, 105, 1–13.

Suckling, D.M., Kean, J.M., Stringer, L.D., Cáceres-Barrios, C., Hendrichs, J., Reyes-Flores, J.
and Dominiak, B.C. (2016) Eradication of tephritid fruit fly pest populations: outcomes and
prospects. *Pest Management Science*, 72, 456–465.

Wan, X.W., Liu, Y.H. and Zhang, B. (2012) Invasion history of the oriental fruit fly, *Bactrocera
dorsalis*, in the Pacific-Asia region: two main invasion routes. *PLoS One*, 7, e36176.

8   Natural Enemies

In natural systems, the *Bactrocera* suffer natural-enemy mortality from insect parasitoids, vertebrate and invertebrate predators, and fungal and bacterial diseases. Enhancing natural-enemy mortality may be as simple as allowing chickens to forage under a fruit tree, or as complex as the introduction and liberation of exotic natural enemies for classical biological control. This chapter introduces those natural enemies and discusses how they have been, or may be, used to enhance fruit fly control. Entomopathogens, while technically biological control agents, are introduced in this chapter but are discussed more extensively in Chapter 9, as their application and usage is more akin to pesticide use than to what most researchers would generally consider as biological control.

8.1   Parasitoids

8.1.1   Introduction to parasitoids

Insect parasitoids are a special feeding guild that are neither true parasites nor predators, but share attributes of both. Like parasites, parasitoids feed upon a single host, but unlike most parasites they inevitably kill the host. However, unlike predators that kill and consume many prey items over their life, parasitoids only ever consume and kill one host, which are normally other insects or spiders. The parasitoid life style is entirely restricted to the insects, and within the insects most commonly (but not exclusively) to the wasps, true flies and beetles. Approximately 100 000 insect species are parasitoids, meaning that just under 10% of all described insects have this life history.

　　The general parasitoid life cycle is that the immature stage of the parasitoid lives and feeds within the host (i.e. endoparasitoids), but parasitoids that live outside the host (i.e. ectoparasitoids) also occur (Fig. 8.1). For endoparasitoids,

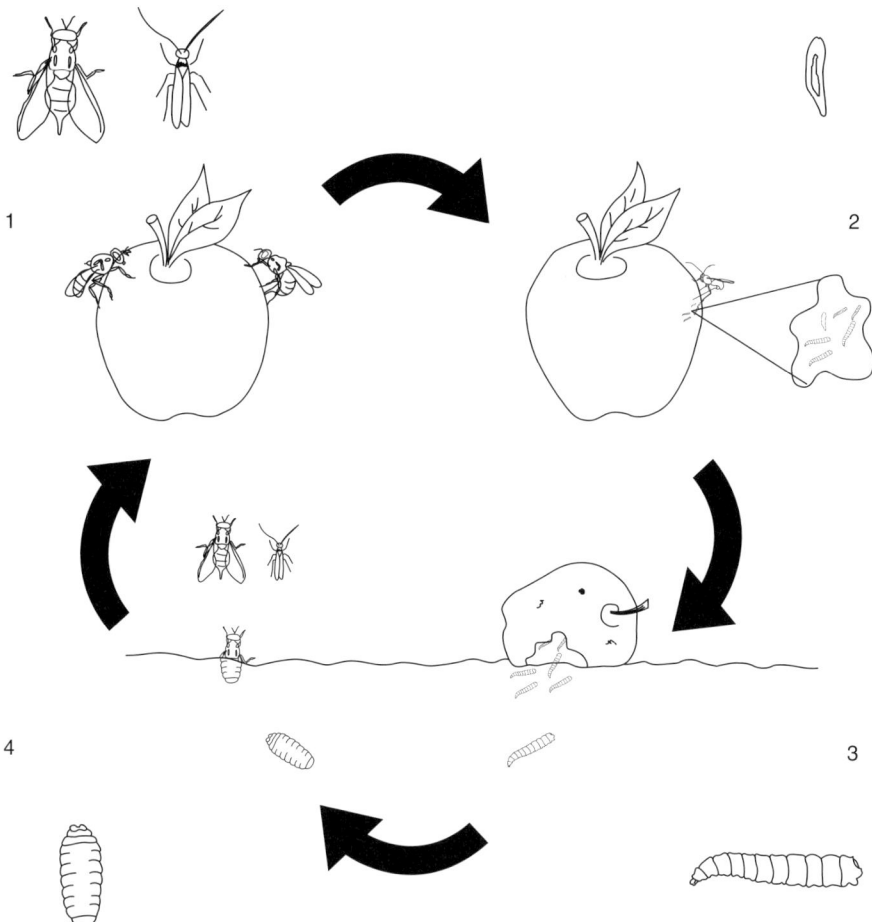

Fig. 8.1. The generalised parasitoid life cycle. (1) The adult female fruit fly lays eggs into fruit; (2) the parasitoid female lays her eggs into the fruit fly eggs or larvae (depending on parasitoid species), after which the fruit fly larva continue to feed and develop as normal, but now with a parasitoid larva internally feeding and growing within them; (3) the mature fruit fly larvae leaves the fruit, still carrying the immature parasitoid; (4) the fruit fly larva pupates in the soil. If unparasitised the fly pupates normally and a new fly subsequently emerges. If parasitised, the wasp larvae pupates within the pupal case of its now dead fruit fly host, eventually emerging as a new generation wasp.

the parasite can be directly inserted as an egg into the host by the ovipositing parasitoid female; the parasitoid egg may be placed on foliage that is subsequently consumed by the host, or the parasitoid egg may hatch external to the host and then the first instar parasitoid larva actively burrows into the host. Hosts of parasitoids may be killed or paralysed when first parasitised (i.e. idiobiont parasitoids), or may continue to develop and grow while hosting the parasitoid (i.e. koinobiont parasitoids). Koinobionts develop without initially damaging the host by feeding on non-essential tissue, such as the haemolymph and fat bodies, but the host is invariably killed at the point when the immature parasitoid is ready to pupate. Parasitoids are also commonly described based on the life stage(s) of the host that they affect. Thus there are egg parasitoids, larval and nymphal parasitoids, pupal parasitoids, and adult parasitoids, as well as combinations of these. The parasitoids most commonly affecting *Bactrocera* are egg/larval/pupal, larval/pupal, and pupal koinobiont endoparasitoids.

All parasitoids of *Bactrocera* are hymenopteran wasps. The most commonly studied are the opine braconids (Hymenoptera: Braconidae: Opiinae) that are a large group of egg/larval/pupal and larval/pupal koinobiont endoparasitoids that attack the dipteran families Tephritidae, Agromyzidae and Anthomyiidae. The Opiinae contains almost 1500 species and is one of the largest sub-families of the Braconidae. Species from other braconid sub-families, as well as species from the families Chalcidoidea, Eulophidae, Pteromalidae and others have also been recorded from frugivorous tephritids, but the number of species associated with tephritids, and their incidence in field collections, is significantly less than the opiines.

8.1.2   Opiine braconids

In classical and inundative biological control, as well as natural attack in endemic areas, the opiine braconids are the most common parasitoids of *Bactrocera*. The opiines are by no means restricted to *Bactrocera*, but attack many frugivorous tephritid genera, including *Dacus*, *Zeugodacus*, *Ceratitis* and *Anastrepha*. Opiines are medium-sized insects of around 1 cm in length, orange to brown to black in colour, and with surface sculpturing that ranges from smooth to highly reticulate. The females have long ovipositors (often as long, or longer, than the rest of the body) that are used to penetrate fruit and find the fruit fly hosts within. Many studies have demonstrated a negative correlation between percentage parasitism and increasing fruit size, as there is a physical limit to how deep into the fruit a female parasitoid can insert her ovipositor to find hosts.

The major genera of Opiinae recorded as attacking *Bactrocera* are *Diachasmimorpha* Viereck, *Fopius* Wharton, *Opius* Wesmael, *Psytallia* Walker and *Utetes* Foerster (Table 8.1). Some of the better-known species are *F. arisanus*, *D. longicaudata* and *D. kraussii*, but over 100 opiine species have been formally associated with frugivorous tephritids. The percentage mortality caused by opiines varies widely between studies. Field parasitism rates have been recorded from less than 10%, to as high as 80%: 10–25% appears most common. The relatively low percentage mortality caused by opiines has seen them treated dismissively by some practitioners, as does the fact they only control the fly after the crop damage has occurred. Nevertheless, as fruit fly controls move away from synthetic pesticides they have an increasingly important role to play not only in classical biological control, which is how they have been most used in the past, but as part of area-wide management for total population suppression and in conjunction with other control approaches, such as the sterile insect technique (SIT).

Table 8.1. Hymenopteran parasitoids as associated with Dacini fruit flies.

Family (sub-family)	Species	Host species	Biology
Chalcididae	*Dirhinus giffardii* Silvestri	*Z. cucurbitae*, *B. demmerezi*, *B. dorsalis*, *B. oleae*, *B. passiflorae*, *B. tryoni*	Pupal ectoparasitoid
Eulopidae	*Aceratoneuromyia indica* (Silvestri),	*Bactrocera*, *Zeugodacus* and *Dacus* species	A gregarious, koinobiont larval/pupal endoparasitoid
	Neochrysocharis formosa (Westwood)	*B. oleae*	
	Pnigalio mediterraneus Ferriere and Delucchi	*B. oleae*	
Eupelmidae	*Eupelmus urozonus* Dalman	*B. oleae*	A koinobiont larval/pupal ectoparasitoid. Can also be a facultative hyperparasitoid
Eurytomidae	*Eurytoma rosae* Nees	*B. oleae*	Larval ectoparasitoid
	Eurytoma martelli	*B. oleae*	Larval ectoparasitoid
Pteromalidae	*Cyrtoptyx latipes* (Rondani)	*B. oleae*	Larval parasitoid
	Pachycrepoideus vindemmiae (Rondani)	*D. ciliatus*, *Z. cucurbitae*, *B. dorsalis*, *B. oleae*, *B. passiflorae*	Polyphagous pupal idiobiont ectoparasitoid and a facultative hyperparasitoid
	Spalangia cameroni	*Numerous species*	Polyphagous pupal parasitoid
Figitidae (Eucoilinae)	*Aganaspis daci* (Weld)	*B. dorsalis*	Solitary, koinobiont larval/ pupal endoparasitoid.
Braconidae (Braconinae)	*Bracon celer* Szépligeti	*B. oleae*	Idiobiont larval ectoparasitoid
Braconidae (Opiinae)	*Diachasmimorpha albobalteata* (Cameron)	*Z. cucurbitae*, *B. dorsalis*	Solitary, koinobiont larval/ pupal endoparasitoid
	D. brevistyli (Paoli)	*Dacus ciliatus*	Solitary, koinobiont larval/ pupal endoparasitoid
	D. carinata (Szépligeti)	*Dacine tephritids*	Solitary, koinobiont larval/ pupal endoparasitoid
	D. dacusii (Cameron)	*'tephritids breeding in cucurbits'*	Solitary, koinobiont larval/ pupal endoparasitoid

Continued

Table 8.1. Continued.

Family (sub-family)	Species	Host species	Biology
	D. feijeni van Achterberg	*B. minax*	Solitary, koinobiont larval/pupal endoparasitoid
	D. hageni (Fullaway)	*B. passiflorae, B. dorsalis, Z. cucurbitae*	Solitary, koinobiont larval/pupal endoparasitoid
	D. kraussii (Fullaway)	*B. tryoni, B. neohumeralis, B. aquilonis, B. halfordiae, B. melas, B. visenda, B. cacuminata, B. kraussi, B. jarvisi, B. murrayi*	Solitary, koinobiont larval/pupal endoparasitoid
	D. longicaudata (Ashmead)	*B. caryeae, B. curvipennis, B. dorsalis, B. frauenfeldi, B. latifrons, B. passiflorae, B. pedestris, B. psidii, B. tryoni, B. umbrosa, B. xanthodes, B. zonata, D. ciliatus, Z. cucurbitae*	Solitary, koinobiont larval/pupal endoparasitoid
	D. tryoni (Cameron)	*B. dorsalis, B. halfordiae, B. passiflorae, B. tryoni, B. xanthodes*	Solitary, koinobiont larval/pupal endoparasitoid
	Fopius arisanus (Sonan)		Solitary, koinobiont egg/larval/prepupal endoparasitoid
	Fopius ceratitivorus Wharton	*B. oleae*	
	Fopius deeralensis (Fullaway)	*B. barringtoniae, B. endiandrae, B. fagraea, B. frauenfeldi, B. jarvisi, B. kraussi, B. laticaudus, B. musae, B. pallidus, B. tryoni*	
	Fopius niger (Szépligeti)	*D. humeralis*	
	Fopius ottotomoanus Fullaway	*Dacus* spp.	
	Fopius persulcatus (Silvestri)	*B. dorsalis, B. latifrons* (Hendel)	
	Fopius schlingeri Wharton		
	Fopius silvestrii (Wharton)	*D. bivittatus*	
	Fopius skinneri (Fullaway)	*B. limbifera, B. pedestris*	
	Fopius vandenboschi (Fullaway)	*B. dorsalis*	
	Opius froggatti Fullaway	*B. kraussi, B. laticaudus, B. psidii*	

Species	Host(s)
Psytallia concolor (Szépligeti)	B. oleae
P. dacicida (Silvestri)	B. oleae
P. dexter (Silvestri)	D. longistylus
P. fijiensis (Fullaway)	B. barringtoniae, B. curvipennis, B. endiandrae, B. fagraea, B. frauenfeldi, B. jarvisi, B. kraussi, B. laticaudus, B. murrayi, B. musae, B. pallida, B. passiflorae, B. psidii, B. tryoni, B. xanthodes
P. fletcheri (Silvestri)	Z. cucurbitae
P. incisi (Silvestri)	B. dorsalis
P. lounsburyi (Silvestri)	B. oleae
P. makii (Sonan)	B. dorsalis, Z. tau
P. muesebecki (Fischer)	B. psidii
P. novaguineensis (Szépligeti)	B. calophylli, B. endiandrae, B. fagraea, B. fuscata, B. halfordiae, B. pallida, B. tryoni, B. kraussi, B. jarvisi
P. perproxima (Silvestri)	D. ciliatus, D. bivittatus
P. phaeostigma (Wilkinson)	D. ciliates, D. demmerezi
P. ponerophaga (Silvestri)	B. oleae
P. walkeri (Muesebeck)	B. dorsalis
Utetes africanus (Szépligeti)	B. oleae
Utetes bianchii (Fullaway)	
Utetes manii (Fullaway)	B. dorsalis
Utetes perkinsi (Fullaway)	B. cacuminata, B. jarvisi, B. kraussi, B. laticaudus, D. tryoni

This information is abstracted from Wharton R.A. and Yoder, M.J. Parasitoids of fruit-infesting Tephritidae. Available at: http://parofit.org. Accessed 22/May/2018.

Many (most?) of the opiine wasps are not specific in the fruit fly species they attack, but will attack multiple species, including species against which they have no evolutionary history, e.g. the originally South-east Asian endemic *Fopius arisanus* will attack the originally African *C. capitata*. This can make their use as classical biological control agents a potential environmental risk if endemic, non-pest fruit fly species occur within a release region and appropriate host testing is not done. However, while referred to in the literature as 'polyphagous' parasitoids, and despite their ability to attack flies across multiple genera, not all fruit fly species are physiologically suitable hosts to all parasitoids. The fruit fly species *B. tryoni*, *B. jarvisi*, *Z. cucumis* and *B. cacuminata* all occur sympatrically in the endemic range of *D. krausii*, but not all are suitable hosts to the wasp. Wasp eggs laid into *B. tryoni* and *B. jarvisi* hatch and develop normally (these flies are physiologically suitable hosts), but wasp eggs laid into *Z. cucumis* and *B. cacuminata* are encapsulated by haemocyte cells of the hosts' haemolymph systems and are killed before hatching (these flies are physiologically non-suitable hosts).

Opiines are day-active on fruit, where they forage for fruit fly eggs or larvae (depending on species), and as such they are easily seen in the field if present. The stage of host first attacked varies with species, for example, *F. arisanus* attacks eggs whereas *D. krausii* preferentially attacks second instar larvae but will also lay eggs into third instars. However, all opiines pupate within the fly puparium. While well recognised, but poorly researched, it appears than many opiines have an adult diapausing stage within the host pupal case. Thus when rearing opiines in the laboratory an initial flush of wasp emergence will happen soon after fly pupation, and then there will be gradual emergence of individual wasps over a period that may last several months. If these parasitised fly pupae are dissected, fully developed, pharate adult wasps will be found inside. It is presumed that this is a 'bet-hedging' strategy, to increase the chances of at least some wasps emerging into a favourable environment. How common this diapause is across opiine species, what percentage of a population diapauses, or if diapause varies in length and the percentage of individuals involved depending on season, are all unknown.

The Wharton Lab have developed a comprehensive and authoritative web site to the parasitoids attacking frugivorous tephritids (including fruit fly genera additional to *Bactrocera*, and parasitoids additional to opiines) (Wharton and Yoder, n.d.). This site includes a key to species, and genus and species pages for all taxa that include descriptions, illustrations and biological information.

8.1.3   Opiine host location and host use

The host location behaviours of several opiine braconid species have been studied in detail and there is significant commonality across species. The parasitoids first locate host fruits using visual and chemical cues, probably in combination. Infested fruit are significantly more attractive to female parasitoids than uninfested fruit, although fruit volatiles of uninfested fruit will still attract females of some species, but other species will only respond to infested fruit.

Volatiles emitted from damaged fruit are more attractive, or only attractive, to female parasitoids when the damage is caused by fruit fly larval feeding, rather than mechanical damage. In choice tests, parasitoids show preference rankings for different fruit types, with at least some measures of wasp offspring performance being correlated with adult wasp preference for fruit. Like many parasitoids, opiines have good learning capacity, and can learn to positively associate certain fruit or odours with larval infestation. In two species of wasp, orientation is modified by adult fruit flies. In *P. concolor* both males and females showed positive orientation to high levels of (Z)-9-tricosene, a *B. oleae* male produced sex pheromone; whereas *D. krausii* females showed greater searching time on fruit where an adult fruit fly was present, even if the fruit was uninfested. How common adult fruit fly derived cues are to modifying opiine searching warrants further attention.

An important aspect of opiine host location associated with the host plant is differential response to different host fruits. In-field parasitism rates of the same fruit fly species can vary significantly between adjacent host plant species, and there is direct experimental evidence to show that this is related to the species of host plant. In a study of foraging by *D. krausii*, where physiologically suitable and unsuitable fruit fly maggot hosts were experimentally switched between three different species of host fruit, the wasp showed consistent preference ranking of the three host fruits, regardless of the maggot species within each. Several field studies have shown that infested native fruit often receives higher parasitism than do exotic agricultural crops, and this may reflect the wasps' evolved searching mechanisms for native rather than exotic plants.

Having found a fruit fly infested piece of fruit, how parasitoids identify larvae within the fruit is less studied, although it is known to include the use of larval vibrational cues and ovipositor probing. It is a common pattern that wasps cannot discriminate between physiologically suitable and unsuitable hosts at the probing stage, and it is considered that they use the host plant location part of searching to find physiologically suitable hosts. In the field, parasitoids are often found where adult flies are present, but within an orchard no correlation was found between *B. tryoni* and *D. krausii* numbers at the individual tree level.

8.2   Other Parasitoid Species

Wharton and Yoder list numerous other parasitoids associated with *Bactrocera* from a range of parasitoid families, including the Chalcididae, Eulopidae, Eupelmidae, Eurytomidae and Pteromalidae. Some of these species are locally important as parasitoids, for example *Spalangia endius* (Pteromalidae) is reported as a major pupal mortality agent of *B. dorsalis* in China and other Asian countries, with this solitary endoparasitoid preferentially attacking pupae of 2–4 days old. However, many of the records of non-opiine parasitoids are rare, association with *Bactrocera* is assumed rather than confirmed (e.g. reared from olive, and so presumed associated with olive fly), and at least some species are facultative hyperparasitoids (parasitoids of parasitoids) of opiines. *Pachycrepoideus vindemmiae* (Pteromalidae), for example, was deliberately introduced into Hawaii for Medfly control, but was subsequently shown to readily hyperparasitise the opiine parasitoids also released. A number of the species recorded (e.g. *Spalangia cameroni* and *S. endius*) are also highly polyphagous on dipteran pupae and so are a risk of attacking non-target organisms. Because of their often casual association with tephritids, and possible hyper-parasitoid activity, usage of these species should be carefully considered prior to active manipulation.

8.3   Manipulation of Parasitoids

8.3.1   Classical biological control

Classical biological control involves moving natural enemies of pests into an area that the pest has invaded, but to which the natural enemies have not themselves followed. The underlying assumption of classical biological control is that the introduction of the natural enemies will reduce the pest population in a density-dependent fashion – in older literature it is referred to as 'restoring the balance of nature'. Classical biological control was first carried out in the late 1880s when the vedalia ladybird beetle, *Rodolia cardinalis*, was moved to California to control cottony cushion scale, *Icerya purchasi,* on citrus. Classical biological control of pests and weeds continues around the world, but increasing environmental and biodiversity concerns over non-target impacts has greatly limited the practice in many countries. This is particularly the case for insect biological control where demonstrating host specificity is much more difficult than it is for weed biological control.

Parasitoids of *Bactrocera* have been moved extensively for biological control. While some such movement has been against invasive fruit flies (e.g. against *B. dorsalis* in Hawaii), parasitoids have also been liberated against endemic pest species with their own endemic parasitoid fauna, (e.g. against *B. tryoni* in Australia or against *Anastrepha* species in South America).

The first fruit fly biological control programmes started at the turn of the 1900s, targeting Medfly in Western Australia and olive fruit fly in Italy, but it is the better-known Hawaiian programme that is credited with starting fruit fly biological control. Research on discovering parasitoids for the introduced pest tephritids in Hawaii (*Z. cucurbitae*, introduced 1895; *C. capitata* 1910, *B. dorsalis* 1946; *B. latifrons* 1983) began in 1912 with a major focus in *C. capitata*, and then exploded again in the late 1940s following the introduction of Oriental fruit fly. Over 30 species of parasitoid were introduced during these projects (as reviewed by Clausen, 1978; Waterhouse, 1993), with every combination of results. Some species were never recovered after liberation, some were recovered for a few years and then disappeared, some established at low levels, some became a dominant parasitoid for several years until apparently out-competed by a later introduction, and others established as important parasitoids in some host plants, but not others. The success of parasitoids also varied with fruit fly species. The egg/larval/pupal parasitoid *F. arisanus* eventually became the dominant parasitoid of *B. dorsalis*, causing up to 70% parasitism on guava, whereas *P. fletcheri* became the dominant parasitoid on *Z. cucurbitae*, although the control exerted on *Z. cucurbitae* in commercial cucurbit crops was limited.

Many of the parasitoids initially established in Hawaii have subsequently been spread from there. For example, in the late 1950s/early 1960s, *F. arisanus, D. longicaudata, F. vandenboschi, P. incisi, D. anthracia, A. indica* and *T. giffardianus* were all introduced and liberated in Australia against *B. tryoni*, despite a native fruit fly parasitoid already occurring there (including *D. tryoni* that had been previously sent to Hawaii!) and the country having a large, non-pest dacine fauna. Of these introductions only *F. arisanus* established, where it now has a wide host range on both pest and non-pest *Bactrocera*, causing approximately 10% population mortality.

Introductions and liberations of parasitoids against *Bactrocera* have continued into the 21st Century. Recent examples include the liberation of *F. arisanus* and *D. longicaudata* into Tahiti to control Oriental fruit fly, introduced in 1996. *Fopius arisanus* was introduced in 2002 and by 2009 was causing 65% mean parasitism on the target, although *D. longicaudata* caused less than 5% control. This has been considered a highly successful programme. Classical biological control of olive fruit fly was commenced in California in 2003, with new foreign exploration, host testing and release, and confirmation of establishment of two agents, *P. humilis* and *P. lounsburyi*, in 2015. The long-term success of this project is still to be determined.

The history of classical biological control against dacines is not great reading. Up until the 1960s significant effort was put into foreign exploration and release, but with little apparent effort to consider the co-evolved relationships between parasitoids and hosts, how placing new parasitoids into a system might impact on existing parasitoids, and with little thought (as was quite common at the time) for non-target impacts on native, non-pest fauna. The overall success rate for the financial investment made in several countries was also low.

More recent examples of classical biological control against *Bactrocera* have addressed these problems, with much greater emphasis on searching for target-specific parasitoids (e.g. as in the case of olive fly), much more consideration to the non-target risks, and better strategising of the parasitoids released so they are likely to complement, rather than compete, with each other. A general perception among many entomologists around the world today is that the age of classical biological control against insect pests is largely passed, because of the environmental risk it poses. Nevertheless, it remains a valuable technique in selected circumstances and the existing strong base of opiine taxonomy, biology and ecology makes the success of any future projects much more likely.

8.3.2 Augmentative, inundative and inoculative biological control

Augmentative biological control involves the release of natural enemies that are already present in the field, so as to increase their numbers and make them more effective over time. Inundative releases are similar, except that for inundative release a specific generation, stage or timing of the pest organism is normally targeted and there is an expectation that many, or most, of the mass-released beneficials will die within one generation. Inoculative biological control involves the release of agents at a key part of the cropping cycle (often early), with an expectation that their numbers will increase over multiple generations during the season. Only a few of these approaches have been tried with opiines and, indeed, not all approaches are relevant.

Augmentation of opiines has been tried experimentally against *Bactrocera*, and operationally against *C. capitata*. In Hawaii, augmentative releases of *Psytallia fletcheri* were made against *Z. cucurbitae*, as parasitism on refugia plants was low and these provided a source of flies in area-wide management. Under experimental conditions the augmentative releases of parasitoids significantly increased parasitoid emergence, but paradoxically did not decrease fly emergence. However, a combination of wasp release and release of sterilised male flies did provide a greater level of protection than did either approach on its own. This experimental data supports theoretical modelling predictions, which also suggest that SIT male releases and parasitoid releases are synergistic in their effects, with the combined approach giving better control than the simple additive effects of both. Operationally, a SIT mass-rearing facility is required to make mass rearing of opiines viable, as their host specificity means that it is difficult or impossible to rear them in the absence of fruit flies: that is why there are operational examples against Medfly, for which there are several SIT factories globally.

A second method of opiine augmentation has been developed and operationally used against dacines in several countries. Developed initially as a 'spin-off' from crop hygiene, 'augmentoria' are essentially portable tent-like/cage-like structures that can be placed in the field into which culled or dropped fruit can be placed. The walls and/or roof of the tent are made of mesh and, if appropriately selected, are of a size that allows emergent parasitoids to escape but retain (and kill) emergent flies. The selection of mesh size and structure is critical to success. Experimental evidence that augmentoria increase local parasitism rates is ambivalent, but research has shown that growers like augmentoria as an additional benefit of crop hygiene activities, while they also serve a valuable extension role in educating growers of the value of natural controls.

Inundative biological control using opiines has not been used to my knowledge. This is not surprising as the approach is probably not suited to agents that only bring about control once damage has occurred. By this I mean that a grower trying to target fruit fly control in their crop will gain no immediate reduction in damage through inundative releases of parasitoids, as the maggots already have to be in the fruit for the wasps to work. It is possible that well-timed, inundative releases of parasitoids to target key wild/native hosts or 'hot-spots' as part of an area-wide management programme would have benefits, but significant ecological data would be needed to justify this approach, as would the economics. Similarly, inoculative releases of opiines against *Bactrocera* have not been made to my knowledge. Such an approach is largely irrelevant in tropical areas where flies are nearly always present, but the approach may have value in temperate regions where fly populations build up slowly after winter, or as part of large-scale protected cropping. In such cases wasps could be released early in the season (i.e. in spring) so they are controlling flies before they become pestiferous in summer. As for inundative releases, significant background biological and economic research would need to be done before such approaches were tried.

8.3.3 Conservation biological control

Conservation biological control is the manipulation of the local environment to provide resources, including alternative prey items, nectar sources and shelter, to encourage and conserve natural enemies within a local environment. Modern research into the field was pioneered by German researchers in the 1980s, who demonstrated that strip mowing (i.e. alternately mowing or not mowing) between tree rows increased the ratio of beneficials to pests within orchards and resulted in higher levels of natural pest control, while still achieving the agronomic outcome of reducing weed competition with trees. Actively seeding flowering plants that provide nectar sources for hymenopteran parasitoids is another well-researched and commercially implemented conservation biological control approach.

Only a small amount of research has been carried out on conservation biological control for opiines. *Diachasmimorpha longicaudata* was found to have good survival on orange pulp and juice, homopteran honey dew and extrafloral nectary secretions, all of which are food resources that might be found in a fruiting plant at which they were foraging for fruit fly hosts. Flowering plants of the type that might grow as weeds, or be seeded within orchards (e.g. *Bidens alba, Lantana camara, Lobularia maritima*), did not provide food resources for the wasp. However, subsequent research showed poor survival of *D. longicaudata* and several other opiines on guava juice and pulp, while the earlier orange juice and pulp results were confirmed. This infers that it is not safe to simply assume that foraging opiines can find the food resources they need within a fruiting orchard, but the relationship between fruit fly host fruit and parasitoid survival needs to be tested for different systems. If the fruit type within an orchard does not provide nutritional resources for parasitoids, mechanisms to provide them (e.g. by occasional planting of suitable fruiting trees) may need to be considered.

8.4 Predators

8.4.1 Ants

Ants (Hymenoptera: Formicidae) are known to prey on all life stages of *Bactrocera*, although quantified data is poor. The best work available is for the role played by weaver ants in deterring ovipositing female flies. In both Africa and Australia the related weaver ants *Oecophylla longinoda* and *O. smaragdina*, respectively, have been shown to significantly decrease *Bactrocera* infestation levels in fruit crops (notably mango). In well-replicated trials on both continents, increased ant colonies, or increased counts of foraging trails, are correlated with decreased fruit infestation. The decreases in infestation rates are equivalent to, or better, than the benefits gained from pesticide usage. Based on direct observation this decreased damage does not seem to be related to direct predation of ovipositing female flies: indeed observations are that ants rarely capture flies. However, flies are less abundant in trees with ants than without, and in the presence of ants flies are more 'skittish' and less likely to settle and oviposit, and if ovipositing, are more likely to be interrupted before completing.

The mechanism by which weaver ants disturb flies seems complex. Based on laboratory work, there is strong evidence that an ant-derived pheromone may play an important role in deterring flies. However, field research suggests that visual cues (i.e. 'angry ants') are also needed. Weaver ants have a usage in Chinese horticultural pest control going back to antiquity, and their benefits are seen in the reduction of many horticultural pests additional to fruit fly. Weaver ants can be manipulated by physically moving nests (ideally at the time of day when foraging is most limited), placing bamboo or similar between tree canopies to make foraging runs, and even by placing small insect traps within trees to supplement ant diet and so increase ant numbers over time. However, there are downsides in using weaver ants that need to be managed. Large numbers of ants in trees can make manual fruit picking uncomfortable, and the ants can farm sessile bugs (e.g. scale insects). Nevertheless, in developing countries, particularly, manipulation of native weaver ants can be an efficient control strategy.

Ground dwelling ants are also known to impact *Bactrocera* fruit flies, although there is significantly less quantified data. Records from Hawaii record the ants *Pheidole megacephala* and *Solenopsis geminate* occasionally predating on *Z. cucurbitae* eggs, larvae and pupae. Egg and larval predation inside a fruit was uncommon and restricted to 'less watery fruit' such as cucurbits, but predation was thought to be common on prepupal larvae prior to burrowing, although this was not quantified. In another Hawaiian study, predation (predominantly ants) was estimated to contribute to 40–60% mortality of the combined prepupal/pupal/teneral adult phase. In two studies on Mediterranean fruit fly where prepupal larvae were dropped under a host tree, approximately 20% (North Africa) and 3% (Hawaii) of larvae were predated by ants, with up to 40% mortality of teneral adults (a figure probably inflated due to the experimental protocol, as acknowledged by the authors). Other studies simply refer to 'significant predation' by ants. Given the level of mortality estimated to be caused by ants in the few studies to have examined them, it is surprising that more work has not been done, particularly on how ants might be encouraged in crop ecosystems.

The early studies from Hawaii for *Z. cucurbitae* report other miscellaneous predators of larvae in fallen fruit, including rove beetles and earwigs. It was acknowledged that these were casual and irregular mortality agents.

8.4.2 Spiders

The impacts of spiders on dacine mortality has been largely ignored, as is the case in many natural enemy studies. Nevertheless, when they are studied, spiders are commonly found to be extremely important predators in natural and agroecosystems. In olive groves, the abundance of olive fruit fly was negatively correlated with cursorial spiders, Linyphiidae and web builders, with the strength of the negative correlation varying with the abundance of different spider groups across season.

8.4.3 Vertebrate frugivores

Drew has made a case that in endemic rainforest systems, vertebrate frugivores are the major mortality agents of *Bactrocera*. Working in south-east Queensland, his argument is based on data for *B. cacuminata* that showed that 66–77% of ripe wild tobacco fruit, the stage infested by *B. cacuminata* larvae, was consumed by brown pigeons (*Macropygia phasianella*). In the same forest, 78% of fallen *Planchonella australis* fruit, the major host for *B. halfordiae*, was consumed by small rodents. In later studies in far north Queensland rainforests by other workers, from 10% to 60% of experimentally placed fruit was consumed overnight by small rodents, and 100% of fruit was consumed if left exposed to larger fruit consumers, such as possums, cassowaries and scrub turkeys (*Alectura lathami*). These latter experiments showed that small rodents preferred fruit fly infested fruits over uninfested fruit (for some fruit species), although the effect was most likely the result of fruit fly larvae softening the fruit, rather than an active hunt for the maggots. Although the evidence is limited, like ants it is likely that vertebrate predation causes much greater fruit fly mortality than is recognised.

While encouraging birds to consume fruit on trees is hardly a useful fruit fly control when trying to grow fruit, the role of vertebrates in predation or destruction of soil dwelling stages of fruit fly is a potentially useful control strategy. Many cultures traditionally run scavenging animals such as pigs and chickens in orchards, and these animals will be causing high local fruit fly mortality. Their use should be encouraged if the local system is suitable.

8.5 Entomopathogens in Natural Systems

Relatively little work has been done on naturally occurring entomopathogens (fungi, bacteria, nematodes) of *Bactrocera* and related genera. In a Portuguese study of olive fruit fly in 'traditional' olive groves (low input, ancient trees, high biodiversity value), 15 fungal species were associated with pupae, of which three (*Cordyceps bassiana*, *Penicillium corylophilum* and *Mucor hiemalis*) were considered entomopathogenic. Subsequent laboratory tests confirmed high pathogenicity of these fungi to olive fruit fly pupae, but they had low or no pathogenicity to late stage larvae or adults. In soil samples taken from a range of commercial production and natural vegetation types on the island of Mauritius, a number of known insect pathogenic fungi were collected. Of these, multiple isolates of *Metarhizium anisopliae*, *Beauveria bassiana* and *Isaria fumosorosea* (= *Paecilomyces fumosoroseus*) were recovered. Subsequent laboratory tests showed that topical application of conidial suspensions of each of these three fungi were pathogenic to adults of *B. zonata* and *Z. cucurbitae*. In the same study, however, the incubated corpses of over 14 000 field collected adult *B. zonata* and nearly 16 000 *Z. cucurbitae* yielded no entomopathogenic fungi, strongly inferring that these fungi are not associated with free flying fruit flies.

A cytoplasmic inclusion virus has been reported from adult *B. tryoni*, causing up to 50% mortality of laboratory flies by two to three weeks after emergence. The virus has been well described and its pathology is to cause fragility and break down of the midgut, with non-lethal effects including sluggishness and a temporary reduction in egg laying. The virus was reported from a long-term laboratory culture and its originating source was unknown. If this virus occurs in field populations of *B. tryoni* or other species, it has not been reported.

8.6 *Wolbachia*

Wolbachia are endosymbiotic bacteria of the family Rickettsiaceae that infect approximately 50% of arthropod species. *Wolbachia* are common reproductive parasites and are found frequently in the testes and ovaries of their hosts. However, while they infect eggs they do not infect mature sperm, and so their transfer is through the female line, or via horizontal transfer. Depending on the *Wolbachia* phenotype, infection can significantly alter host reproduction, particularly in a manner that increases the percentage of infested females in a population. This is achieved by: (i) killing males during larval development, leading to female biased sex ratios; (ii) altering male development so they become females, or infertile pseudo females; (iii) creating parthenogenesis (reproduction without males); and (iv) causing cytoplasmic incompatibility, where infected males are unable to successfully reproduce with uninfected females or females infected with another *Wolbachia* strain. Because of their impact on reproduction, which has implications for everything from speciation to pest management, *Wolbachia* are widely researched by entomologists.

Wolbachia are recorded from both *Bactrocera* and *Dacus* species, but they are rare. For example, in China, only 19 out of 15 000 individual flies sampled were detected as being positive for *Wolbachia*. In Australia, 38 flies from a sample of 592 individuals were detected as *Wolbachia* positive, with *Wolbachia* infestation being largely restricted to tropical collection sites. As many as five genetically distinct *Wolbachia* strains have been recorded from a single individual fly, and in one study it was found that these strains were common across many *Bactrocera* species, strongly suggesting the existence of horizontal (i.e. ecological) transfer of the *Wolbachia*. The low incidence of natural infestation of *Wolbachia* suggests that they do not play an important role in the ecology and biology of *Bactrocera* and, indeed, there is some evidence that *Wolbachia* infestation in *Bactrocera* species may be transitory within a species and the infestation is lost over time.

In the laboratory, a strain of cytoplasmic incompatibility inducing *Wolbachia* from *C. capitata* has been successfully transferred to olive fruit fly, where it was subsequently shown to cause male sterility if infected females were mated with non-infected males. If a genetic sexing line is available for mass rearing, then this approach offers a mechanism for producing sterile males for SIT without needing to irradiate the flies. I am not aware if this approach, or other forms of *Wolbachia* manipulation, is being pursued for commercial scale usage against *Bactrocera* anywhere in the world.

8.7 Strepsiptera

A single strepsipteran has been recorded as parasitising several *Bactrocera* species in Australia and the nearby Solomon Islands. The Strepsiptera are an endoptery-gote order of approximately 600 species worldwide that are thought to be a sister order to the Coleoptera, and their life cycle is unique within the insects. Female strepsipterans are primarily endoparasites of their hosts with most of the female's body imbedded within the host, but with the anterior portion emerging externally from the host. Females never leave the host and have a highly reduced morphology, with no wings, legs or eyes. However, the non-feeding and short-lived winged males (approximately 5-hour lifespan) are mobile and actively seek females for mating, cuing in on a female produced pheromone. Mated females may produce many thousands of triungulinid, which are a legged and mobile first instar larval stage. Depending on species these may live within the parental female's haemolymph system and leave the mother without damage, whereas in other species the triungulinid may eventually consume the mother before leaving the host insect to find new hosts. Male triungulinid larvae invading a new host eventually pupate within that host before emerging as adult males, whereas female triungulinid develop to maturity without a pupal stage and remain within the host. It is not clear what damage Strepsiptera do to their hosts. Mature females may occupy up to 90% of the abdominal space of their hosts, presumably causing fitness loss; and male insects that are parasitised are often sterilised.

The family Dipterophagidae (or sub-family Dipterophaginae of the family Halictophagidae depending on author) contains only the one species, *Dipterophagus daci*, and this is the only strepsipteran known to parasitise Diptera. The biology of *D. daci* is not well known, but 17 species of *Bactrocera* and two species of *Dacus* have been recorded as hosts. The population dynamics of *D. daci* appear to cycle with, but lag after, host population peaks, and parasitism rates range from under 1% to a maximum of 7%. The parasites were far more common in tropical than sub-tropical Australia. It is thought that the hosts are parasitised as teneral adults, while the testes of male flies showed no obvious signs of parasitism. What, if any, effects *D. daci* has on the abundance of host populations is unknown.

8.8 Further Reading and References Cited

Allwood, A.J. and Drew, R.A.I. (1996) Seasonal abundance, distribution, hosts and taxonomic placement of *Dipterophagus daci* Drew and Allwood (Strepsiptera: Dipterophagidae). *Australian Entomologist*, 23, 61–72.

Apostolaki, A., Livadaras, I., Saridaki, A., Chrysargyris, A., Savakis, C. and Bourtzis, K. (2011) Transinfection of the olive fruit fly *Bactrocera oleae* with *Wolbachia*: towards a symbiont-based population control strategy. *Journal of Applied Entomology*, 135, 546–553.

Carmichael, A.E., Wharton, R.A. and Clarke, A.R. (2005) Opiine (Hymenoptera: Braconidae: Opiinae) parasitoids of tropical fruit flies (Diptera: Tephritidae: Dacinae) of the Australian and South Pacific region. *Bulletin of Entomological Research*, 85, 545–569.

Clausen, C.P. (1978). Tephritidae. In Clausen C.P. (ed.) *Introduced Parasites and Predators of Arthropod Pests and Weeds: a World Review*. USDA Agriculture Handbook, Washington, DC, USA, pp. 320–335.

Daane, K.M., Wang, X., Nieto, D.J., Pickett, C.H., Hoelmer, K.A., Blanchet, A. and Johnson, M.W. (2015) Classic biological control of olive fruit fly in California, USA: Release and recovery of introduced parasitoids. *BioControl*, 60, 317–330.

Deguine, J.-P., Atiama-Nurbel, T. and Quilici, S. (2011) Net choice is key to the augmentorium technique of fruit fly sequestration and parasitoid release. *Crop Protection*, 30, 198–202.

Drew, R.A.I. (1987) Reduction in fruit fly (Tephritidae: Dacinae) populations in their endemic rainforest habitat by frugivorous vertebrates. *Australian Journal of Zoology*, 35, 283–288.

Ero, M.M. and Clarke, A.R. (2012) Host location by the fruit fly parasitoid *Diachasmimorpha krausii*: role of fruit fly species, life stage and host plant. *Agricultural and Forest Entomology*, 14, 101–110.

Ero, M.M., Hamacek, E. and Clarke, A.R. (2011) Foraging behaviours of *Diachasmimorpha krausii* (Fullaway) (Hymenoptera: Braconidae) and its host *Bactrocera tryoni* (Froggatt) (Diptera: Tephritidae) in a nectarine *(Prunus persica* (L.) Batsch var. *nectarina* (Aiton) Maxim) orchard. *Australian Journal of Entomology*, 50, 234–240.

Gonçalves, F.M., Rodrigues, M.C., Pereira, J.A., Thistlewood, H. and Torres, L.M. (2012) Natural mortality of immature stages of *Bactrocera oleae* (Diptera: Tephritidae) in traditional olive groves from north-eastern Portugal. *Biocontrol Science and Technology*, 22, 837–854.

Jenkins, C., Chapman, T.A., Micallef, J.L. and Reynolds, O.L. (2012) Molecular techniques for the detection and differentiation of host and parasitoid species and the implications for fruit fly management. *Insects*, 3, 763–788.

Merkel, K., Migani, V., Ekesi, S. and Hoffmeister, T.S. (2016) From behavioural studies to field application: improving biological control strategies by integrating laboratory results into field experiments. In: Ekesi, S., Mohamed, S.A. and De Meyer, M. (eds.) *Fruit Fly Research and Development in Africa. Towards a sustainable management strategy to improve horticulture*. Springer, Basel, Switzerland, pp. 369–388.

Messing, R.H. and Wang, X. G. (2009) Competitor-free space mediates non-target impact of an introduced biological control agent. *Ecological Entomology*, 34, 107–113.

Mohamed, S.A., Ekesi, S. and Hanna, R. (2010) Old and new host-parasitoid associations: parasitism of the invasive fruit fly *Bactrocera invadens* (Diptera: Tephritidae) and five African fruit fly species by *Fopius arisanus*, an Asian opiine parasitoid. *Biocontrol Science and Technology*, 20, 183–196.

Mohamed, S.A., Ramadan, M.M. and Ekesi, S. (2016) In and out of Africa: parasitoids used for biological control of fruit flies. In: Ekesi, S., Mohamed, S.A. and De Meyer, M. (eds.) *Fruit Fly Research and Development in Africa. Towards a sustainable management strategy to improve horticulture*. Springer, Basel, Switzerland, pp. 325–368.

Morrow, J.L., Frommer, M., Shearman, D.C.A. and Riegler, M. (2014) Tropical tephritid fruit fly community with high incidence of shared *Wolbachia* strains as platform for horizontal transmission of endosymbionts. *Environmental Microbiology*, 16, 3622–3637.

Moussa, A.Y. (1978) A new cytoplasmic inclusion virus from Diptera in the Queensland fruit fly, *Dacus tryoni* (Frogg) (Diptera: Tephritidae). *Journal of Invertebrate Pathology*, 32, 77–87.

Newell, I.M. and Haramoto, F.H. (1968) Biotic factors influencing populations of *Dacus dorsalis* in Hawaii. *Proceedings of the Hawaiian Entomological Society*, 20, 81–139.

Nishida, T. (1955) Natural enemies of the Melon fly, *Dacus cucurbitae* Coq. in Hawaii. *Annals of the Entomological Society of America*, 48, 171–178.

Peng, R.K. and Christian, K. (2006). Effective control of Jarvis's fruit fly *Bactrocera jarvisi* (Diptera: Tephritidae), by the weaver ant *Oecophylla smaragdina* (Hymenoptera: Formicidae), in mango orchards in the Northern Territory of Australia. *International Journal of Pest Management*, 52, 275–282.

Picchi, M.S., Bocci, G., Petacchi, R. and Entling, M.H. (2016) Effects of local and landscape factors on spiders and olive fruit flies. *Agriculture, Ecosystems and Environment*, 222, 138–147.

Quilici, S. and Rousse, P. (2012) Location of host and host habitat by fruit fly parasitoids. *Insects*, 3, 1220–1235.

Sivinski, J. and Aluja, M. (2012). The roles of parasitoid foraging for hosts, food and mates in the augmentative control of Tephritidae. *Insects*, 3, 668–691.

Sivinski, J., Aluja, M. and Holler, T. (2006) Food sources for adult *Diachasmimorpha longicaudata*, a parasitoid of tephritid fruit flies: effects on longevity and fecundity. *Entomologia Experimentalis et Applicata*, 118, 193–202.

Vargas, R.I., Long, J., Miller, N.W., Delate, K., Jackson, C.G., Uchida, G.K., Bautista, R.C. and Harris, E.J. (2004) Releases of *Psytallia fletcheri* (Hymenoptera: Braconidae) and sterile flies to suppress Melon Fly (Diptera: Tephritidae) in Hawaii. *Journal of Economic Entomology*, 97, 1531–1539.

Vargas, R.I., Leblanc, L., Putoa, R. and Piñero, J.C. (2012) Population dynamics of three *Bactrocera* spp. fruit flies (Diptera: Tephritidae) and two introduced natural enemies, *Fopius arisanus* (Sonan) and *Diachasmimorpha longicaudata* (Ashmead) (Hymenoptera: Braconidae), after an invasion by *Bactrocera dorsalis* (Hendel) in Tahiti. *Biological Control*, 60, 199–206.

Vayssières, J.-F., Offenberg, J., Sinzogan, A., Adandonon, A., Wargui, R., Anato, F., Houngbo, H.Y., Ouagoussounon, I., Diamé, L., Quilici, S., Rey, J.-Y., Goergen, G., De Meyer, M. and Van Mele, P. (2016) The use of Weaver ants in the management of fruit flies in Africa. In: Ekesi, S., Mohamed, S.A. and De Meyer, M. (eds) *Fruit Fly Research and Development in Africa. Towards a sustainable management strategy to improve horticulture*. Springer, Basel, Switzerland, pp. 389–434.

Waterhouse, D.F. (1993). Pest fruit flies in the Oceanic Pacific, Chapter 2. In: *Biological Control: Pacific* Prospects*, Supplement 2*. Australian Centre for International Agricultural Research, Canberra, Australia, pp. 4–47.

Wharton, R.A. and Gilstrap, F.E. (1983) Key to and status of opiine braconid (Hymenoptera) parasitoids used in biological control of *Ceratitis* and *Dacus s. l.* (Diptera: Tephritidae). *Annals of the Entomological Society of America*, 76, 721–742.

Wharton, R.A. and Yoder, M.J. Parasitoids of fruit-infesting Tephritidae. Available at http://paroffit.org. (accessed 25 May 2018).

9 Pre-harvest Management

In countries where pest fruit flies are endemic they are commonly the most serious horticultural pest insects. Dacini fruit flies negatively affect growers through direct crop loss, the cost of implementing control treatments, and lost market-access opportunities if trying to sell into areas where fruit flies are absent or regulated. *Bactrocera* and other Dacini can be problematic to control because: (i) they can occur at high population levels; (ii) the flies are mobile and commonly polyphagous so they can move into a crop from adjoining non-crop/non-managed areas, or simply move from crop to crop sequentially; (iii) a single 'sting' may make the fruit inedible (or at least unmarketable); (iv) an individual female may sting many fruits; and (v) the most effective pesticides for fruit fly control are no longer available or are being withdrawn. Despite these problems, good fruit fly control can still be gained through treatment combinations, particularly where neighbours work together to reduce the total pest pressure. The following chapter deals first with individual control options, and then discusses how they can be best implemented in an integrated fashion.

9.1 Insecticides

Several classes of pesticide have historically given excellent fruit fly control. Of these, the systemic organophosphates (e.g. dimethoate and fenthion) are good examples. One or two well-timed late-season sprays provided high levels of fruit fly control. However, older-generation, broad-acting insecticides are increasingly being banned around the world and there are no automatic replacements for them appearing for fruit fly management. General orchard cover sprays to control adult fruit flies can also negatively affect beneficial insects and disrupt management of other pest groups, and so should be used with caution. Even if the chemicals used are not banned at the place of use, major fruit importers such as the European Union and the United States are becoming increasingly critical of pesticide residues and so buyer demand, even if not local legislation, is likely to see the demise of the widespread use of synthetic pesticides over the medium term.

The relatively novel pesticide spinosad, a compound of two products spinosyn A and spinosyn D derived from the soil bacteria *Saccharopolyspora spinose*, does have use for fruit fly control and is being widely used as an organophosphate replacement. Spinosad is registered as an organic product in many countries, has very low mammalian toxicity and breaks down rapidly upon exposure to sunlight. For fruit flies it is used as the toxicant within the proprietary protein-bait product GF-120.

As for other insects, repeated application of insecticides against Dacini can lead to resistance. Low to high resistance to organophosphates (trichlorophon), pyrethroids (beta-cypermethrin) and avermictin have been reported for *B. dorsalis* in China; to organophosphates (fenthion, malathion, trichlorophon), pyrethroids (beta-cypermethrin), avermictin, spinosad and fipronil for *Z. cucurbitae* in China and Taiwan; to spinosad for *Z. cucurbitae* in Hawaii; and to organophosphates for *B. oleae* across the Mediterranean basin. The resistance to spinosad is particularly troublesome. As the insecticidal agent in the globally used Dow Agrochemicals protein bait spray GF-120, the chemical is exposed frequently (up to weekly) to target flies, who actively search for the protein spray. In small crops (melons, tomatoes, etc.), particularly, there will be no seasonal break in application, creating exactly the right environment for resistance to rapidly develop. Chinese researchers have demonstrated the potential for 40-fold resistance to spinosad in selected lines of *Z. cucurbitae* versus control lines. Sixteen-fold resistance of *Z. cucurbitae* to spinosad was reported in Hawaiian melon crops 10 years after weekly sprays of GF-120 commenced.

Despite their likely general decline in global agriculture, and the limited chemistries well suited for fruit fly control, pesticides have a place in the fruit fly toolbox. Synthetic pesticides offer confidence of immediate efficacy and, even if not used routinely, can still be used as an emergency treatment to suppress damaging populations when other controls fail. The chemistries available, be they organophosphates, neonicotinoids or pyrethroids, will vary with country and market regulations, and in all cases should be used with caution with respect to the applicator health, environmental health, and the disruption of other integrated pest management (IPM) practices.

9.2   Sterile Insect Technique

9.2.1   Introduction and concept

The sterile insect technique (most commonly referred to as SIT) is a biological pest control approach based on mating disruption that was developed over 60 years ago by US scientists. SIT is a control strategy that works best for insects where the female mates only once, or a few times, in her life – such insects are thought to include *Bactrocera* fruit flies. In such insects, if the male the female mates with is sterile, then the female will lay infertile eggs. The SIT operates by mass rearing huge numbers (tens to hundreds of millions) of males of the target insect, sterilising them (commonly with a radioactive cobalt-60 source or X-ray), and then releasing them. If enough sterile males are released that they can out-compete the wild males for partners, then most of the wild females will also then become sterile and the pest population will collapse within a few generations. This approach, when used properly, can drive local populations to extinction. SIT is used operationally for fruit fly management around the world, although currently with limited applications against dacines.

There are several benefits of SIT that makes it a preferred control tool. These include that it is environmentally safe, as only the pest species is targeted and there is no potential for non-target effects. Additionally, if for any reason negative aspects did arise, the releases can simply be stopped and as the released insects are sterile they will simply die out. SIT works really well within an area-wide IPM (A-W IPM) programme as releases can target pest insects anywhere in the environment, be it on-farm, in scrub or forest, or in towns; for this reason SIT is regarded as a key element of A-W IPM. The long history of SIT and the strong international support it has received from multinational organisations means that SIT is well imbedded in international protocols as a recognised component of pest risk reduction and market access.

On the downside, SIT is expensive and operationally complex. The key element of SIT is getting enough mass-reared and sterilised males into the field so they can out-compete wild males for wild females. This requires many things to work. The 'over-flooding' ratio is the multiplier of how many more sterile males you need to release than there are wild males in the environment. Methods are available to calculate this in some detail based on local population size, but operationally around the world the ratio is anywhere from 20:1 to 100:1 and general practice is to release from 1000–5000 males per hectare per week. Thus the number of flies needing to be reared to treat a production area is large, from the tens to hundreds of millions per week. A moderate-sized Mediterranean fruit fly SIT factory in Valencia, producing 500–600 million sterile flies/week, cost €8 million to build in 2007. The start-up cost and ongoing production costs make SIT an expensive control option. Further, just because male flies are released, it does not mean they are competitive against wild flies. To be competitive the released flies need to survive after release, they need to be able to find females, the females need to choose them as partners, and the mating needs to physiologically inhibit the female from mating again (this is discussed further below). Getting all of this right requires a great deal of scientific and technical expertise, and so quality maintenance of mass-reared flies is a constant issue for SIT. Finally, SIT works best within an already established area-wide programme, as the lower the wild fly population the smaller the release numbers need to be and hence the cheaper and more effective the programme. SIT therefore rarely operates as a stand-alone programme, unless it is being used as an eradication tool for incursion management.

More than most parts of this book, the following sections on different aspects of SIT are introductory only, designed to identify the various steps in the SIT process. Huge amounts of technical literature and international expertise are available to support any attempt to develop a SIT programme commercially. If interested in doing so, the first point of contact is the Insect Pest Control Section of the Joint FAO/IAEA Programme Nuclear Techniques in Food and Agriculture, located in Vienna, Austria. References and web-links to FAO/IAEA operations manuals, which are largely focused on SIT, but also include aspects of sampling and area-wide management, are provided in the reading list at the end of this chapter.

9.2.2 Mass rearing, pre-release and release

The mass rearing of tens to hundreds of millions of flies weekly is the first step of the SIT process. The mass rearing of flies is a factory operation and has similar operational issues to any factory: running at a profit, quality maintenance, input supply and output distribution, worker training and care, etc. These issues are critical to the success of a SIT factory, but are not discussed here. Rather, this section focuses on biological aspects to be considered when mass rearing flies.

Male-only lines
The mass rearing and release of males may also mean the mass rearing and releasing of females. Releasing females potentially causes two problems: (i) it increases the factory and release costs as 50% of flies produced are useless; and (ii) released males can be 'distracted' from mating with wild females through mating with sterile females. For *C. capitata*, this last point has been theoretically predicted to decrease the efficiency of SIT to a point where the technique may not be operationally viable.

To overcome these problems, strategies are required to kill or remove females, preferably as early as possible in the production cycle. Genetic sexing strains (GSS) for *B. dorsalis* and *B. carambolae* exist where the sexes may be separated based on pupal colour. Males exhibit the wild-type brown pupal colour, while females are mutant and exhibit a white colour. For *C. capitata* male-sexing lines are also available through temperature-dependent lethality: male eggs are more heat tolerant than female eggs and in the factory the eggs are floated through a temperature-controlled water bath to kill the females. A temperature-linked GSS for a *Bactrocera* species is not yet developed. The development of GSS lines is not trivial, even using modern gene editing techniques such as CRISPR-Cas9. A current major initiative targeting *B. tryoni* SIT in Australia is struggling with developing a male-only line.

Diets
Suitable and cheap ingredients for diet, successful utilisation of the diet by maggots, and then disposal of the spent diet, is a major factor for consideration in factory scale production of flies, as the diet may run into tonnes of material per week. The development of artificial diets for fruit fly maggots largely stems from SIT rearing work. A novel methodology that builds on the liquid diet

approach, but sets the liquid diet into a gel, shows promise as an operational replacement for traditional solid and liquid diets.

Competitiveness of flies and pre-release conditioning

Mass-reared males need to be sexually competitive against wild males if they are to gain access to wild females. The nature of laboratory mass rearing rapidly selects for traits that may act against the competitiveness of mass-reared males. Initial attributes are non-sexual, and relate simply to a fly's ability to survive in the wild, forage for food resources and water, handle extremes of fluctuating temperatures, and general activity patterns. Traits directly attributed to mating include the ability of the fly to find a mating site, release pheromones and carry out appropriate courtship leading to mating. The males also then need to be able to repress the female's re-mating propensity. Poor-quality males can range from flies that are physically incapable of surviving in the wild, through to physically robust males that achieve mating but do not suppress female re-mating.

International standards exist for quality control of mass-reared fruit flies (FAO/IAEA/USDA, 2014), and include measures of pupal weight, pupal emergence, flight ability, survival under stress, and sex ratio after irradiation. Unfortunately many of the protocols and recommendations are based on research for *C. capitata*, with significantly less basic science done for *Bactrocera* species. Mating competitiveness tests, where mass-reared males and wild/wildish males compete for females in walk-in field cages offer a good measure of reared-fly quality at the mating stage, but do not account for the ability of the reared flies to survive and find mating sites in nature.

The quality of mass-reared males can be significantly enhanced if they are held in pupal emergence facilities and provided with resource supplements prior to release. The appropriate supplements can both increase the rate of sexual maturation and make males more competitive. For example, *Z. cucurbitae* treated with methoprene (an insect growth regulatory) had increased mating competitiveness and longevity. Similarly, immature *B. tryoni* males fed with raspberry ketone and protein matured faster than control males, while their matings were similar in suppressing female re-mating activity.

Little information is available on SIT release strategies specifically for Dacini (FAO/IAEA, 2017). As for other tephritids, aerial releases of adult have been made, including *B. tryoni* in Australia and *B. carambolae* in South America. In Australia, small-scale releases have also involved the blower ejection of adult flies from a vehicle driving along suburban streets, and the liberation of flies in the pupal form by letting adults emerge from pupae held in insulated boxes or from sand beds. Pupal-release strategies are likely to result in significant adult mortality of SIT flies before they are sexually mature and capable of mating, but are cheap compared to aerial releases as there are no post-emergence holding costs, and distribution from the rearing facility to point of release is much easier. A compromise being currently considered in Australia is to ship sterilised pupae to farmers, and then allow the pupae to emerge into small field cages where they can be supplied with adult diet supplements and allowed to become sexually mature before release.

9.2.3   Considerations before embarking

International experience has demonstrated that SIT can be a highly effective pest management strategy against tephritids: however, there are few examples of SIT being used effectively against Dacini. SIT has been successfully used against *B. tryoni* for outbreak eradication, and has been trialled successfully against *B. dorsalis* in Thailand. However, there are no current examples where SIT is used routinely for *Bactrocera* pest management, as it is for example against *C. capitata*, although Australia is working to develop this capacity. Why is this?

First, SIT is expensive. The minimum investment to start a sterile insect facility is several million dollars. The new Australian factory in Port Augusta, South Australia, cost AU$3 million (not counting the land), and it is a small to medium facility with a maximum capacity of 50 million flies per week. The Australian investment does not include a separate rear-out facility. Once a factory is built and fitted, costs include management, technical and operating staff, consumable supplies and semi-permanent fittings (such as cages). Then once flies are being produced they need to be shipped to release sites, and then released. The question is, who covers this cost? Throughout the world, SIT costs are commonly heavily subsidised through governments, and occasionally through local rates. If such support is not possible, then it is unlikely that SIT will be viable in the long term as growers, even wealthy growers in countries such as Australia, are unlikely to have the funds to use SIT.

A second reason why SIT is perhaps less frequently used against Dacini than for other pests is the problem of multiple pests. SIT, by its nature, is species specific. Where there is only one dominant fruit fly pest species then SIT, targeting that pest, may be practical. This, for example, is the case with *C. capitata* in Spanish citrus orchards. However, for much of the native ranges of the Dacini, there are at least two, commonly many more, pest flies in any given production system. Controlling just one fly through SIT, for example *B. tryoni* in Australia, or *B. dorsalis* in South-east Asia or Africa, is just likely to see the emergence of secondary pest fruit flies, or the return of what were the primary pest fruit flies in Africa before displacement by *B. dorsalis*.

For these reasons, many of the other fruit fly control approaches (dealt with in the following sections), which are cheaper for an individual farmer and effective against multiple fly species simultaneously, are likely to remain preferred for Dacini pest management.

9.3   Lure-and-kill Technologies

Lure-and-kill techniques, as the name implies, work by attracting the target pest to a particular location and there killing it. In the wider pest management literature there are subtle variations in how authors define different related techniques, whether they are lure-and-kill, versus trapping, versus lure-and-infect. However, as they all work on the same basic principle (i.e. attract the target and then in some way kill it), and are commonly used together in fruit fly pest management, I am treating them here within the one larger section. For a broader review of lure-and-kill in insect pest management see El-Sayed *et al.* (2009).

9.3.1 Male annihilation technique

The male annihilation technique, commonly referred to as MAT, uses as its basis the strong male *Bactrocera* response to the male lures methyl eugenol and cue-lure (see Section 5.7). Approximately 2–3 ml of lure, mixed with an equal part of contact pesticide, applied to an absorbent substrate, becomes an exceptionally effective device that can trap-out male flies from the local area for up to 6 weeks. The carrier is dependent on what is available and can range from 'low tech' (such as bits of coconut husk), to high tech (such as SPLAT technology). The density of MAT stations per area varies with usage and country, but as reviewed by Vargas *et al.* (2014) is commonly around 400–600 units per square kilometre for eradication programmes, and 2–4/hectare for methyl eugenol and 4–8/hectare for cue-lure as part of ongoing pest management. Counterintuitively, very high densities of MAT devices are not as effective as lower densities, as high concentrations of the male lure in the environment makes it difficult for a lure source to be located and so flies are not poisoned.

Methyl eugenol is considered a more powerful attractant than cue-lure, and for this reason MAT against cue-lure-responsive species is considered less effective. Early work by Steiner in the Pacific, and Japanese researchers on the Okinawan islands, demonstrated that it was possible to drive the methyl eugenol-responsive *B. dorsalis* to extinction using MAT alone, but full eradication using MAT only for cue-lure responsive species is much rarer. For the cue-lure flies, MAT in combination with protein bait sprays (see Section 9.3.2) is required to achieve high levels of population suppression or eradication. Recent development of new, species-specific cue-lure derivatives, such as raspberry ketone formate (i.e. melolure) for *Z. cucurbitae*, and raspberry ketone trifluoroacetate for *B. tryoni*, offer promise for greater MAT efficacy for cue-lure-responsive species. When combined with the regular application of protein bait sprays, the MAT/protein combination has been shown to provide 90–95% in-field *Bactrocera* control in a wide range of usages, from small-scale mixed semi-commercial production in Africa, Asia and the Pacific, to large-scale commercial monoculture production in Australia. Several theoretical and life-table studies have demonstrated that removing males alone from a fruit fly population is an inefficient means of gaining control, and so MAT for *Bactrocera* pest management is nearly always recommended to be done in conjunction with protein bait spray.

While the lure and toxicant for MAT have traditionally been applied in the field using impregnated fibre-board blocks, or more recently fibre cards, this technology is limiting. Fibre-board blocks need to be nailed onto a tree or post, or the cards hung, and in both cases collection again is required to avoid environmental pollution. As an alternative, SPLAT (specialised pheromone and lure application technology) is a proprietary product used for the controlled release of semiochemicals. Research from Hawaii and Australia has demonstrated that SPLAT, as a carrier for methyl eugenol or cue-lure and spinosad, is a highly viable alternative to fibre blocks or cards, while being quicker to apply and with natural, non-toxic breakdown in the environment over time. SPLAT-ME and SPLAT-CL is now commercially available.

9.3.2 Protein bait spray

Protein bait spray is a lure-and-kill technique that targets both male and female flies, although it is mainly directed at females. It is commonly used in conjunction with MAT to have a joint male/female-targeted programme. In protein bait spray control, an insecticide-baited protein liquid is splashed, spot sprayed or strip sprayed onto host plants or used in traps to attract flies that come and feed upon it, thereby poisoning themselves. While not an organic approach (depending on the insecticidal agent), insecticide usage within an orchard using protein bait spray is reduced to only a few per cent of that required for insecticide cover sprays because of the restricted and targeted application. The biological basis of the control is that both male and female dacine flies are known to need protein in order to become sexually mature. Further, it has been assumed that females (at least) will need a life-long source of protein because the continuous production and laying of eggs is 'protein expensive'. It is therefore assumed that the flies will continuously forage for protein sources and feed upon them if found.

Protein bait sprays themselves are highly variable in content and creation, but are most commonly based upon hydrolysed proteins. As reviewed by Epsky *et al.* (2014), protein baits developed from early observations that ammonia was attractive to various genera of fruit flies, and when added to sugar/fruit-based lures increased lure attractancy. This led to the development of the hydrolysed protein baits, where acid hydrolysis or enzymatic hydrolysis is used to break down protein, commonly brewer's yeast, to peptides, free amino acids and smaller breakdown products such as ammonia. The original source of protein, the hydrolysis type used, intermediate processing, the product's final pH, and the addition (or otherwise) of extra ammonia have all been shown to affect attractancy of protein bait to flies. While there are many commercial bait sprays available, relatively simple processing of spent beer yeast (*Saccharomyces* spp.) from commercial breweries has been demonstrated in both Asia and the Pacific to make an effective and cheap bait spray product. The Dow Agrochemical product GF-120, sold under the trade names GF-120™ NF Naturalyte fruit fly bait and Naturalure™ Naturalyte fruit fly bait, are arguably the most widely available commercial fruit fly protein baits at the present time.

Although widely used, and with good success, the effective use of protein bait sprays is not without issue. The literature is clear that different fly species respond differently to protein sources; for example, two studies have demonstrated that female *Z. cucurbitae* respond more strongly to GF-120 than do *B. dorsalis*. Sexually mature, gravid female *B. tryoni* were rarely found to be

protein hungry in the field, providing a physiological basis as to why their protein response was limited to a few metres. Sexually mature male *B. tryoni* rarely
responded to protein at all. In contrast, sexually immature flies of this species
are protein responsive and so bait sprays, or protein-baited traps, are effective
when there is a large percentage of young flies in the population.

Application of protein bait needs to be done with caution. Protein-bait
formulations (especially those with high salt or pH levels) have the capacity
to cause phyotoxicity to the leaves of some plants, while the skin of some
fruit types (notably mango) can be permanently marked by even a few drops of
bait spray. Different flies also forage for protein in different places and different
heights within the crop. Based on foraging behaviour, protein for *B. dorsalis* is
recommended to be sprayed at mid-crop canopy, whereas *B. tryoni* recommendations are for the upper canopy. Many of the cucurbit infesting *Zeugodacus*
and *Dacus* only enter the cucurbit crop to oviposit, and forage for food resources
elsewhere, and so for these pests it is recommended that protein be applied to
border plantings, rather than within the crop itself. Because of the differences
between and interactions among lure attractiveness, fly species and crop type,
local research should be done before providing recommendations on how to
implement a protein-bait programme.

One limitation of protein bait sprays is their need for frequent application, commonly recommended at 7–10 days. The sprays are susceptible to
being washed off the plant, and repeat application after rain is a normal recommendation. On the other hand, when it is not raining, the bait sprays dry to a
crust and lose their attractiveness. The use of gelling agents, which delay dehy-
dration, increase rain fastness and allow re-wetting during rain, help to address
these problems, but do not fully solve them. The development of bait stations,
which retain the protein baits in a container, with or without a visual attractant,
are a preferred means to overcome these problems. By definition, bait stations
differ from protein-baited traps, in that the dead insects are not retained. Bait
stations utilising protein as the attractant are still in their infancy for *Bactrocera*,
but one Hawaiian study comparing the efficacy of GF-120 sprays directly onto
foliage, or onto the undersides of inverted yellow plastic plates attached at
right angles to a pole or similar, thus providing a visual attractant cue and rain
protection to the protein, found equal efficacy of both. Israeli scientists are
currently developing methodologies for slow protein-release stations that show
good promise.

9.3.3   Mass trapping

As the name suggests, mass trapping involves the utilisation of high densities of
trapping devices to reduce a pest population. MAT is a type of mass trapping,
but is generally considered separate to it in the Dacini literature, if not in some
of the wider tephritid and pest management literature. Because of the effectiveness of MAT and protein bait spray, mass trapping has not been extensively
developed for most of the Dacini pest species, and most tephritid mass-trapping
literature targets the non-lure responsive *Anastrepha* and *Ceratitis*.

A recent review of tephritid mass trapping divided traps into two types – wet and dry. Wet traps include the classical MacPhail traps, bell-shaped with an invaginated opening in the bottom that allows flies to enter and simultaneously creates a water reservoir in which the trapped flies drown. MacPhail traps and their precursors have been used for Dacini management since the early 20th century, holding fruit pulp/juice and some form of nitrogen/ammonia source to attract the flies. The following recipe from a 1944 Australian paper is typical of many of these early attractants (May and Caldwell, 1944):

> Formula – Rind and rag of one ripe or ripening orange, about 2 ½ inches in diameter; concentrated aqueous ammonia, 18 per cent., 6 teaspoonfuls; tank water, quarter pint. The orange rind is shredded with a sharp knife. The ammonia and the water are added and the mixture is kept in a tightly stoppered bottle for at least 24 hours before using; this stock mixture can be stored under these conditions for periods up to one month. To make the lure ready for the [McPhail] traps, two tablespoonfuls, i.e. eight teaspoonfuls, of the liquid are added to 3 ½ pints of tank water. This lure catches practically no insects other than fruit flies and the fouling of the traps is therefore negligible.

Modern versions of wet fruit fly traps do effectively the same thing, commonly using modified versions of PET soft-drink bottles as the trap body, and often with a protein attractant. Liquid-lure traps for dacines suffer from poor efficiency, often large amounts of bye-catch which foul the trap, and the logistic difficulties of needing to clean and refill on a regular basis. For these reasons they are rarely recommended for commercial purposes, although the ability to make them cheaply from recycled and/or common household materials means they are often recommended for subsistence and backyard growers.

In contrast to wet traps, dry traps are used commercially for *Bactrocera* pest management. In the Mediterranean basin, a dry trap that combines a food attractant (ammonium bicarbonate salt), a pheromone component (1,7-dioxaspiro [5.5] undecane), and a knock-down insecticide has been found to be an effective, stand-alone method for the control of *B. oleae*. Pilot trials demonstrated that while cost and efficacy of the trap was similar to the protein sprays previously used, the amount of insecticide used to control flies in an orchard was dramatically reduced, and contained within the trap. Recent experience in temperate Australia is showing that a dry, protein-based trap is catching large numbers of female *B. tryoni* early in the active season (i.e. spring), but efficacy declines as the season progresses. In temperate Australia the early season flies are seeking protein for sexual maturation, explaining the seasonal efficacy of the traps.

Sticky traps, with or without an additional chemical attractant, are occasionally sold for *Bactrocera* control. In China, a green-coloured trap is commercially available for *B. minax* control; in Australia a blue trap is available for *B. tryoni* control; whereas the Ladd trap (a yellow panel with a red, semi-spherical centre), has been shown to be effective in trapping both *B. dorsalis* and *B. tryoni*. However, sticky traps are rarely recommended for control, although they do have a role in population monitoring. Sticky traps are logistically difficult to work with, need to be replaced or cleaned frequently, need to be set at high densities in order to reduce populations, and have large bye-catch, including beneficials.

A truly effective, mature-female lure is the 'holy grail' of trap development for the polyphagous Dacini, as eliminating gravid females is the surest method for reducing fruit infestation levels. A female lure based on the odours of cucumber has been developed for *Z. cucurbitae*, and has also been shown to be attractive to the Australian melon specialist *Z. cucumis*. No effective lure has, however, yet been developed for the females of highly polyphagous species such as *B. dorsalis* and *B. tryoni*.

9.3.4 Lure-and-infect

See the discussion on auto-dissemination in Section 9.4.

9.4 Entomopathogens

Entomopathogens have been tested against *Bactrocera* for use as pesticide-replacement technologies for in-field control, although much of this work is still laboratory based. Conidial suspensions of *Metarhizium brunneum* have been shown to cause 60% mortality of adult *B. oleae*, 82% mortality of pre-pupal third instar larvae, but only 33% mortality of pupae. Two strains of *Aspergillus flavus* and one strain of *A. tamarii* had high toxicity against adults of *Z. cucurbitae*, but low insecticidal activity against eggs, larvae and pupae. Similarly, *Metarhizium anisopliae*, *Beauveria bassiana* and *Isaria fumosorosea* have been shown to be toxic to adults of *B. zonata*, but to a much lesser extent the juvenile stages. In a two-season trial, *Metarhizium anisopliae* applied as a soil inoculation reduced *B. dorsalis* fruit infestation in mango orchards to 45% and 38% of that experienced in an untreated control orchard over years 1 and 2, respectively. When combined with a GF-120 spinosad bait spray treatment, the fruit infestation rates were 16% and 11% of the control for year 1 and year 2, respectively.

Additional to their use as prophylactic soil drenches, fungi have the potential to be used directly against adults in a technique known as auto-dissemination, or by some authors as 'lure-and-infect'. In this approach male flies are attracted to a feeding or male-lure trap that is contaminated with fungal conidia, where they pick up a high conidial load. Because it takes up to a week for the fungi to kill the adults, and not all conidia will immediately hatch, the fungal spores then have the potential to be horizontally transferred to females during mating, or to other males during male–male interactions. Participation in mating behaviour by infected males declines within 3 days of initial infection, and so the temporal window for horizontal transfer is short. In Kenya, using a Lynfield trap lined with velvet that carried the conidia of *M. anisopliae*, *B. dorsalis* infestation in mangoes was reduced from 61% (control) to 6% in ripening fruit, and male flies in traps fell from 56 flies/trap/day to less than 2 flies/trap/day. As part of IPM programmes, lure-and-infect should be more widely researched and potentially adopted.

Entomopathogenic nematodes have also been trialled for *Bactrocera* control. In laboratory trials, *Steinernema feltiae* caused dose-dependent mortality to *B. dorsalis* and *Z. cucurbitae* when applied against pre-pupal third instar larvae, with 85% mortality being the highest level reached. Similarly, *S. feltiae, S. carpocapsae* and *Heterorhabditis bacteriophora* all caused significant mortality (20–90%) of *B. tryoni* late third instars, but had no effect against pupae. In the latter study, substrate moisture and temperature significantly affected the results, but surprisingly, nematode density had a much lesser effect except for *S. carpocapsae* where *B. tryoni* mortality acted in a positive dose-dependent fashion. *Steinernema feltiae, S. carpocapsae, S. riobrave, S. glaseri, H. bacteriophora*, and *H. marelatus* all caused significant mortality of *B. oleae*, although the mortality caused by *S. feltiae* (68%) was significantly greater than all other nematode species. Substrate moisture and temperature were again confirmed in this study as key impactors of nematode efficacy.

Despite the large amount of laboratory data showing the potential value of entomopathogens for *Bactrocera* control, field evaluations and commercial uptake is still minimal except in Africa, where Dr Sunday Ekesi and ICIPE scientists have driven research in this field. This situation is likely to change as entomopathogens are better researched, more thought is given to how they can be incorporated into IPM and A-W IPM programmes, and ready access to pesticides declines.

9.5 Crop Hygiene

9.5.1 General concept

A relatively simple concept, the idea of crop hygiene is based on a premise that a significant percentage of a local pest fruit fly population breeds in the crop. If fallen infested fruit is disposed of before larvae leave the fallen fruit, then the local pest population will be suppressed. Crop hygiene was one of the first 'modern' pest control strategies put in place for fruit fly, being formally recommended and practised in many tropical crop production areas from the late 1800s.

Numerous studies around Asia, the South Pacific and Hawaii have demonstrated benefits of crop hygiene. So widespread are the assumptions that crop hygiene is beneficial, that it is automatically included in any summary of fruit fly management tools, and many export systems demand evidence of crop hygiene as a key part of a systems approach. Nevertheless, while widely accepted and often automatically mandated, the practicalities of crop hygiene are not straightforward and its overall benefit should be considered on a crop-by-crop and location-by-location basis. Some of the issues to be considered before automatically incorporating a crop hygiene schedule are as follows.

- *Biological need:* The likely benefits of crop hygiene in reducing a local population depends on two things: how many flies are actually in the crop, and what other local sources of flies are there? In well-protected orchards being grown for commercial production, infestation levels may only be a few per cent or less. If the crop is heavily picked, then the few remaining fruits per plant are unlikely to be producing many flies, even at the whole-orchard level. Or alternatively, the crop may just be a poor host for the fly. For example, lemons are extraordinarily poor hosts for Queensland fruit fly, even though the fly will sting them. In this case carrying out crop hygiene in a lemon orchard is likely to be simply wasted energy, as successful pupation of *B. tryoni* from lemon is less than 1%. In contrast, some tropical fruit orchards may have high background levels of infestation, an orchard may not have been commercially managed in a given season and so may be heavily infested, or significant amounts of late-season fruit may be left in the orchard after picking and fruit fly management has ceased. In such cases the orchard may well be producing large numbers of flies for the next generation. In such cases, removing or destroying fallen fruit should certainly be considered.

- *Biological usefulness:* Crop hygiene is only likely to be useful if a significant proportion of the next season's or next crop's pest population is coming from the orchard being managed. In physically isolated production areas where the commercial crop(s) are the major likely source of the flies, then hygiene may be warranted. Alternatively, in area-wide management schemes where an entire local pest population is being targeted then again hygiene may be warranted. However, if the orchards are relatively small in scale and being grown in matrix of many alternative native and commercial hosts that may not be being managed, then local hygiene may have no significant impact on the size of the next population to re-infest the crop.

- *Operational practicalities:* Whatever the biological benefits of crop hygiene, in most situations it is difficult to implement well. The pre-pupal larvae of most major *Bactrocera* pest species leave the fruit within a few hours, and 48 hours is considered the longest that larvae will stay in the fruit after it has fallen. This means that fallen fruit needs to be collected and destroyed every 24 hours to be effective, and even that will not give 100% control. For most commercial growers this is simply not operationally feasible or economic. A marked exception to this is Chinese citrus fruit fly, where larvae can stay in the fallen fruit for up to several weeks before leaving to pupate. In this case even 7–10 day fruit removal cycles will give excellent local control.

Physical removal and destroying of the fallen fruit poses further challenges. In orchards with large amounts of fallen fruit, physically destroying it can be a challenge. Burial requires deep burial at depths of >50 cm between the uppermost layer of fruit and the soil surface. Burial at depths less than this and all that is being done is to increase the likelihood of fly survival through pupation. Fruit can be heaped, covered in plastic sheet and be allowed to 'cook' and this can be effective for small-to-medium–sized fruit loads. In mechanised countries, mulching of fruit in-field is reported to have mixed benefits. Fruit needs to be mulched to a fine level to destroy larvae, and commercial mulchers attached to tractors may not achieve sufficient fruit breakdown. Mulching in the orchard inter-row will also miss fallen fruit directly in the tree row, unless this fruit is deliberately raked into the inter-row.

While there are many difficulties with effective hygiene, in some production systems there are simple answers. Running chickens, pigs or other livestock within an orchard will provide good natural orchard hygiene and this can provide an excellent means of local fruit fly control in organic, backyard and subsistence-level production. As for any agricultural production, whether or not to practice crop hygiene for fruit fly control may also depend on whether it achieves wider benefits, for example, by controlling other orchard pests or diseases.

9.5.2 Packing-shed fruit dumps

Directly related to crop hygiene, but often considered independent to it, are commercial fruit dumps. Any commercial packing shed at full harvest will sort and reject potentially tonnes of fruit. Sometimes this waste fruit is put to other direct commercial uses (e.g. juicing), or to indirect commercial uses (e.g. livestock feed). However, in some situations it is taken to a convenient local gully where it is dumped and left to naturally break down. The author has first-hand experience of seeing such dumped fruit directly laid into by Queensland fruit fly (even though this species rarely attacks fallen fruit), and larvae developing in the fruit. Given how big some of these dumps can be, they should be regarded as a potentially major source of flies and treated accordingly – for example, by daily burial of any new fruit added.

9.5.3 A role for protein bait sprays as an alternative to crop hygiene

Where managing the flies emerging from late-season fruit is considered necessary for population suppression, but crop hygiene is not an operationally realistic option, then late-season protein bait sprays should be considered. *Bactrocera* adults, both females and males, are most attracted to protein baits when they first emerge. Even though an orchard may be picked, or all remaining fruit dropped, weekly bait spray applications for 2–3 weeks after the crop is over will effectively target the new generation of emerging flies. If done properly, post-season protein bait sprays may well give better fly control than crop hygiene, although I am not aware of any direct comparative tests.

9.6 Abandoned Orchards, Feral Fruit Trees and Garden Fruit Trees

Directly related to the issue of crop hygiene is the importance of managing abandoned orchards and feral or garden fruit trees. The aim of such control is again the depression of a local population of flies, so making control in commercial crops easier.

9.6.1 Abandoned orchards

There seems to be little argument against the need to control abandoned fruit orchards. Any orchard, even an abandoned one, will produce large amounts of fruit and so potentially be a tremendous source of pest fruit flies. Further to flies, an abandoned orchard in the middle of a horticultural landscape may be a refuge and subsequent source of numerous other pests and diseases and so should be destroyed. However, how this is done may cause issues. Some countries have governmental controls that allow the prompt removal of abandoned orchards, but others do not, and yet others have the regulatory power to enforce removal but for various reasons may not use them. There is no easy answer to these problems, other than to say they should be addressed.

9.6.2 Feral fruit trees

In any attempt to implement area-wide management, or instigate some form of systems approach for market access, the removal of feral fruit trees is considered an important element. However, as for my comments on crop hygiene, I think feral trees also need to be looked at on a case-by-case basis.

In many tropical countries feral guavas may grow in small to large groves and can be largely unmanaged with respect to fruit fly. Guava is a good larval host of most major *Bactrocera* pests and if trying to implement an area-wide management programme then this is a clear example where feral trees should be managed. Management can be through tree destruction, fruit stripping, or some other control depending on the value of the plants for other uses (e.g. for soil conservation, shade, amenity use, etc.). At the other extreme is a situation often seen in inland or sub-coastal Australia. Here individual fruit trees (apples, stone fruit, etc.) may establish from bird dispersed seed in the middle of open grasslands or shrubby eucalypt forest. Such trees struggle to survive and if they flower they set minimal fruit, nearly all of which is consumed by birds and small mammals before it ripens. Do these feral trees pose a fruit fly threat to neighbouring horticultural production districts? Most likely not, but international protocols still suggest that such trees should be removed if attempting to achieve areas of freedom or low-pest prevalence. As for crop hygiene, I firmly believe that local input of feral fruit sources to the total local fly population needs to be assessed before it is automatically assumed that all feral fruit trees in a landscape need to be managed. In some cases they certainly will, but in others cases they may not.

9.6.3 Backyard fruit trees

In all agricultural and urban landscapes there is a source of fruit trees other than commercial and feral – these are the 'backyard' or home-garden trees. Most farmers will have a few individual trees for personal use in their home gardens, as do most people living in urban towns and also many living in larger cities if the land is available. Such trees can be a significant source of fruit fly as the trees may be fed and watered regularly and so produce heavy crops, but paradoxically pest control is often poor and much of the fruit may be routinely infested. In Australia it is well documented that B. tryoni is much more abundant in major cities such as Brisbane and Sydney than it is in agricultural areas. This is because the constant supply and high diversity of fruiting home-garden trees means that it has a largely continuous larval food resource. This problem is no less in rural towns, and managing the breeding of flies in garden trees must be a key element in any area-wide management programme. Where this has been successfully achieved it has required extensive public awareness and education attained through the use of local media, presentation at local fairs and fetes, and individual 'door knocking' (i.e. individual, house-to-house contact). Growers also need to be included in this message. It is surprising how growers may give excellent fruit fly control in their orchards, but ignore the large mango tree behind the house that the children climb in.

9.7 Mechanical Protection

Mechanical protection covers a number of different crop protection activities, but all involve placing a physical barrier between the fruit and the fruit fly. Physical barriers can be highly efficacious, giving 100% control, but the costs or mechanics of different types of barriers make them only suitable for some types of crop or production system.

9.7.1 Fruit bagging

Fruit bagging is a simple concept and involves placing a physical barrier (i.e. the bag) between the fly and the fruit. As the name suggests, bagging involves a bag being placed over the fruit with the bag then closed with string or wire around the fruit stem or branch. Bags can be ordinary brown paper bags, made from folded and stapled newspaper, or can be purchased commercially as a waxed-paper product made explicitly for fruit fly control. Experience shows that bagging not only protects fruit from fruit flies, but also other pests and mechanical damage and so the fruit tends to be of higher visual quality. Depending on the fruit types, bagged fruit can also be sweeter than unbagged fruit, a physiological response of the fruit to shading. In an informal trial in Papua New Guinea, plastic bags (that were cheap and locally available) also gave good control, although plastic bags are not recommended for bagging because the fruit can 'sweat' and high moisture levels can build up inside the bag. This can lead to fruit rot and other diseases.

Bagging is a highly efficacious control technique, as a fully bagged piece of fruit is 100% protected from fruit fly. The down side of fruit bagging is that it is obviously highly labour intensive and is not suitable for all fruit types, particularly those that do not have an obvious stem. Nevertheless, fruit bagging is used commercially in several countries for premium tree fruit. For such fruit, bags may be uniquely marked on the day of bagging, so that when fruit is suitable for harvest all bags with the same marking can be picked without opening the bag. By cutting the stem above the bag, the still bagged fruit is then sent to market in its own packaging. Despite this example, in most situations individual fruit bagging is not suitable for commercial production because of the labour costs involved, but it is suitable for non-commercial garden production. In Papua New Guinea, where banana fruit fly is endemic, bagging of whole banana bunches using banana leaves and 'bush rope' (twine made from local vines or peeled bark) is a traditional management practice. This is not done for all banana varieties, but only for sweet varieties (versus cooking varieties) that are left to ripen on the plant.

9.7.2   Whole-tree netting, orchard netting and protected cropping

These are all just larger versions of fruit bagging, and act to separate the crop from the fly. Individual tree netting is a viable technique for small-scale commercial and non-commercial production. It is important not to net before pollination has occurred, but after that whole-tree netting can be applied at any point. Nets that touch the fruit provide no protection, as flies can and will lay through the netting into the fruit. The netting thus needs to be held away from the plant, or an acceptance made that some fruit may be stung.

   Whole-orchard netting is infrequent, but in experiments in Australia it has been shown to work effectively. It may be a commercially viable option if the orchard has other reasons to be netted, for example as protection from hail, or from fruit bats or birds. Protected cropping, generally large-scale glass house production, can give outstanding fruit fly control, especially if fruit fly reduction strategies (e.g. trapping, baiting) are initiated in the surrounding area to reduce external fruit fly pressure and thus chances of fruit fly entry. Examples exist of large-scale glasshouse production of crops such as tomato gaining market access through the glasshouses being recognised as pest-free places of production, or the protected area being incorporated into a systems approach for market access.

9.8   Barrier and Repellent Sprays

Not widely used in *Bactrocera* control, barrier sprays are suspensions or emulsions that, when sprayed onto the crop, provide a physical, deterrent and/or repellent barrier to ovipositing flies. Barrier treatments are used commercially for *B. oleae* (see Section 9.8.1), and have been experimentally demonstrated as effective against other species. To be effective, barrier sprays need to cover the entire fruit, not be easily lost due to rain or other weathering, nor have their effectiveness lessened if the crop is still physically growing (i.e. the film is 'thinned' by the fruit getting bigger).

9.8.1 Kaolin particle films

Kaolin particle films are highly refined suspensions of kaolinic clay ($Al_4Si_4O_{10}(OH_8)$). When sprayed as a liquid onto a crop they dry to a white, hydrophilic layer over the fruit and leaves. These types of agricultural clays have been used for a long time to prevent fruit 'burn', but in different countries they are now registered for insect control. Depending on the insect, kaolin particle films may give protection by being repellent, having anti-ovipositional qualities, and by strongly reflecting light. Kaolin sprays are used commercially to control *B. oleae*, for which they have been shown to give better season-long control than organophosphate cover sprays. The clays can cause detrimental effects to natural enemies, and this needs to be kept in mind during use.

9.8.2 Agricultural and vegetable oils

Horticultural mineral oils are widely used against some groups of horticultural pests (e.g. scale insects) but they are not commercially used against *Bactrocera*. In the laboratory, an nC_{20-22} horticultural oil significantly reduced landings, fruit exploration, probing and oviposition by *B. tryoni* on tomato fruit that had been dipped in an oil emulsion. True repellence was detected, as well as possible blocking of fruit volatile cues that may have led to flies landing on fruit. Similarly, vegetable oils (especially safflower oil) reduced *B. tryoni* oviposition in laboratory trials, but these oils did not appear to have a repellent effect, rather the authors suggested a mechanical effect of the oils making a slippery surface which impeded female oviposition. In the same trial, essential oils of peppermint and some Australian native trees repelled female flies, but the oils evaporated rapidly and the effect was lost within a few hours. When translated to the field in small tomato plots, mineral oils provided variable levels of control, ranging from 0% to 50–60% to nearly 90% in three different seasons. The difference in results appeared to be due to the background fly pressure. In the year when greatest protection was obtained, fly pressure was low.

9.8.3 Neem

Extracts of the seeds and ground leaf of the neem tree (*Azadirachta indica* A. Juss. [Meliaceae]) are known to have insecticidal and anti-feedant properties for many insects. The insecticidal properties are commonly associated with the disruption of normal growth and moulting, whereas the anti-feedant properties are associated with repellence. In *B. dorsalis*, neem seed extracts significantly decreased the number of flies landing on treated guava and the number of eggs laid into fruit, inferring a repellent effect. Similar laboratory results have been reported for *B. zonata* and *Z. cucurbitae*. Reports of use or efficacy of neem sprays in the field are not available.

9.9   Production During Periods of Low-Pest Pressure

Even in tropical conditions, fruit flies may have seasonally low abundance because of dry conditions, or because the production area is in a continental inland area with cold winters. If there are suitable crops to be harvested during such time then they get a natural advantage of being produced during a period of low-pest pressure. In addition to the immediate benefits of reduced pest problems, a period of low-pest abundance can also have market-access benefits.

The phrase 'low-pest prevalence' has a specific meaning in regulatory entomology (see Chapter 11) and can be included formally as part of a systems approach for market access. In Australia, domestic quarantine is facilitated by the interstate certification assurance (ICA) scheme, in which states research and propose operational treatments that, when approved and implemented, greatly reduce the phytosanitary risk posed by the interstate movement of plant and plant products. Two such ICA agreements from the state of Queensland, while not formally incorporating production during a period of low-pest prevalence, do tie the applicable dates under which the agreements may operate to dates during which it is known that local fruit fly numbers are low, in this case the end of March to mid-August for the flies *B. tryoni* and *B. neohumeralis*. The crops these ICAs are relevant to are citrus (ICA-28) and strawberries (ICA-34), and in both cases growers get a significant operational advantage by being able to grow and market their crops while flies are largely inactive.

9.10   Pupal Controls

Control of *Bactrocera* at the pupal stage has not received a great deal of attention. Chinese workers, particularly, have focused on the value of inundative releases of pupal parasitoids, such as pteromalids of the genus *Spalangia*. These parasitoids show effectiveness in the laboratory, occur naturally as fruit fly parasitoids in the field, and can be mass reared easily in the laboratory on alternative and more easily produced hosts (such as *Musca*), which makes their usage as inundative biological control agents much more feasible. However, these wasps also show low or no host specificity to *Bactrocera* and may also be facultative hyper-parasitoids of opiine braconids. For these reasons, any plan to use these parasitoids specifically to target *Bactrocera* pupal control should first check about possible non-target effects and disruption to existing biological controls.

Entomopathogenic fungi, such as *Metarhizium anisopliae*, and entomopathogenic nematodes such as *Steinernema riobrave* have been laboratory tested as *Bactrocera* pre-pupal larval and pupal controls, often with good results. However, examples of field validation and usage are much fewer. Soil inoculation of *M. anisopliae* significantly reduced populations of *B. dorsalis* in African mango orchards compared to untreated controls, but the level of control obtained was not as great as that obtained through the use of GF120 bait sprays. The combined usage of both treatments reduced damage to 10–15% of that seen in controls. More field testing of entomopathogens against *Bactrocera* is clearly warranted. Based on laboratory studies, issues that warrant further study include the efficacy of treatments against pre-pupal maggots versus the pupae themselves, longevity of inoculum in the soil under different dryness and ultraviolet regimes, and optimum soil moisture regimes to maximise control. Mechanisms and rate of soil application also clearly need more research.

Direct mechanical destruction of pupae gives excellent control if it is operationally feasible. Observational data suggests that it is untargeted, traditional practices that may provide the best examples of such control. In many subsistence and non-commercial orchards, livestock (chickens, pigs) are allowed to forage within orchards. Animals such a pigs can consume large amounts of fallen fruit, so destroying pre-pupal larvae. In Bhutan, the author has observed the winter tethering of water buffaloes under citrus orchards in the sub-tropical Himalayan foothills region. No emergent *B. minax* were recovered from such orchards in the spring, as the physical disturbance and dunging of the orchard soil presumably killed over-wintering pupae. Similarly, scratching chickens are well documented as general predators of soil insects and pupae. In village situations, or non-commercial organic farms, the use of animals to control *Bactrocera* pupae has merit.

9.11 Mixed Controls

With the exception of insecticide cover sprays, fruit fly controls are most often applied as a mixture of strategies. This is rarely through choice; growers in every country prefer simple pest management tools for cost and time savings. Unfortunately, in the absence of pesticides, none of the currently available tools have sufficient efficacy to deliver high-level control on their own. The plus side is that, if done properly, a mix of alternative controls can give high levels of pest suppression. Because Dacini are mobile, and the polyphagous species have the capacity to breed on native, feral or unmanaged hosts away from the crop, it is considered that the best control is gained when control across a district is organised. However, good on-farm control can still achieve significant pest management benefits, and the failure of neighbours to engage in coordinated control efforts should not be used as a reason to not carry out any controls.

9.11.1 Integrated pest management or on-farm best management practice

For the pest *Bactrocera*, the most commonly recommended best management practice involves the use of MAT, regular (weekly to two weekly) protein bait sprays, and crop hygiene. When done properly, this combination of controls has been demonstrated to give >95% uninfested fruit. MAT and bait sprays target males and females, respectively, while the hygiene acts to decrease total population size. Growers need to be mindful that not just their commercial crops need to be managed – individual trees, non-commercial orchards (e.g. young orchards not into production, old orchards waiting to be pulled) also need to be managed to give good on-farm suppression. For small, subsistence-level gardens the fly 'spill-over' from non-controlled plots will negatively impact many of the benefits of these controls, but research has shown that production gains are still made. For large farms, such as those found in Australia, on-farm best management practice can be an effective replacement for insecticide-only control strategies.

9.11.2   Area-wide integrated pest management

A-W IPM is the term used to describe what is currently considered to be the best management practice for polyphagous tephritids, such as *B. dorsalis* and *B. tryoni*. A-W IPM involves the coordinated action of numerous stakeholders across a production district so as to manage the entire local fruit fly population, not just that component breeding in commercial crops. The strategy is based on the explicit assumptions that: (i) pest flies can breed in non-commercial fruit resources (garden plants, feral trees, wild hosts, etc.) and so move into commercial crops; and (ii) best control is gained when these off-farm flies are managed in parallel with best-practice on-farm management. Thus A-W IPM involves the coordinated management of fruit flies both in and away from places of commercial production. The benefits are reduced pest pressure and lower risk for commercial growers, and increased yields of uninfested fruit for backyard and non-commercial growers. Experience from around the world, involving both Dacini and other tephritid pests, is that good A-W IPM does deliver excellent fruit fly control. However, implementing A-W IPM is not a casual undertaking and several issues need to be considered.

What area?
It is far from clear how big the 'area' has to be in area-wide management. *Bactrocera* spp. have a reputation as highly mobile insects and this is one of the key drivers for beginning an A-W IPM programme. However, as covered in the section on movement, for the most part *Bactrocera* only undertake short-distance local foraging movements and lifetime movement may be, at most, between 500–5000 m. At this scale, a production area of even a few farms, if buffered by non-fruit fly suitable country around, may be sufficient to gain the benefits of area-wide control. Experience from parts of South-east Asia is that coordinated control for a village and the farm lands tended by members of that village will lead to substantially improved controls. Involving a few neighbouring villages and their farm lands gives excellent control. So, in summary, the area involved in A-W IPM does not have to be huge to start seeing benefits, and this is important. In any community effort it is easier to engage in a common practice with your nearest neighbours, than it is to try to engage in a new practice with strangers, or at least people a long way off. There are distinct social and operational benefits of starting local, and then after seeing benefits begin to engage with those neighbours a bit further away, and so on.

Stakeholders and stakeholder engagement
For 'normal' pest management, the only stakeholders that need to be routinely engaged are the growers. In A-W IPM, the entire community are stakeholders

and so the majority need to be engaged for the programme to work (this is one major benefit of starting small). Stakeholders include growers of susceptible crops, growers of crops that may breed flies but for which the flies are not a commercial pest, people living in rural towns, people living on small rural blocks who have crops for non-commercial sale, governments of different levels (local, state), private and public crop consultants, etc. Not surprisingly, these different stakeholders will vary in their willingness to engage. Local councils may see the value for helping coordinate A-W IPM and undertake some control activities on public land, but their resources may be limited and investment in other areas of public good may be seen to be more valuable. Growers of crops that may breed fruit flies, but which do not suffer significantly from fruit fly damage, may also be difficult to engage. In south-eastern Australia, *B. tryoni* breeds at low levels in wine grapes, but the fly is not considered a commercially damaging pest to that crop. In a region where wine grapes and stone fruit grow largely intermixed, there is a need to get wine-grape growers engaged in AW-IPM programmes to attain fly suppression for stone fruit pest management, but not surprisingly this has proven a difficult case to make to the grape growers. The Australian experience is that people in rural towns will engage in area-wide management, but need awareness and education programmes. In summary, an increasing amount of social-science literature, and practical experience, is demonstrating the critical importance of engaging the wider community early in the process when considering an A-W IPM programme.

Who pays?

An obvious question about A-W IPM is who pays? It is reasonable to expect growers to pay to control pests on their own properties, but should they also pay to control a fly in a rural town or on public common land, to fund a local coordinator, etc.? Growers may be willing to fund the entire costs of A-W IPM, but logically only if they get the same or better level of on-farm control for a reduced cost. If this scenario does not occur, then other mechanism are needed to help meet costs. However, people not involved in fruit growing, be it commercially or in a backyard, have little incentive to pay to control a pest that causes no direct damage to them. The capacity of larger society to meet such costs is known as 'willingness to pay'. Willingness to pay varies between cultures and populations within cultures, depending on acceptance or otherwise of public-good expenditure and local capacity to pay. If a local community is willing and able to pay to meet the costs of A-W IPM, then some mechanism needs to be put in place to gather those payments.

While the question of 'who pays?' may seem an obvious point, it is one that is often not asked until too late. Several high-profile AW-IPM programmes against *Bactrocera* have been found to be successful in the start-up phase when governments and researchers were subsidising costs, including operational control costs, public awareness and education costs, and coordination costs. However, when these public subsidies were withdrawn, farmers fell back to on-farm management. For this reason, commonly seen around the world is that ongoing AW-IPM strategies (or some component of it) are at least partially subsidised by government.

Pest management tools

A-W IPM utilises additional tools, and a different mindset, to on-farm management. On-farm management is largely about reducing the immediate pest population, hence the focus on MAT and protein bait sprays, or insecticides if they
are available. In contrast, A-W IPM is about reducing an entire fruit fly population in a region for the entire year, and so removing that fly as an on-farm pest
simply because there are not enough flies remaining in an area to be damaging.
For *Bactrocera*, MAT and protein baiting remain invaluable tools within the
A-W IPM 'tool box', but they are supplemented by an increased importance
on crop hygiene, the removal of fruit-bearing feral and street trees, increased
management or removal of backyard and garden breeding sources, the use of
supplementary release of parasitoids, and where available the use of SIT. SIT
has a key role in A-W IPM by being able to target flies breeding in off-farm
areas, such as natural forest and urban areas where other forms of fruit fly control may be logistically limited, impossible to apply or socially unacceptable. It
is important to note, however, that many successful A-W IPM programmes
have been run without SIT.

The mindset change for A-W IPM is about recognising that each control
does not have to be a 'silver-bullet', giving 100% control on its own, but operates as part of a larger, integrated whole. Parasitoids are a good example of this.
In regions where *Bactrocera* are endemic, the role parasitoids play in fruit fly
pest management has been largely ignored because parasitism levels are often
low, around 20%, and parasitoids do not protect the current crop from being
damaged. But if as part of an A-W IPM package it is realised that relatively
little extra work, perhaps through supplementary releases or habitat manipulation, might push that figure up to 30–35% parasitism, then already one in
every three fruit flies in the area is being controlled. Thirty per cent population
mortality will not protect a current crop from heavy damage, but in an area-
wide approach it is an important mortality component of the total population.
Growers, politicians and administrators like 'silver bullets' because it makes
decision making and the selling of an idea easy. Unfortunately, silver bullets
rarely work in sustainable pest management, particularly so in A-W IPM.

9.11.3   Regulatory controls

Fruit fly management, when being done to prevent or minimise infestation in
crops that are for local sale or consumption, can operate quite independently
of any larger regulatory control or management. However, fruit and vegetables
being grown for international (and occasional domestic) trade often have to
meet strict regulatory controls around the management of risk posed by the
movement of fruit fly infested fruit. These controls are managed by the country's
National Plant Protection Office, or their delegated officers, operating under the
regulatory framework of the International Plant Protection Convention. Such
regulatory fruit fly controls, which include systems approaches, pest-free places
of production, and areas of low-pest prevalence are the focus of Chapter 11.

9.12 Further Reading and References Cited

Akter, H. and Taylor, P.W. (2018) Sexual inhibition of female Queensland fruit flies mated by males treated with raspberry ketone supplements as immature adults. *Journal of Applied Entomology*, 142, 380–387.

Akter, H., Mendez, V., Morelli, R., Pérez, J. and Taylor, P.W. (2017) Raspberry ketone supplement promotes early sexual maturation in male Queensland fruit fly, *Bactrocera tryoni* (Diptera: Tephritidae). *Pest Management Science*, 73, 1764–1770.

Anon (2018) Adaptive area wide management for Queensland fruit fly using Sterile Insect Technique. Available at https://area-wide-management.com.au/ (accessed 23 May 2018).

Barclay, H.J. and Hendrichs, J. (2014) Models for assessing the male annihilation of *Bactrocera* spp. with methyl eugenol baits. *Annals of the Entomological Society of America*, 107, 81–96.

Barry, J.D., Miller, N.W., Piñero, J.C., Tuttle, A., Mau, R.F.L. and Vargas, R.I. (2006) Effectiveness of protein baits on melon fly and oriental fruit fly (Diptera: Tephritidae): attraction and feeding. *Journal of Economic Entomology*, 99, 1161–1167.

Broumas, T., Haniotakis, G., Liaropoulos, C., Tomazou, T. and Ragoussis, N. (2002) The efficacy of an improved form of the mass-trapping method, for the control of the olive fruit fly, *Bactrocera oleae* (Gremin) (Dipt., Tephritidae): Pilot scale feasibility studies. *Journal of Applied Entomology*, 126, 217–223.

Cáceres, C., Rendon, P. and Jessup, A. (2012) The FAO/IAEA Spreadsheet for Designing and Operating Insect Mass-rearing Facilities. Procedures Manual. Available at http://www-naweb.iaea.org/nafa/ipc/public/Spreadsheet-insect-mass-rearing.pdf (accessed 23 May 2018).

Cunningham, R.T. (1989) Male annihilation. In: Robinson, A.S. and Hooper, G. (eds.) *Fruit Flies, Their Biology, Natural Enemies and Control, vol 3B*. Elsevier, Amsterdam, The Netherlands, pp. 345–351.

Deguine, J.-P., Atiama-Nurbel, T., Aubertot, J.-N., Augusseau, X., Atiama, M., Jacquot, M. and Reynaud, B. (2015) Agroecological management of cucurbit-infesting fruit fly: a review. *Agronomy for Sustainable Development*, 35, 937–965.

Dominiak, B.C., McLeod, L.J. and Landon, R. (2003) Further development of a low-cost release method for sterile Queensland fruit fly *Bactrocera tryoni* (Froggatt) in rural New South Wales. *Australian Journal of Experimental Agriculture*, 43, 407–417.

Duyck, V.A., Hendrichs, J. and Robinson, A.S. (eds.) (2005) *Sterile Insect Technique. Principles and practice in area-wide integrated pest management*. Springer, Dordrecht, The Netherlands.

Ekesi, S. (2016) Baiting and male annihilation technique for fruit fly suppression in Africa. In: Ekesi, S., Mohamed, S.A. and De Meyer, M. (eds.) *Fruit Fly Research and Development in Africa. Towards a sustainable management strategy to improve horticulture*. Springer, Basel, Switzerland, pp. 275–292.

Ekesi, S. and Billah, M.K. (2007) *A Field Guide to the Management of Economically Important Tephritid Fruit Flies in Africa*. ICIPE Science Press, Nairobi, Kenya.

Ekesi, S., Maniania, N.K. and Mohamed, S.A. (2011) Efficacy of soil application of *Metarhizium anisopliae* and the use of GF-120 spinosad bait spray for suppression of *Bactrocera invadens* (Diptera: Tephritidae) in mango orchards. *Biocontrol Science and Technology*, 21, 299–316.

El-Sayed, A.M., Suckling, D.M., Wearing, C.H. and Byers, J.A. (2006) Potential of mass trapping for long-term pest management and eradication of invasive species. *Journal of Economic Entomology*, 99, 1550–1564.

El-Sayed, A.M., Suckling, D.M., Byers, J.A., Jang, E.B. and Wearing, C.H. (2009) Potential of 'lure and kill' in long-term pest management and eradication of invasive species. *Journal of Economic Entomology*, 102, 815–835.

Epsky, N.D., Kendra, P.E. and Schnell, E.Q. (2014) History and development of food-based attractants. In: Shelly, T.E., Epsky, N., Jang, E.B.; Reyes-Flores, J. and Vargas, R.I. (eds.) *Trapping and the Detection, Control, and Regulation of Tephritid Fruit Flies: Lures, Area-Wide Programs, and Trade Implications*. Springer, Dordrecht, The Netherlands, pp. 75–118.

FAO/IAEA (2013) *Trapping Guidelines for Area-wide Fruit Fly Programmes*. International Atomic Energy Agency, Vienna, Austria. 58pp. Available at http://www-naweb.iaea.org/nafa/ipc/public/Trapping-Guideline-september2013-final.pdf (accessed 23 May 2018).

FAO/IAEA (2016) In: Barclay, H.L., Enkerlin, W.R., Manoukis, N.C. and Reyes-Flores, J. (eds.), *Guidelines for the use of mathematics in operational area-wide integrated pest management programmes using the Sterile Insect Technique with a special focus on tephritid fruit flies*. Food and Agriculture Organization of the United Nations. Rome, Italy. 95 pp. Available at http://www-naweb.iaea.org/nafa/ipc/public/tephritid-fruit-flies-manual.pdf (accessed 23 May 2018).

FAO/IAEA (2017a) In Enkerlin, W.R., Reyes, J. and Ortiz, G. (eds), *Fruit Sampling Guidelines for Area-Wide Fruit Fly Programmes*. Food and Agriculture Organization of the United Nations, Vienna, Austria. Available at http://www-naweb.iaea.org/nafa/ipc/public/fruit-sampling-guidelines.pdf (accessed 23 May 2018).

FAO/IAEA (2017b) In: Zavala-López J.L. and Enkerlin W.R. (eds.), *Guideline for Packing, Shipping, Holding and Release of Sterile Flies in Area-wide Fruit Fly Control Programmes*, Second edition. FAO/IAEA, Rome, Italy, 140 pp. Available at http://www-naweb.iaea.org/nafa/ipc/public/Guideline-for-Packing-Sept2017.pdf (accessed 23 May 2018).

FAO/IAEA/USDA. 2014. *Product Quality Control for Sterile Mass-Reared and Released Tephritid Fruit Flies, Version 6.0*. International Atomic Energy Agency, Vienna, Austria. 164 pp. Available at http://www-naweb.iaea.org/nafa/ipc/public/QualityControl.pdf (accessed 23 May 2018).

Gurr, G.M. and Kvedaras, O.L. (2010) Synergizing biological control: Scope for sterile insect technique, induced plant defences and cultural techniques to enhance natural enemy impact? *Biological Control*, 52, 198–207.

Haq, I. and Hendrichs, J. (2013) Pre-release feeding on hydrolysed yeast and methoprene treatment enhances male *Bactrocera cucurbitae* Coquillett (Diptera: Tephritidae) longevity. *Journal of Applied Entomology*, 137 (s1), 99–102.

Hsu, J.-C., Haymer, D.S., Chou, M.-Y., Feng, H.-T., Chen, H.-H., Huang, Y.-B. and Mau, R.F.L. (2012) Monitoring resistance to spinosad in the Melon fly (*Bactrocera cucurbitae*) in Hawaii and Taiwan. *The Scientific World Journal*, Volume 2012, Article ID 750576, doi:10.1100/2012/750576.

IAEA (2009) *Development of bait stations for fruit fly suppression in support of SIT*. Vienna: International Atomic Energy Agency. Available at https://www.iaea.org/resources/technical-report/development-of-bait-stations-for-fruit-fly-suppression-in-support-of-sit (accessed 25 April 2019).

Isasawin, S., Aketarawong, N., Lertsiri, S. and Thanaphum, S. (2014) Development of a genetic sexing strain in *Bactrocera carambolae* (Diptera: Tephritidae) by introgression of sex sorting components from *B. dorsalis*, Salaya1 strain. *BMC Genetics*, 15 (Suppl 2). doi:10.1186/1471-2156-15-S2-S2.

Jang, E.B., Casana-Giner, V. and Oliver, J.E. (2007) Field captures of wild melon fly (Diptera: Tephritidae) with an improved male attractant, raspberry ketone formate. *Journal of Economic Entomology*, 100, 1124–1128.

Jang, E.B., McQuate, G.T., McInnis, D.O., Harris, E.J., Vargas, R.I., Bautista, R.C. and Mau, R.F. (2008) Targeted trapping, bait-spray, sanitation, sterile-male, and parasitoid releases in an areawide integrated melon fly (Diptera: Tephritidae) control program in Hawaii. *American Entomologist*, 54, 240–250.

Jin, T., Lin, Y.-Y., Jin, Q.-A., Wen, H.-B. and Peng, Z.-Q. (2016) Population susceptibility to insecticides and the development of resistance in *Bactrocera cucurbitae* (Diptera: Tephritidae). *Journal of Economic Entomology*, 109, 837–846.

Klungness, L.M., Jang, E.B., Mau, R.F.L., Vargas, R.I., Sugano, J.S. and Fujitani, E. (2005) New sanitation techniques for controlling tephritid fruit flies (Diptera: Tephritidae) in Hawaii. *Journal of Applied Science and Environmental Management*, 9, 5–14.

Korir, J.K., Affognon, H.D., Ritho, C.N., Kingori, W.S., Irungu, P., Mohamed, S.A. and Ekesi, S. (2015) Grower adoption of an integrated pest management package for management of mango-infesting fruit flies (Diptera: Tephritidae) in Embu, Kenya. *International Journal of Tropical Insect Science*, 35, 80–89.

Langford, E.A., Nielsen, U.N., Johnson, S.N. and Riegler, M. (2014) Susceptibility of Queensland fruit fly, *Bactrocera tryoni* (Froggatt) (Diptera: Tephritidae), to entomopathogenic nematodes. *Biological Control*, 69, 34–39.

Lloyd, A.C., Hamacek, E.L., Kopittke, R.A., Peek, T., Wyatt, P.M., Neale, C.J., Eelkema, M. and Gu, H. (2010) Area-wide management of fruit flies (Diptera: Tephritidae) in the Central Burnett district of Queensland, Australia. *Crop Protection*, 29, 462–469.

Maniania, J.N.K. and Ekesi, S. (2016) Development and application of mycoinsecticides for the management of fruit flies in Africa. In: Ekesi, S., Mohamed, S.A. and De Meyer, M. (eds.) *Fruit Fly Research and Development in Africa. Towards a sustainable management strategy to improve horticulture*. Springer, Basel, Switzerland, pp. 307–324.

Mau, R.F.L., Jang, E.B., Vargas, R.I. (2007) The Hawaii Area-Wide Fruit Fly Pest Management Programme: Influence of partnerships and a good education programme. In: Vreysen, M.J.B., Robinson, A.S. and Hendrichs, J. (eds.) *Area-Wide Control of Insect Pests*. Springer, Dordrecht, The Netherlands, pp. 671–683.

May, A.W.S. and Caldwell, N.E.H. (1944) Fruit fly control. *Queensland Agricultural Journal*, 58, 224–229.

Moadeli, T., Taylor, P.W. and Ponton, F. (2017) High productivity gel diets for rearing of Queensland fruit fly, *Bactrocera tryoni*. *Journal of Pest Science*, 90, 507–520.

Mwatawala, M.W., Mziray, H., Malebo, H. and De Meyer, M. (2015) Guiding farmers' choice for an integrated pest management program against the invasive *Bactrocera dorsalis* Hendel (Diptera: Tephritidae) in mango orchards in Tanzania. *Crop Protection*, 76, 103–107.

Navarro-Llopis, V. and Vacas, S. (2014) Mass trapping for fruit fly control. In: Shelly, T.E., Epsky, N., Jang, E.B.; Reyes-Flores, J. and Vargas, R.I. (eds.) *Trapping and the Detection, Control, and Regulation of Tephritid Fruit Flies: Lures, Area-Wide Programs, and Trade Implications*. Springer, Dordrecht, The Netherlands, pp. 513–555.

Nguyen, V.L., Meats, A., Beattie, G.A.C., Spooner-Hart, R., Liu, Z.M. and Jiang, L. (2007) Behavioural responses of female Queensland fruit fly, *Bactrocera tryoni*, to mineral oil deposits. *Entomologia Experimentalis et Applicata*, 122, 215–221.

Perez-Staples, D., Harmer, A.M.T., Collins, S.R. and Taylor, P.W. (2008) Potential for pre-release diet supplements to increase the sexual performance and longevity of male Queensland fruit flies. *Agricultural and Forest Entomology*, 10, 255–262.

Petacchi, R., Rizzi, I. and Guidotti, D. (2003) The 'lure and kill' technique in *Bactrocera oleae* (Gmel.) control: effectiveness indices and suitability of the technique in area-wide experimental trials. *International Journal of Pest Management*, 49, 305–311.

Piñero, J.C., Mau, R.F.L. and Vargas, R.I. (2009a) Managing Oriental fruit fly (Diptera: Tephritidae), with spinosad-based protein bait sprays and sanitation in papaya orchards in Hawaii. *Journal of Economic Entomology*, 102, 1123–1132.

Piñero, J.C., Mau, R.F.L., McQuate, G.T. and Vargas, R.I. (2009b) Novel bait stations for attract-and-kill of pestiferous fruit flies. *Entomologia Experimentalis et Applicata*, 133, 208–216.

Piñero, J.C., May, R.F.L. and Vargas, R.I. (2010). Comparison of rain-fast stations versus foliar bait sprays for control of oriental fruit fly, *Bactrocera dorsalis*, in papaya orchards in Hawaii. *Journal of Insect Science*, 10, 157.

Piñero, J.C., Enkerlin, W. and Epsky, N.D. (2014) Recent developments and applications of bait stations for integrated pest management of tephritid fruit flies. In: Shelly, T.E., Epsky, N., Jang, E.B., Reyes-Flores, J. and Vargas, R.I. (eds.) *Trapping and the Detection, Control, and Regulation of Tephritid Fruit Flies: Lures, Area-Wide Programs, and Trade Implications.* Springer, Dordrecht, The Netherlands, pp. 457–492.

Prokopy, R.J., Miller, N.W., Piñero, J.C., Barry, J.D., Tran, L.C., Oride, L. and Vargas, R.I. (2003) Effectiveness of GF-120 fruit fly bait spray applied to border area plants for control of melon flies. *Journal of Economic Entomology*, 96, 1485–1493.

Reynolds, O.L., Osborne, T., Crisp, P. and Barchia, I.M. (2016) Specialized pheromone and lure application technology as an alternative male annihilation technique to manage *Bactrocera tryoni* (Diptera: Tephritidae). *Journal of Economic Entomology*, 109, 1254–1260.

Ryckewaert, P., Deguine, J.-P., Brévault, T. and Vayssières, J.-F. (2010) Fruit flies (Diptera: Tephritidae) on vegetable crops in Reunion Island (Indian Ocean): state of knowledge, control methods and prospects for management. *Fruits*, 65, 113–130.

Siderhurst, M. and Jang, E. (2010) Cucumber volatile blend attractive to female melon fly, *Bactrocera cucurbitae* (Coquillet). *Journal of Chemical Ecology*, 36, 699–708.

Siderhurst, M.S., Park, S.J., Buller, C.N., Jamie, I.M., Manoukis, N.C., Jang, E.B. and Taylor, P.W. (2016) Raspberry Ketone Trifluoroacetate, a new attractant for the Queensland fruit fly, *Bactrocera Tryoni* (Froggatt). *Journal of Chemical Ecology*, 42, 156–162.

Sirjani, F.O., Lewis, E.E. and Kaya, H.K. (2009) Evaluation of entomopathogenic nematodes against the olive fruit fly, *Bactrocera oleae* (Diptera: Tephritidae). *Biological Control*, 48, 274–280.

Steiner, L.F. and Lee, R.K.S. (1955) Large-area tests of a male annihilation method for oriental fruit fly control. *Journal of Economic Entomology*, 48, 311–317.

Thaochan, N. and Ngampongsai, A. (2015) Effects of autodisseminated *Metarhizium guizhouense* PSUM02 on mating propensity and mating competitiveness of *Bactrocera cucurbitae* (Diptera: Tephritidae). *Biocontrol Science and Technology*, 25, 629–644.

Vargas, R.I., Stark, J.D., Hertlein, M., Neto, A.M., Coler, R. and Piñero, J.C. (2008) Evaluation of SPLAT with spinosad and methyl eugenol or cue-lure for 'attract-and-kill' of Oriental and Melon fruit flies (Diptera: Tephritidae) in Hawaii. *Journal of Economic Entomology*, 101, 759–768.

Vargas, R.I., Leblanc, L., Piñero, J.C. and Hoffman, K.M. (2014) Male annihilation, past, present, and future. In: Shelly, T.E., Epsky, N., Jang, E.B.; Reyes-Flores, J. and Vargas, R.I. (eds.) *Trapping and the Detection, Control, and Regulation of Tephritid Fruit Flies: Lures, Area-Wide Programs, and Trade Implications.* Springer, Dordrecht, The Netherlands, pp. 493–511.

Vargas, R.I., Piñero, J.C. and Leblanc, L. (2015) An overview of pest species of *Bactrocera* fruit flies (Diptera: Tephritidae) and the integration of biopesticides with other biological approaches for their management with a focus on the Pacific region. *Insects*, 6, 297–318.

Vayssieres, J.F., Sinzogan, A., Korie, S., Ouagoussounon, I. and Thomas-Odjo, A. (2009) Effectiveness of spinosad bait sprays (GF-120) in controlling mango-infesting fruit flies (Diptera: Tephritidae) in Benin. *Journal of Economic Entomology*, 102, 515–521.

Vijaysegaran, S. (2016) Bait manufactured from beer yeast waste and its use for fruit fly management. In: Sabater-Muñoz, B., Vera, T., Pereira, R. and Orankanok, W. (eds.) *Proceedings of the Ninth International Symposium of Fruit Flies of Economic Importance.* Published by the Editors, pp. 227–248.

Vreysen, M.J.B., Robinson, A.S. and Hendrichs, J. (2007) *Area-Wide Control of Insect Pests: From research to field implementation.* Springer, Dordrecht, The Netherlands.

Weldon, C.W., Preter, J. and Talyor, P.W. (2010) Activity pattern of Queensland fruit fly (*Bactrocera tryoni*) are affected by both mass-rearing and sterilization. *Physiological Entomology*, 35, 148–153.

10 Phytosanitary Measures

PETER LEACH

The movement of fresh fruit and vegetables through trade can pose quarantine risk as fruit fly eggs and larvae may be moved in those products. For this reason, importing countries may request that phytosanitary measures are applied to traded commodities to reduce that risk to an acceptable level. This chapter introduces phytosanitary measures, with a particular focus on disinfestation measures and potential issues that researchers and industry should consider when undertaking disinfestation research. Many of the examples provided are drawn from research conducted in Australia, and many of the issues are not new – but nevertheless they do not go away. Issues such as the use of laboratory-reared flies, assessing the host status of commodities, natural versus artificial infesting techniques, and appropriate efficacy levels have been addressed by many researchers in their individual research projects and reviews.

The work of developing phytosanitary treatments is a highly specialised area carried out by researchers who understand and appreciate the trade ramifications of even the slightest variations in protocol, and the sensitivities of how information is reported. The author of this book does not work in this field and has only superficial knowledge of the discipline: as such I could make grave errors if writing about it. For this reason this chapter is authored by Peter Leach, a research leader in the field and at the time of publication Chairman of the Phytosanitary Measures Research Group, an international technical group operating under the umbrella of the International Plant Protection Convention (IPPC) (see Chapter 11), and a member of the IPPC Standards Committee Technical Panel for Phytosanitary Treatments. The chapter utilises, more than elsewhere in the book, case examples from Peter's work to illustrate subtleties to consider when carrying out research into phytosanitary measures.

10.1   What Treatment to Use?

One of the most common questions industry will ask researchers working in phytosanitary treatment research, is what treatment should they use? The answer is simple. The treatment that meets your trading partner's requirements, results in the least damage to fruit, and is economical. The points must be addressed in that order. If the plant health and food safety legislation in the intended export market does not permit the use of irradiation, then there is no point trying to negotiate a treatment using this technology. The best technology may be cold or heat treatments, or even methyl bromide. Similarly, the most economical treatment may be post-harvest treatment with insecticide (approximate treatment cost of US$0.01 per kilogram), but if there is no maximum residue level (MRL) established or the MRL cannot be met, then the treatment will never be approved despite the ease and cost of the treatment.

The great majority of phytosanitary measures fall into the broad category of disinfestation treatments. These treatments are applied post-harvest to ensure that if infested fruit has been packed, then there is little or no risk of viable eggs or larvae being transported. For this reason, the bulk of this chapter is focused on disinfestation. However, two other forms of phytosanitary measures can be applied: host status determination and systems approaches. These topics are discussed in greater detail in other sections of this book (host status, Sections 6.6.1 and 11.3.4; systems approaches, Section 11.3.3), but are also introduced here.

10.2   Systems Approach

Systems approaches integrate two or more independent risk-reduction steps to reduce the total risk of transporting a pest or disease to below an acceptable level. In this way they are quite different to non-host status or disinfestation treatments, which are single-step reduction tools. The concept of systems approaches is discussed fully in Section 11.3.3, while the individual risk-reduction steps that can (at least theoretically) be used within a system are any of the single-step pre-harvest or phytosanitary treatments listed in Chapters 9 and 10.

10.3   Host Status

10.3.1   Background

The use of non-host status is an internationally accepted phytosanitary measure as, by definition, there is no risk associated with the movement of fruit and vegetables if they are not capable of hosting a fruit fly. The relevant regulatory aspects of non-host status are covered in Section 11.3.4, while biological aspects of Dacini host use are discussed in Chapter 6, and particularly Section 6.6.1. However, this topic is also addressed in this chapter because the complexities of determining host status from an operational phytosanitary measures perspective are worth developing, and also because the host status of a commodity can have a dramatic effect on the methodology employed when carrying out disinfestation research. While there are a limited number of commodities that meet the criteria of non-hosts, it has been used successfully for decades by large industries such as bananas and avocados.

10.3.2 Conflicting terminology and standards

There are various terms that have been used to describe the host status of fruit. In addition to standard terms such as host, non-host, conditional non-host and conditional host, there are also references to potential hosts, field host, natural host, primary host, commercial host, secondary host, and major, good, moderate, poor or occasional hosts. Some of these definitions are based on host collection records (Hancock *et al.*, 2000), while others are based on laboratory trials to determine the susceptibility to fruit fly infestation (Lloyd *et al.*, 2013). While numerous labels have been used to categorise host status there is no international acceptance for many of these terms.

The three IPPC standards currently approved for host status testing are:

- International Plant Protection Convention International Standard for Phytosanitary Measures (ISPM) No. 37 *Determination of host status of fruit to fruit flies (Tephritidae)* (see Section 11.3.5);
- Asia Pacific Plant Protection Commission (APPPC) Regional Standards for Phytosanitary Measures (RSPM) No. 4 *Guidelines for the confirmation of non-host status of fruit and vegetables to Tephritid fruit flies*; and
- North American Plant Protection Organization (NAPPO) *Specifications for a Standard on Host Status*.

While the definitions for a non-host are similar in the three standards, the definitions for a conditional non-host or conditional host are quite different. In the APPPC standard the definition of a conditional non-host is 'Fruit and vegetables at a specified maturity and specified physical condition that cannot support the development of viable adults of a fruit fly species'. An example of a conditional non-host to *B. tryoni* is mangosteen. If damaged fruit are exposed to gravid females, they will lay eggs that will develop into viable adults. However, if mangosteens are not damaged then *B. tryoni* is unable to oviposit in the fruit due to the thick pericarp and so is considered a 'conditional non-host'. Using the NAPPO standard, mangosteens would be classified as a 'conditional host'. ISPM 37 would classify undamaged mangosteen as a 'non-host'.

Another difference between the methodologies recommended in each guidelines is the age of the flies used in the trials. The APPPC guidelines recommend keeping colonies for 12 months. The NAPPO standards recommend colonies be no older than three generations, while ISPM 37 recommends colonies be no more than five generations and reared on a natural host to ensure natural oviposition behaviour.

10.3.3   A role for small-cage laboratory tests in assessing host status

ISPM 37 states that host status testing should be done in large cages on whole plants. This is quite different to the APPPC standard that supports the carrying out of host status testing in small cages. During the Papaya Fruit Fly (PFF) campaign in the 1990s, host status testing was undertaken on a wide range of crops using the New Zealand Standard (Anon., 2001), the forerunner of the APPPC standard. Results were positive and found that three varieties of pumpkin and certain varieties of longan, lychee, rambutan, mangosteen, tahitian lime, rosella and angled loofah were all conditional non-hosts to the then *B. papayae* (now *B. dorsalis*): the condition being they were not damaged. Since then additional testing with *B. tryoni* has been undertaken and conditional non-host status has been accepted for domestic trade for more than 20 years using small-cage trials.

The major criticism of the New Zealand Standard is that the test is an unrealistic no-choice, forced test and flies will lay eggs into fruit that they would not normally infest in the field (Jessup and McCarthy, 1993). Using this test you may have 50 gravid females exposed to 500 g of fruit for 24–48 hours. For a tree that produces 50 kg of fruit per year this is equivalent to 5000 flies per tree. This is extreme, completely unrealistic and does not represent population pressures that would ever be experienced in field conditions. The authors of the methodology were aware of this flaw and stated that fruit that failed this test was not considered a host, but considered a 'potential' host. The plus side is that if crops can pass this test, you are fairly confident that they are a non-host to fruit fly in the field: host status testing using laboratory-reared flies presents a worst-case scenario.

The record of a potential host, through the New Zealand Standard, still means that fruit would need to be treated to access regulated markets, but the failure does not automatically list the commodity as a host and industry has the option to undertake additional testing (i.e. large-scale field collections) to try and accurately identify the host status of the commodity. For very small tropical fruit industries in Queensland, with limited research budgets, laboratory testing has provided a very quick and cheap option for undertaking testing. Additionally, while the New Zealand Standard was considered to be a harsh test, it is a dramatic improvement on some of the early host status trials in Queensland where fruit were placed in cages with 7500 female flies for several hours.

While the New Zealand Standard provided market access for commodities that were considered conditional non-hosts, the data was also used to develop systems approaches for varieties of lychee and grapes that did not pass the test but produced low numbers of pupae and were considered poor hosts. These were the first systems approaches published in the national Interstate Certification Assurance Scheme in Australia and incorporated four elements: monitoring, field treatments, inspections for damaged fruit and the low host susceptibility of the commodities. A systems approach was required for grapes because post-harvest insecticide treatments were not meeting MRL requirements and methyl bromide fumigation was reducing fruit quality. For varieties of lychee that did not pass the host status test, a systems approach was implemented because a number of growers were organic and did not want to use chemical treatments.

10.3.4 How much 'pressure' to apply and efficacy assessment of host status testing

One criticism of the use of small-scale laboratory trials for host status testing is that they utilise much lower numbers of test insects than are typically used to develop post-harvest treatments. Another difference between the two measures is that disinfestation trials usually report efficacy rates and confidence limits whereas host status trials typically do not (Follet and Hennessy, 2007). Follet and Hennessy have suggested that non-host status is a valid measure if there is a high efficacy and statistical confidence, but if the level of confidence is low then non-host status may require additional measures and then form part of a systems approach. While their calculations of statistical confidence are appropriate for standard disinfestation treatments (heat, cold, etc.) it may not be an appropriate formula for host status testing trials. Laboratory-based host status testing is not aiming at biological reality, but is deliberately skewed towards a worst-case scenario of flies laying into fruit that they may not attack in the field. To improve the confidence interval around host status testing results you would need to add some 'forced infestation' weighting to Follet and Hennessy's formula. But exactly what weighting you would provide is unknown because we really do not have accurate information on the exact pressure fruit are exposed to in the field and how to replicate this accurately when using laboratory-reared flies.

We know that fruit fly populations vary from season to season, and year to year, but we simply do not know what exact pressure crops are exposed to. In the case of *B. tryoni*, which has a distribution covering the dry tropics, wet tropics, subtropical and temperate areas, you cannot make a generic statement about population pressure. This problem is made worse as most of our monitoring for field numbers comes from male trap catch numbers, and we do not have accurate figures on how male trap catches correlate with female population pressure. In Queensland we have undertaken field studies looking at a range of commodities and from this data we have cases where no flies were caught in traps but infested fruit was recorded; conversely, we have examples where male flies were caught in traps but we have not recorded infestation in fruit. We have never been able to collate infestation rates to male trap catch numbers. Unfortunately, until we get better female traps and can accurately determine what real-world fly pressure is, we have to go with our best guess of what constitutes realistic numbers of test insects for host status testing. One thing we can probably be guaranteed is that field population pressures are going to be very different for each host and each fruit fly species; and for flies with large geographical distributions this pressure will change each season, from region to region, and within regions.

10.4 Disinfestation Methods

A recent international collaboration between Australian, New Zealand and US scientists undertook a review of post-harvest disinfestation techniques. Over 30 fumigants, including 15 major fumigants and 18 minor fumigants, were reviewed. Despite numerous publications stating alternatives to methyl bromide fumigation were being developed, the reality is that there have been no major recent advances in the field of fumigant development for control of fruit fly. Methyl bromide is still the most widely used and approved fumigant for international exports. It also widely used by many national plant protection organisations (NPPOs) as a remedial treatment of commercial consignments where live insects are detected. A similar result was found for non-chemical treatments. There are also many positive publications available on treatments such as controlled energy treatments (electrical, microwave, radio frequency), controlled atmosphere treatments and physical treatments (e.g. pressure, vacuum or physical removal), but the major commercial treatments for fruit fly are still cold, heat and, more recently, irradiation.

10.4.1 Irradiation

Phytosanitary irradiation of fresh fruits and vegetables provides a safe, efficacious and chemical free solution to quarantine restrictions. Many fruits and vegetables can tolerate low doses of irradiation and the treatment is fast and safe and can take less time for treatment compared to other commercial phytosanitary treatments such as cold and heat treatments. In terms of international standards for phytosanitary treatments, irradiation is easily the most advanced technology. ISPM 18 *'Guidelines for the use of irradiation as a phytosanitary measure'* was adopted in 2003 and published in 2016. This standard, in conjunction with ISPM 28 *'Phytosanitary treatments for regulated pests'* (adopted in 2007 and published in 2016) has helped fast track bilateral negotiations. A total of 32 schedules (annexes) have been published in ISPM 28 to date, and 15 of these are for irradiation treatments (see Section 11.3.1). In fact, the first 14 annexes adopted were for irradiation.

Irradiation is the only treatment that has a generic fruit fly treatment (150 Gy) approved by the IPPC. This means no more research on fruit fly is required and the treatment can be applied to all fruit and vegetables except those stored in modified atmosphere. It is also the only treatment that has a generic approval (400 Gy) for all insect pests (except Lepidoptera that pupate internally) in countries such as Australia, India, Indonesia, Mexico, New Zealand, Pakistan, South Africa, Thailand, United States and Vietnam. The 150 Gy generic treatment for fruit fly was originally proposed by the International Consultative Group on Food irradiation in 1986, but USDA/APHIS was the first NPPO to adopt generic irradiation treatments (both the 150 Gy and 400 Gy generic doses) and these treatments have been the major catalyst in fast tracking new irradiation protocols in the last decade. The fact that no more funding is required for efficacy trials against fruit fly is a major benefit to small industries who simply cannot afford to fund efficacy packages that meet international standards. The cost of conducting efficacy trials can be millions of dollars to complete, especially for cold treatments.

One very important aspect of irradiation research is dosimetry, which should only be undertaken by a specialist in the field. Dosimetry is the calculation of the absorbed dose in the target tissue (i.e. fly egg or larva). Most entomologists do not have the skills required, or access to, research-scale facilities to undertake very accurate doses mapping. A major impediment to the adoption of the generic dose for fruit fly was early research results indicating the treatment required to control fruit fly was between 200–300 Gy. These results have since been reviewed and several issues such as potential contamination of the trials, or incorrect reporting of dosimetry results, have been identified as reasons why doses of >150 Gy were initially recommended.

One of the most important concepts with dosimetry is the DUR or dose uniformity ratio. This is the ratio between the maximum absorbed dose a batch received, and the minimum absorbed dose in the batch. In commercial facilities treating whole pallets (approximately 1 tonne of fruit per pallet), the minimum dose is usually in the centre of the pallet whereas the maximum dose is in the outer cartons on the pallet. It is very difficult to undertake efficacy research in large-scale commercial irradiation facilities as the DUR in these facilities is usually in the order of 2:1 or 3:1. So fruit may have received a minimum dose of 150 Gy and a maximum dose of 452 Gy (note: average doses are not important in irradiation trials). Because some of the test insects did receive the maximum absorbed dose of 452 Gy, this then becomes the minimum dose that could be recommended.

10.4.2   Heat

Heat treatment research has been undertaken for a wide range of species using technologies such as vapour heat treatment (VHT), high temperature forced air and hot water treatment. There are now five heat treatments approved in the annexes of ISPM 28, but there is no generic treatment for fruit fly. One of the main reasons for the lack of a generic treatment is that there have been very large differences in the heat tolerance of the flies studied, with flies such as *B. latifrons* recording very high heat tolerance (Jang *et al.*, 1999). The main problem with heat treatments is that the temperatures required to control fruit fly are very close to the upper limit of tolerance for most fruits and vegetables. As such you could recommend a generic treatment, but if the treatment is not applicable to most fruit and vegetables then there is no real benefit, as researchers will still be developing data sets for every fly species/commodity combination in an attempt to reduce the treatment conditions and minimise damage to the fruit.

The only industry that uses heat treatment for phytosanitary disinfestation in Australia is the mango industry, who currently use VHT for exports to Japan, Korea and China. Research on VHT of tomato, melons, zucchini and squash has been completed, but the treatment has not been adopted by industry as it is not viewed as cost competitive compared to alternative treatments such as methyl bromide, insecticide treatments, irradiation or systems approaches. One of the major disadvantages of heat treatment is the volume of produce that can be treated at a time. The majority of VHT units in operation are 5-tonne units. With the treatment taking 6–8 hours (including warm up and cool down) it is very difficult to conduct more than two treatments per day. As such, throughput of the VHT facility is only 10 tonnes per day. A recently constructed facility in Brisbane, Queensland, is a 10-tonne unit that will improve throughput, but the volume of fruit that can be treated per day is still very low compared to treatments like irradiation and methyl bromide, which can potentially treat hundreds of tonnes per day at a lower cost.

10.4.3 Cold

Cold treatments are the most commonly used disinfestation treatments worldwide, especially for temperate fruit. There are now nine cold treatments approved in the annexes of ISPM 28, but like heat treatment there is no generic treatment for fruit fly despite decades of research and hundreds of studies on a wide range of fruit fly species and commodities. There is also no internationally accepted in vitro technique.

Cold treatments have a similar problem to heat treatments in the fact that many treatments are close to the lower tolerance of the commodity. The approval of a generic treatment may not provide many benefits as industry adoption for longer treatments and the associated costs is likely to be very low. Additionally, the debate over the cold tolerance between species and even the tolerance of the same species from different regions is still quite contentious. For example, Annex 24 of ISPM 28 'Cold treatment for *Ceratitis capitata* on *Citrus sinensis*' took nearly a decade to be reviewed and approved for publication. One of the major objections was that the efficacy of the cold treatment could be influenced by population differences or by the variety/cultivar of citrus treated.

While the treatment was eventually approved, research was undertaken to address the issue of potential differences in cold tolerance between different populations of *C. capitata* (Hallman et al., 2018). Colonies were collected from three continents, reared under identical conditions and were less than a year old when dose-response trials were undertaken. Hallman *et al.* found that the three populations did not differ significantly in cold tolerance at 1.1°C and classified their research as the first step in a quest for generic phytosanitary cold treatments.

If a conservative generic treatment for cold can be established, it would help a lot of industries because cold research is extremely expensive and time consuming. A good example of how resource-intensive cold research can be is work by De Lima et al. (2017). Research was undertaken to gain access to the Japanese market. Data packages developed at three temperatures in nine grape varieties took over a decade to complete, and at peak periods over 100 staff were employed. The cost of the verification trials alone was AU\$1.5 million. While the research was originally for Japan, it has resulted in market access to several other international markets including China and Korea.

10.4.4 Fumigation

Methyl bromide is still the most widely used fumigant for phytosanitary use for fresh fruit and vegetables. Research on alternative fumigants to replace methyl bromide for internal feeding pests such as fruit fly has not been very positive, but there is some interest in the use of phosphine. While phosphine has high efficacy levels, at low temperatures it is a very slow treatment (several days to a week) (Li *et al.*, 2014).

In Australia we have recommended research on methyl bromide, but we are looking at ways to maximise the efficacy of the treatment using lower doses and longer time periods. While methyl bromide fumigation is a contentious issue, the use of higher temperatures and longer time periods (i.e. low-dose usage) is endorsed by the IPPC as a strategy to reduce methyl bromide use (FAO/IPPC 2017). Our research is focusing on the use of temperatures at 18°C (rather than 11°C) or above and using treatment periods of 4–6 hours rather than the traditional 2-hour treatment. It should be noted that all our research is funded by industry groups. Australian exporters wanted a quick treatment so that they could use air freight rather than sea freight, because Australia has some of the cheapest airfreight worldwide and some of the most expensive and erratic sea freight options. An obvious alternative to methyl bromide would have been the use of irradiation, but many of our major trading partners do not approve the use of irradiation for phytosanitary purposes. This was a major factor in industry funding low-dose methyl bromide research. The low-dose methyl bromide research has been very successful to date and China has approved protocols for nectarines and peaches using a treatment of 18 g/m^3 at 18°C for 5 hours and 30 minutes.

10.5 Disinfestation Methodology

When undertaking post-harvest disinfestation research there are several general procedures that researchers have followed for many decades. The first step is to undertake a review of the host records to determine for what species efficacy research is required. The next steps are to establish colonies and undertake life-history trials to determine the development rates of each juvenile stage of each species of concern. Results from the life-history trials can then be used to determine infestation time, so that when most tolerant stage testing is undertaken you are actually treating the correct stage (e.g. eggs, first, second, and third instar larvae). Once the most tolerant life stage of the most tolerant species is determined, you can then conduct large-scale trials against this stage and the treatment should be effective against all species of concern in the commodity being tested.

This procedure sounds straightforward and numerous protocols for a range of different treatment technologies (e.g. heat, cold, methyl bromide and irradiation) have been successfully developed and operated commercially for decades. However, the reality is that while methodologies are generally similar, different researchers have employed different techniques (often for legitimate reasons) and this has led to a situation where it is very difficult to directly compare results between different laboratories. This is has resulted in a situation where, after decades of research, there are still no internationally agreed testing procedures for post-harvest research and, in the case of heat and cold treatments, no generic treatments for fruit flies. One strategy that is vitally important when undertaking trials is that researchers should always consult with their NPPO, who in turn can negotiate with the NPPO of the importing country to ensure the methodology employed in the trials meets their requirements.

One aspect of research that is consistent between the different technologies is that research methodology is typically very conservative. For irradiation treatments the maximum dose received in trials will be the minimum dose recommended in potential protocols. For fumigation treatments the highest temperature and highest fumigant concentration will become the minimum parameters used for commercial consignments. For cold and heat treatments the treatment may start when half of the temperature probes have reached the target temperature. Monitoring of fruit temperatures in trials may use the smallest fruit in the trials, whereas commercial consignments use the largest fruit in the treatment batch. For cold treatment the treatment will also start when half the probes are within 0.5°C of the target temperature (e.g. start at 3.5°C for a 3°C treatment). In addition to this, you are treating the most tolerant stage as opposed to the most common stage found in commercial consignments and treating them in very large numbers that in no way represents the level of infestation found in export-quality fruit. Once again it is always a good strategy to consult with your NPPO before undertaking efficacy trials.

The use of systems approaches is not covered in this chapter (see Section 10.2), but it should be noted that 'stand-alone' post-harvest treatments are, in fact, unregulated systems approaches. The vast majority of export-quality fruit has very low or nil infestation rates because growers have insect monitoring or in-field control programmes, and damaged fruit may be removed at harvest or in the packing line. The application of the post-harvest treatment is simply required because we do not currently have inspection systems that can reliably detect infested fruit if it is there. If rapid, accurate detection technology was available it would replace the need for post-harvest treatments for exports of fresh fruit and vegetables. If a female trap is the holy grail for pre-harvest researchers, then a rapid, non-invasive inspection technique is the holy grail for post-harvest researchers

10.5.1   Review of host records

The first process in undertaking disinfestation research, or undertaking a pest risk analysis (PRA), is developing a host list. If there are multiple, reliable positive infestation records clearly stating the condition of the fruit when sampled then the commodity is regarded as a host and phytosanitary measures are required to develop export protocols. Host records may also indicate that fruit in an undamaged, hard-mature state (e.g. certain varieties of bananas or avocado) may be a conditional non-host to fruit fly. But the absence of host records does not automatically mean that a commodity is not a host. The 'absence of evidence for a specific host/quarantine pest association may be the result of a recent pest invasion into the range of the host, or the fact that the association has not been investigated' (Follet and Hennessy, 2007). Another scenario for an 'absence of evidence' is that new crops have been introduced to an area, or that traditional crops have expanded their production range into new areas due to advances in breeding programmes (e.g. highbush and lowbush blueberries), and have not yet been subjected to pest pressure.

Host records are crucial to any PRA but it is important that the veracity of the records be confirmed and unfortunately this is not always possible, especially for old records. An example is a host record of *B. tryoni* in pumpkin recorded during the Queensland PFF eradication campaign of the 1990s. During the PFF campaign approximately 85 000 fruit samples were collected and over 21 300 host records recorded (Hancock *et al.*, 2000). Hancock *et al.* acknowledge that the large volume of fruit sampled led to a number of suspected errors, especially where records were based on a single record. Other potential sources of error included misidentification of the fly (especially for teneral specimens), misidentification of the plant samples, contamination of samples during transportation, transcription errors, or misreading sample numbers. Nevertheless, these potential sources of error did not stop the record of *B. tryoni* from pumpkin being listed.

Bactrocera tryoni has never been reported as a pest of pumpkins despite that crop being grown in endemic *B. tryoni* areas for more than a century. However, during the PFF eradication campaign a single host record was reported in the PFF database. Because we have retained the original data sheets we were able to review the record and believe it is incorrect. The original datasheets show the sample contained seven pumpkins with a total weight of 100 g. The sampling protocol for large fruit like pumpkins was to store fruit separately with a unique identification code for each sample. We suggest that the identification of the host plant was incorrect as the weight of seven pumpkins should have been several kilograms and split amongst seven samples. No further detection of *B. tryoni* has been found in pumpkin since this initial record.

Another example where host records can potentially cause problems is the record of *B. halfordiae* as a pest of citrus by the early Queensland fruit fly researcher Alan May in 1953. This species is generally regarded as a non-economic rainforest species that does not attack commercial hosts, and the early records were likely misidentified. No further infestations have been reported despite large-scale assessments of citrus fruit throughout Queensland for the last 70 years. This supports our assertion that *B. halfordiae* should be regarded as a non-economic fruit fly. It is also an excellent example of why voucher specimens are so important if we are to ascertain the veracity of host records, and also that voucher specimens of flies used in post-harvest research projects should be retained.

10.5.2 The use of laboratory flies

Disinfestation research on fruit flies is normally conducted in a laboratory environment that allows access to large numbers of insects with tight control (e.g. through constant temperature regimes, artificial media) over the development rates of juvenile stages, which is a stringent requirement of current testing procedures.

In early disinfestation studies there were no guidelines on how long a colony could be maintained. There are several examples of studies where the colony flies used had been maintained for several hundred generations, with no supplementation with wild stock being undertaken. Some colonies have been maintained for 20–30 years (i.e. 200–400 generations) with occasional infusion of wild flies. But it is more common that colonies are maintained for 3–5 years before either being replaced completely, or supplemented with wild flies.

There are concerns over the use of laboratory colonies and if they represent the true tolerance of wild flies, especially for heat and cold tolerance studies. In the laboratory environment, the genetic bottlenecking and intense selection laboratory flies undergo when being brought into mass culture selects for flies that do best at a constant temperature. That could potentially have negative consequences by reducing or removing original high and low temperature tolerances. The main assumption is that colony insects may be less tolerant than wild flies, which experience diurnal temperature fluctuations.

To address this issue, various NPPOs and the IPPC's ISPM and RSPM have recommended that colonies be replaced or supplemented on a regular basis to maintain genetic diversity. However, the guidelines are very broad and do not address issues such as the minimum number of insects required to establish a colony, or how colonies should be supplemented. The exception to this is the APPPC RSPM 4 where the authors state the laboratory colonies should be founded from an appropriate number of individuals (100 to 1000). Heather and Hallman (2008) also recommend that cohorts of 200 mated females should be collected to ensure an acceptable probability of sampling the genetic variation in a contiguous population.

While most guidelines for efficacy studies recommend keeping colonies for 12 months, the guidelines for host status testing recommend much shorter periods. The NAPPO standard recommends colonies be no older than three generations, while ISPM 37 recommends colonies be no more than five generations. The NAPPO standard also states 'Colony should be no older than three generations at the initiation of the trials, without re-stocking, and maintained on natural hosts to ensure normal oviposition behaviour'. Rearing early generations on natural hosts rather than artificial diets does not ensure 'normal oviposition behaviour'. The flies are still laboratory flies that are adjusting to being reared in an artificial condition and being fed artificial food sources. The major problem with this is that we know that when wild flies are brought into the lab they take time to adjust to the new conditions. Egg hatch, pupal viability and fecundity is always variable in early generations, but normally increases and stabilises as colonies get older (e.g. towards F_5 from the wild).

Many guidelines also state that colonies should be regularly supplemented to maintain genetic diversity, but once again they do not provide specific guidance on how to undertake supplementation. Heather and Hallman (2008) recommended that supplementing colonies to maintain genetic diversity can be achieved by the addition of field collected adults or juveniles annually. But once again they did not stipulate how this should be achieved. One option is to simply mix pupae from old and recently established colonies, but there is no guarantee there is cross-mating between the two populations. Another option is to cross-mate wild males with colony females and vice versa. The choice to cross laboratory females with wild males is often chosen to retain the high fecundity of the colony females and it is easier to catch wild males than females. But regardless of how you supplement the colony, all techniques have problems because they are time consuming and it may take several generations to build colonies back up to a size where they are large enough to use for efficacy trials (e.g. 30 000 adult flies).

As mentioned earlier, young fly colonies are placed under enormous stress to adapt to laboratory conditions and we are still not sure if adaptive stress and treatment stress are synergistic or antagonistic. Based on the limited data available, it would appear that if you want to bias your results towards a positive result then you should use young colonies as they are easier to kill. Research on the heat tolerance of *B. tryoni* has shown that heat tolerance increases with the age of the colony. *In vitro* studies of *B. tryoni* comparing the heat tolerance of newly colonised flies (F_1–F_4) and an established colony (220 generations) found that the young colony (which were still adapting to artificial food sources, constant temperatures, etc.) were significantly less tolerant to heat than long term colonies.

10.5.3   Reporting colony establishment and production

While there are general issues related to the use of laboratory flies as described above, one area that can be improved immediately is the reporting of how colonies have been established. To date, many authors simply state that new colonies are established every 12 months but provide no information on the size of the founder population, culturing techniques, identification of the flies, and if voucher specimens were established. While the brevity of reported details may be due to size restrictions in peer reviewed journals, there is no excuse not to report these details when developing technical market access submissions (TMAS). Depending on the reporting requirements of the importing country's NPPO, a TMAS may be 50 pages long or more than 600 pages long (cold treatments generate very large raw data sets).

I would recommend that all researchers should write TMASs with all the information regarding colony establishment and maintenance, detailed methodology and all raw data generated in the project provided. This can be valuable to the interpretation of the results if future research shows that, for example, the age of the colony or the rearing medium influences tolerance to physical treatments. The IAEA has established a database on insect radio tolerance and the IPPC is currently developing databases to collate reports on heat and cold disinfestation studies and this is the perfect place where TMASs could be stored for future reference.

Colony establishment
Examples of fruit collections undertaken to establish a new colony of *B. tryoni* are present in Table 10.1. Standard details to be recorded normally include the species of fruit collected and the number of flies collected. For this particular

colony a total of 1645 *B. tryoni* was used in establishment. The data shows that infestation rates can vary dramatically amongst the same host plant in the same region. For example, a total of 1048 pupae was recovered from guava from collections made on 12 and 13 March 2017, yet only five of the flies were identified as *B. tryoni*. Fruit collected on the same day but in different areas of the same region recorded much higher infestation rates and hundreds of *B. tryoni* were collected. The high variability of infestation rates in the field is another reason why disinfestation researchers usually use laboratory flies rather than undertaking trials with field-infested fruit, even though the use of laboratory flies is far more labour intensive. If trials were undertaken using field-infested fruit it is impossible to control/determine the number or life stages of flies present prior to treatment (see Section 10.5.4 for further discussion).

One thing the data does not show is how time consuming the identification of live flies is, especially in areas that have multiple fruit fly species present. Undertaking identification of live adult flies is a highly specialised skill that requires years of training to do accurately. An alternative to identifying live flies would be to temporally sedate flies (CO_2 or cold), but our experience is that this does cause some mortality no matter how gently or quickly flies are handled.

As mentioned earlier, there are age requirements on colonies in several RSPMs and NPPO standards for disinfestation research. However, there are no clear guidelines on how to define the age of the colony. The example provided in Table 10.1 has samples being collected from 12 March until the 5 April. So does the age of the colony start from when the first collection was made or the last collection? This issue may seem trivial, but market access negotiations can be stalled over very minor issues like this. In this case the choice

Table 10.1. Data sheet retained showing collection details for a new *Bactrocera tryoni* colony established in 2017.

Host plant fruit	Date collected	Collection site	Fruit weight (g)	Number of pupae collected	Number of emerged flies identified as *B. tryoni*
Common guava, *Psidium guajava*	12-13/03/2017	Ashfield Rd, Bundaberg	4757	1172	430
Common guava, *Psidium guajava*	12-13/03/2017	Elliot Heads Rd, Bundaberg	10193	1938	247
Common guava, *Psidium guajava*	12-13/03/2017	Innes Peak Rd, Bundaberg	11854	1048	5
Common guava, *Psidium guajava*	14/03/2017	Elliot Heads Rd, Bundaberg	18969	2893	681
Common guava, *Psidium guajava*	28/03/2017	Elliot Heads Rd, Bundaberg	19634	2935	166
Feijoa, *Acca sellowiana*	05/04/2017	Dayboro Rd, Whiteside	4426	765	116
		Total		**10751**	**1645**

of the first collection or last collection date could be the difference between finishing trials within the designated time period, or having to halt trials and recommence research next season. Once again, this issue may seem trivial, but it does highlight that more prescriptive guidelines are required. But any guidelines developed will probably be plagued by the need to base them on expert advice unless more research on some of the very basic areas of disinfestation research are undertaken.

Rearing method

Other details that should be included in a TMAS are the location of the laboratory and the conditions the flies were held under (Table 10.2). When fruit fly are maintained in laboratory cultures it is standard practice to rear the colonies on artificial media and use artificial oviposition receptacles to collect eggs from the adult cages: once again all these details should be provided in a TMAS (Table 10.3).

In Queensland we have successfully trained multiple *Bactrocera* spp. to adapt to carrot-based media and to lay eggs into perforated plastic cups internally smeared with fruit juice. When young colonies are being established it is very difficult to collect eggs from plastic cups, so a hollowed-out fruit dome of a known host is used. For *B. tryoni* we use a hollowed-out apple dome because insecticide free apples are readily available all year round. The use of whole fruit rather than a fruit dome poses several problems. The first is that fruit may contain pesticide residues (not all organic fruit is organic) or the fruit may

Table 10.2. Illustrative example of information on larval rearing medium and adult diet for *Bactrocera tryoni* colony as would be reported in a technical market access submission.

Larval rearing medium	Adult diet
Ingredients: 500 g dehydrated diced carrot 3 L hot tap water 150 g torula yeast 10 g Nipagin [p-hydroxybenzoic acid methyl ester (methyl paraben)] 10 ml concentrated hydrochloric acid diluted in 100–250 ml of water	*Ingredients:* Water Sucrose (sugar cubes) Autolysed brewer's yeast (dry and as a yeast/sugar paste)
Method: The dehydrated diced carrot was soaked in 3 L of hot tap water for at least 30–45 minutes and then blended with the other ingredients in a food processor. In this diet, torula yeast provided protein and vitamins, hydrochloric acid inhibited bacterial growth and Nipagin inhibited mould growth.	*Method:* Water was placed on top of the cage in a plastic cup with a perforated lid placed upside down on cellulose sponge and a wet cloth. Sugar was placed on top of the cage. Autolysed brewer's yeast and sugar (10:1) was mixed with water to a paste consistency and smeared on the top of the cage. Dry yeast was also provided to the flies by sprinkling the yeast onto the top of small containers covered in cloth placed in the cage.

Table 10.3. Illustrative example of culturing procedure for *Bactrocera tryoni* colony as would be reported in a technical market access submission.

Life stage	Details
Eggs	Collected over ~3–6-hour period and cultured on carrot media
Pupae	Sieved 13–14 days after culturing
Adults	Eclose 3–6 days after pupal collection
Adults	Fertile ~ 10 days after eclosion
Adults	Cages discontinued approximately 5 weeks after eclosion

already be infested and contain another fruit fly species that would contaminate the colony. Another issue with the use of infested fruit is that it may contain parasitised larvae or eggs. While parasitoids should not survive in well-maintained colonies, their presence is always concerning (we have occasionally recorded *Encarsia* spp. in our controlled temperature rooms).

As colonies adapt to laboratory conditions it is possible to get them to adapt to the plastic perforated cups by placing them in the cages beside the hollowed-out fruit domes. Most species will eventually adapt to the plastic cups and the use of fruit domes can be discontinued (an exception is *Z. cucumis*). The main advantage of plastic cups is they are very quick to set up and mould levels in the rearing media are generally much lower than when fruit domes are used (even fruit domes washed in a mild algaecide still introduce higher levels of mould to the culturing medium).

Colony quality assurance

Quality assurance records from colonies used in disinfestation research are essential when developing TMASs. Parameters such as egg hatch, pupation rate, adult emergence rate, sex ratio and total viability should be included (Table 10.4). These parameters are recorded every time culturing is undertaken, but NPPOs normally request the quality assurance statistics for the flies that were used in the efficacy trials rather than the complete records. Flight ability is also assessed, but because this criterion is not as crucial to disinfestation trials as it is to SIT programmes, we normally assess flight abilty only once per generation rather than for every batch that is cultured.

While all quality assurance records are important there is an emphasis on egg hatch. The New Zealand Standard for Host Status Testing stated that 'regular egg hatch records must be available for laboratory colonies indicating egg viability of at least 70%'. This figure has been increased to 80% for disinfestation research by several NPPOs. While we regularly record 80% egg hatch for *B. tryoni* colonies, egg hatch rates do vary between species especially when new colonies are being established. If minimum egg hatch rates are going to be imposed, then it should not be an arbitrary figure. It should be an egg hatch rate that is equivalent to the egg hatch rate of wild flies. Unfortunately, we do not have accurate records of egg hatch for field flies. We can tell the number of flies we record per kilogram of fruit from field collections, but this does not tell us what the egg hatch rate of the wild eggs was.

Table 10.4. Illustrative example of quality measurements of *Bactrocera tryoni* colonies utilised in the most tolerant stage trials as would be reported in a technical market access submission.

Generation and batch No.	Egg hatch (%)	Pupation (%)	Adult emergence (%)	Sex ratio male: female (%)	Total viability (%)
F8 (18)	74	97	88	48:52	63
F9 (19)	82	91	97	47:53	72
F9 (20)	83	95	95	48:52	74
Overall quality	**80**	**94**	**93**	**48:52**	**70**

10.5.4 *In vitro*, *in vivo* and infestation techniques

When developing quarantine disinfestation treatments researchers should focus on identifying the most tolerant stage of the most tolerant species known to infest a particular crop. In the case of heat treatments, *in vitro* techniques have been established that provide quick, cheap and accurate indicators of the relative heat tolerance between species. *In vitro* trials utilise 'naked' eggs or larvae, exposed directly to the treatment (e.g. a hot water dip) before being transferred to an artificial diet. Countries such as Japan and New Zealand published standards for heat-tolerance testing in the 1980s and 1990s, which were eventually incorporated into the APPPC RSPM No. 1 - *Guidelines for the development of heat disinfestation treatments of fruit fly host commodities* (FAO/APPPC, 2004). Other trading partners that have endorsed the use of *in vitro* techniques in the development of heat treatments are the US, China and Korea who have all approved import protocols based on data developed following the Japanese approved methodology. Unfortunately for cold tolerance research, no *in vitro* techniques have been developed that are internationally accepted.

The advantage of the use of *in vitro* techniques in heat-tolerance research is that they provide very quick treatments, with uniform treatment regimes, a known number of insects, and the age of the insect treated can be tightly controlled. Typically, insects are exposed directly to immersion in hot water, which allows eggs and larval stages to be exposed to exactly the same treatment conditions. When heat-tolerance studies are undertaken in fruit (i.e. *in vivo*) the insects can be located at different depths within the fruit with eggs deposited just below the skin surface while larval stages can be found at various depths within the fruit. As such they do not receive identical exposure to the treatment. It is important to note that *in vitro* techniques for heat-tolerance research are very accurate at identifying the differences in tolerance between species. They are not always reliable at determining the most tolerant stage of a species found in fruit. Once the most tolerant species is determined using *in vitro* techniques, dose-response trials against the most tolerant species should then be undertaken in fruit.

When efficacy trials are undertaken in fruit there are several accepted methods of infestation available, each with advantages and disadvantages. Fruit can be collected from the wild and is classified as naturally infested. While this technique can be less labour intensive it is normally used as a last resort (at least in fruit flies) as it is impossible to control/determine the number or life stages of flies present prior to treatment.

The use of laboratory-reared flies is the preferred method for NPPOs as it allows the use of specific life stages and very large numbers of insects can be reared relatively quickly in a healthy colony. Fruit can be infested using laboratory-reared flies where the fruit is either exposed directly to caged flies (also referred to as natural infestation) or artificially infested by collecting insects from the colony and placing them inside the fruit. There is a lot of flexibility in the choice of infestation methods, but when fruit is artificially infested by placing laboratory-reared larvae in fruit it is very important that additional trials are undertaken to demonstrate that artificial infestation has not reduced the tolerance of the test insects to the treatment being examined.

One method of artificial infestation that is commonly used is the injection of eggs into fruit. This technique provides even infestation rates and larvae feed and develop in the test fruit. This technique has been used extensively in Australian studies of *C. capitata*. Artificial injection of eggs was required because, unlike *B. tryoni* that will readily oviposit in fruit placed in cages, *C. capitata* does not and it is difficult to collect sufficient eggs to conduct disinfestation trials if trial fruit are exposed directly to adult flies.

10.5.5   Life-history trials

Life-history studies are usually the first trials undertaken in disinfestation research and the results are used to determine infestation times for subsequent trials. Life-history results are also a good indicator of the host status of a commodity. When undertaking life-history trials it is very important to monitor and record the temperature and relative humidity during the trials. Standard storage conditions commonly used in trials is in the range of 25±2°C and > 60% relative humidity.

To determine the development rate of the larval stages, trials are conducted for between 8 and 12 days depending on the fruit fly species and commodity being studied. The minimum number of fruit to be inspected should be enough to ensure that several hundred insects are sampled each day. For a good host, or medium-sized fruit capable of sustaining large numbers of larvae, you may be able to sample three or five fruit per day. For smaller fruit, or poor hosts such as grapes, you may need to sample 50 fruit per day to obtain a sufficient sample that makes sense (i.e. accounts for highly variable results).

If possible, fruit should be assessed immediately after sampling (i.e. not frozen) so that the number of live insects present can be determined. Fruit are destructively sampled and insects removed from the fruit and examined under a dissecting microscope to calculate the proportion of the life stage present. The live larvae in each sample are examined individually and classified as either first (L1) or late first instar (LL1), second (L2) or late second instar (LL2), third instar (L3) or late third instar (LL3), based on two characteristics: (i) shape of the mouth hooks; and (ii) presence or absence of anterior spiracles as described by

Anderson (1962). The number of dead larvae should also be recorded in each sample, but dead larvae are not used to estimate the percentage of each life stage present. Dead larvae may indicate the fruit is a poor host and, if so, you are recording natural mortality. Alternately, it may indicate that fruit used are not insecticide free. We have had results where good egg hatch was recorded but we recorded 100% mortality of first intars in every sample. This research was in ripe mango, which is a very good host to fruit fly and, as such, we knew the fruit had been insecticide treated and so we abandoned the trials.

Identification of larvae to early or late larval stages is very time consuming, so it very important that you have enough trained staff and good quality microscopes to complete the assessments as quickly as possible (e.g. several hours or less). If instarring is taking longer than expected another option is to freeze the samples. The downside with this technique is that if the fruit being tested is a poor host you may have very high natural mortality and not being able to distinguish between live and dead larvae can dramatically skew your results. In a good host, freezing samples may not affect your results at all.

As stated above, this research is labour intensive and does require skilled staff. If there is enough published literature on development rates of the fly species and the host commodity being tested then no research may be required and existing development rates can be used. Efficacy trials (most tolerant stage trial and confirmatory trials) will have extra infested fruit (i.e. 'instar fruit' or 'instar check') that are sampled at the commencement of the trials to confirm the percentage of each life stage treated (see 50% rule, Section 10.5.6).

Another issue that can occur when you are naturally infesting fruit for life-history trials is that fruit may accidently be heavily or over-infested. In this case you may need to take a sub-sample of larvae present in each fruit. As long as sub-samples of the larvae present are randomly selected it is perfectly legitimate to stop sampling when several hundred larvae have been counted. You will get a better indication of development rates by counting 300 larvae in five fruit, rather than counting 1500 larvae in a single fruit.

Because sampling needs to be taken every day and may run for 12 days or more, the resources required can be quite significant. In fact, most activities in disinfestation require work to be undertaken 7 days per week and often after hours. Even routine activities such as colony maintenance require staff to be present even if no disinfestation trials are running (e.g. during holiday periods). While heavy workloads may be sustained for short periods they are very difficult to maintain for long periods, especially in small teams. Any strategy that can help avoid weekend and after hours work should be utilised. In the case of larval development work, it may be better to have several infestation times so that samples can be undertaken during normal working hours (Table 10.5).

To determine embryonic development rates fruit are normally artificially infested with a known numbers of eggs (counted on black filter paper) and the number of unhatched eggs is recorded at half hourly intervals for approximately 12 hours. Samples are then stored again and a final sample undertaken 24 to 48 hours later to determine the total number of eggs hatched. You could continue sampling at regular intervals, but the additional data is of little value compared to the resources required. Based on previous research on *B. tryoni*

Table 10.5. Egg hatch data (number of unhatched eggs) of *Bactrocera tryoni* (at 26 ± 2 °C and 70% RH) to determine 50% egg hatch on carrot media and on cellulose sponge. Note the variation between cohorts within a media type, and the particularly large variation across media types.

Time	Hours	Carrot A	Carrot B	Carrot C	Sponge A	Sponge B	Sponge C
5.30	36.0	100	100	100	100	100	100
6.00	36.5	100	100	100	100	100	100
6.30	37.0	100	100	100	100	100	100
7.00	37.5	100	100	100	100	100	100
7.30	38.0	100	100	100	100	100	100
8.00	38.5	100	100	100	100	97	100
8.30	39.0	100	100	100	95	88	91
9.00	39.5	100	100	100	81	76	79
9.30	40.0	100	100	100	63	56	68
10.00	40.5	100	99	99	51	38	57
10.30	41.0	100	97	99	42	27	35
11.00	41.5	97	95	97	30	21	25
11.30	42.0	88	87	89	22	19	19
12.00	42.5	82	78	85	18	13	16
12.30	43.0	77	72	76	15	11	14
13.00	43.5	69	67	70	12	10	13
13.30	44.0	64	59	59	11	9	12
14.00	44.5	61[a]	54	51	11	9	12
14.30	45.0	57	52	45	11	9	12
15.00	47.5	48	44	38	11	9	12
Friday 20/8/10, 16.00	70.5	24	18	18	8	7	11

[a]50% egg hatch (corrected to include egg mortality) are highlighted.

we know that egg hatch will normally commence at around 36 hours and be completely finished by 72 hours. We also know that the substrate the samples are held on influences development rates. Examples of embryonic and larval development rates are provided, respectively, in Tables 10.5 and 10.6.

Egg hatch of *B. tryoni* follows a normal distribution so we are interested when peak hatch has occurred (i.e. 50% of the viable eggs have hatched). In Table 10.5, carrot sample A had a total of 100 eggs of which 24 did not hatch. This means there were 76 viable eggs in the sample. So peak hatch time will be when 50% of the 76 viable eggs are hatched, i.e. 38 eggs have hatched. Converting this figure back to unhatched eggs we can determine that peak hatch time is when there are less than 62 unhatched eggs in the sample.

In the three samples in carrot media, peak hatch times were 44.5, 44 and 44 hours, respectively. So if we want to treat mature eggs (60% developed) we will treat at 26.4 hours, which is 60% of 44 hours. This information is crucial if we want to make valid comparison between the heat tolerances of different species in a commodity. In heat treatments we know that young eggs (several hours old) are particularly susceptible to heat, while 60% developed eggs (mature eggs) are significantly more tolerant than 80% developed eggs.

Table 10.6. Illustrative life-history data of *Bactrocera tryoni* larval stages in mandarin at 26 ± 2 °C and 70% RH. The data illustrates variation between three cohorts (A-C), and sampling structured to identify two developmental classes within each larval instar.

Sample	Egg	L1	LL1	L2	LL2	L3	LL3	Pupae
Day 2 (A)	11	84	0	0	0	0	0	0
Day 2 (B)	13	82	0	0	0	0	0	0
Day 2 (C)	19	59	0	0	0	0	0	0
Day 3 (A)	15	59	19	5	0	0	0	0
Day 3 (B)	11	24	33	29	0	0	0	0
Day 3 (C)	10	56	25	4	0	0	0	0
Day 4 (A)	0	3	0	95	1	0	0	0
Day 4 (B)	0	0	0	100	0	0	0	0
Day 4 (C)	0	8	1	84	1	0	0	0
Day 5 (A)	0	4	0	23	22	47	0	0
Day 5 (B)	0	1	1	15	26	56	0	0
Day 5 (C)	0	4	0	11	27	62	0	0
Day 6 (A)	0	0	0	0	0	97	0	0
Day 6 (B)	0	0	0	0	1	97	0	0
Day 6 (C)	0	0	0	1	0	95	0	0
Day 7 (A)	0	0	0	0	0	76	23	0
Day 7 (B)	0	0	0	0	1	98	0	0
Day 7 (C)	0	0	0	0	0	93	6	0

The development rates for a range of fruit fly species in a number of substrates is provided in Table 10.7. The development time for mature eggs for many species is approximately 23 to 26 hours, but for *Z. cucumis* it is only 15 hours. Since we know that heat tolerance varies with the stage of development we need to use different infestation times to ensure eggs for each species are treated at the same development stage.

While the heat tolerance of different aged egg stages of several dacine species has been known for several decades, there has been little research on the tolerance variation within larval stages. One of the main reasons for this is that the treatment times for eggs can be easily and accurately manipulated, whereas larval stage development is very difficult to manipulate, especially for second and third instars. Recently studies by Kaneyuki *et al.* (2016) have investigated the tolerance of early, mid and late first, second and third instar larvae of *B. dorsalis* and *Z. cucurbitae* using *in vitro* studies in hot water. They found that heat tolerance of late aged first and second instars decreased for both species, but mixed results were recorded for third instar larvae. As a result, the authors recommended that heat tolerance within each larval instar period should be considered in future disinfestation tests.

Kaneyuki *et al.* should be commended for undertaking research in an area that has been badly neglected, but it should be noted that the differences recorded were found using *in vitro* tests (i.e. completely artificial conditions) and at sub-lethal temperatures. When a significant difference in tolerance is found at sub-lethal levels it does not mean that the treatments are failing and so should be suspended. It simply highlights the next priority area for research.

Table 10.7. Development times for eggs, first, second and third instars of a number of fruit fly species in a range substrates at 26 ± 2 °C and 70% RH.

Fruit fly species	Commodity	Variety	Egg 60%	Age of insect treated (hours[a])		
				L1	L2	L3
B. tryoni	Apple	Red Delicious	23	44–408	96–408	168–648
B. tryoni	Pear	Packham	23	44–240	72–240	144–240
B. tryoni	Nashi	Nijisseiki	23	44–240	72–240	144–240
B. tryoni	Mango[b]	Keitt	23.2	48–72	72–96	120–144
B. tryoni	Mango	Kensington Pride	24.6	48	72	96
B. tryoni	Mango	B74	24.3	48	72	120
B. tryoni	Mandarin		24	48	96	144
B. jarvisi	Mandarin		21	48	96	144
B. neohumeralis	Mandarin		23	48	84	144
B. kraussi	Mandarin		24	48	96	144
B. frauenfeldi	Papaw	Yellow papaya	25	60	84	132
B. musae	Papaw	Yellow papaya	28	72	84–96	132
Z. cucumis	Papaw	Yellow papaya	13.5–15	36	60	84–144
B. bryoniae	Tomato		20	48	96	144
B. tryoni	Carrot media	–	26	48	96	144
B. jarvisi	Carrot media	–	23	48	96	144
B. neohumeralis	Carrot media	–	25	48	96	144
B. frauenfeldi	Carrot media	–	26	48	96	144
B. musae	Carrot media	–	29–31	72	120	168–192
Z. cucumis	Pumpkin media	–	15	36	60	96

[a]The age of the insects is the hours or days post mean egg collection time. [b]All mango results are for ripe, eating-mango, not hard-mature mango which is the current industry standard for harvest.

While life-history results are primarily used to determine infestation times for efficacy trials, they also provide invaluable information on the host status of commodities. This information may be critical for determining if fruit can be naturally infested (refer to host status testing), or if artificial infestation techniques are required.

10.5.6 The 50% rule

While Kaneyuki *et al.* (2016) is the first publication to formally report differences in heat tolerance within larval stages, nearly every previous study on heat tolerance (or any treatment for that matter) has actually tested different larval stages but failed to report it. As mentioned previously, it is very difficult to accurately control larval development rates for second and third instar larvae. This issue is acknowledged in various NPPO guidelines worldwide, by the fact that a treatment for any life stage is considered valid if 50% of the target life stage is present at the time of treatment. When any trials are undertaken you will always use the life-history results to determine your infestation times. However, to double check that the correct larval stage has actually been treated you need to infest extra fruit which can be sampled at the start of the treatment to determine the percentage of each life stage present: this technique is often referred to as instar checks. Instar checks are vital as there are numerous examples where development rates of larval stages in efficacy trails vary slightly from the original life-history trial results.

Often when authors are reporting results from instar checks they state that more than 50% of the target life stage was present or simply omit the results from the instar checks. This issue is probably one of the most important areas that needs to be addressed if we are to accurately report the life stages treated in trials. A very simple example is that if you treated 10 000 third instars and your instar results show that you had 1% early first instar, 1% late first instar, 15% early second instar, 26% late second instar, 56% early third instar and no late third instars then you have met the 50% rule. Typically this would just be reported as you treated 10 000 third instars. But the correct interpretation is that you actually treated 100 early first instar, 100 late first instar, 1500 early second instar, 2600 late second instar and 5600 early third instar.

While incorporating instar results into estimates of mortality has many advantages for research on poor hosts, it also addresses issues such as differences in tolerance within larval stages. It is strongly recommended that you consult your NPPO before adopting this methodology, as it is a new concept that has not been approved internationally.

10.5.7 Most tolerant stage testing

The second step when undertaking efficacy trials is usually most tolerant stage testing. By undertaking dose-response trials the most tolerant stage of the most tolerant species can be determined. Treatments can then be developed by undertaking large-scale trials against one life stage, rather than undertaking trials against all stages of all species of concern. This is the standard testing procedure recommended by NPPOs worldwide and the IPPC. In the case of heat treatments this may be true. *In vitro* techniques using hot water dipping have been established and provide quick, cheap and accurate indicators of the relative heat tolerance between species.

Unfortunately for cold tolerance research, no *in vitro* techniques have been developed that are internationally accepted and trials are normally conducted in fruit (naturally or artificially infested). One publication that has used *in vitro* trials on media is Hallman *et al.* (2011). Comparative trials were undertaken to compare the cold tolerance of invasive African *B. dorsalis* (then under the name of *B. invadens* and presumed a distinct biological species, see Section 2.2.6) to three species that had been well studied and treatment schedules were available. A major assumption with this research was that third instars in all species were the most tolerant stage. Comparative trials were undertaken with *B. invadens*, *Anastrepha ludens*, *B. dorsalis* (non-African) and *C. capitata*. Results showed that third instar *B. invadens* was no more cold tolerant than the other species. As such, the treatment approved for *A. ludens*, *B. dorsalis* and *C. capitata* would also control *B. invadens*. The United States Department of Agriculture (USDA) has approved a treatment schedule based on these results and unpublished data from Kenya.

Research on *in vitro* techniques for cold treatment of *B. tryoni* have been investigated but the results were not encouraging. Similar to Hallman *et al.* (2011), the *in vitro* trials for *B. tryoni* were undertaken by placing eggs on rearing media that was held inside petri dishes. The use of rearing media provided a treatment substrate that was easily replicated and known to be suitable for insect survival. But the results recorded showed that *in vitro* samples dramatically underestimated the treatment time and they could not accurately predict the most tolerant stage (Table 10.8). Because of the labour-intensive nature of the work, and the lack of international standards, we have discontinued research on *in vitro* testing procedures and only undertake *in vivo* trials for cold treatment research.

Table 10.8. Comparison of LD_{99} estimates of *Bactrocera tryoni* eggs and larval stages at 1°C for *in vitro* and *in vivo* treatments.

Treatment	Stage	LD_{99} (days)	LD_{99} (95% fiducial limits)
In vivo (mandarin), 1 ±0.5°C	L2	7.9	7.3–8.7
	L1	5.5	5.0–6.1
	L3	4.8	4.4–5.4
	Eggs	4.0	3.6–4.6
In vitro (carrot-based diet) 1 ± 0.5°C	Eggs	4.4	4.2–4.6
	L3	2.6	2.3–3.2
	L1	2.4	2.1–2.8
	L2	2.2	1.9–2.8

When undertaking analysis of dose-response trials, regression analysis can be undertaken to calculate lethal dosage (LD) values that can be used to determine if there are any significant difference between stages. While LD values are essential to determine if there are significant differences between stages, their use to predict potential doses for large-scale confirmatory trials is notoriously inaccurate. Researchers usually undertake empirical research using a range of times, temperatures and heating rates that result in minimal reductions in fruit quality while providing high efficacy rates to determine the final treatment dose.

10.5.8   An alternative to most tolerant stage testing

In addition to stopping work on *in vitro* techniques for cold research we have also stopped research on dose-response trials. We now undertake large-scale trials against all stages. It takes approximately 30 days to undertake one replicate of a dose-response trial, and at the end of the trial you have treated several hundred insects at each dose. Undertaking a large-scale trial takes a similar amount of time, but you have treated more than 10 000 insects of each stage. The decision to undertake testing against all stages was not made because it was less labour intensive, but simply due to the fact that we were undertaking research on methyl bromide and only had access to one treatment chamber. As such we could only undertake research using a single dose at a time.

For heat and cold treatments, only having access to a single treatment unit (e.g. a cold room) is not a problem as you can open the treatment unit and remove a sample without interfering with or compromising subsequent samples. For methyl bromide, opening the treatment chamber during a treatment is not an option for obvious reasons. To undertake most tolerant stage testing with methyl bromide using seven doses, we would have had to undertake 84 trials (7 doses × 4 stages × 3 replicates). Once the most tolerant stage was identified, we would still then have had to undertake large-scale trials. The adoption of testing all stages was first used for research on capsicums and tomatoes, and as mentioned earlier, China has approved protocols for nectarines and peaches using this methodology.

Another example of large-scale trials 'treating all stages' is research we have recently undertaken on cold treatments. The aim of the project was to treat a minimum of 30 000 insects of each life stage. The research was undertaken in a poor host and recorded highly variable development rates, especially for third instars (Table 10.9). A total of five trials was undertaken, treating an estimated total of 318 076 insects with no survivors recorded. While large numbers of insects were recorded for all stages, the trials could be considered as failed because in all five trials the number of third instars failed to meet the 50% rule.

However, if the instar results are actually used to correct the estimated treated insects the results could be classified as highly successful. For example, in Trial 5, the total number of insects treated was estimated to be 66 599. Using the 50% rule, this trial against third instars would be rejected because 54.8% of the insects present were second instar larvae. However, if the instar results are used to correct the estimated number of insects treated, then the results would be reported as a total of 1931, 36 496 and 28 171, respectively, for first, second

Table 10.9. Numbers of insects treated in confirmatory cold treatment trials. The % of life stages present was based on dissection of extra fruit (i.e. instar fruit, or instar checks) during the actual trials. While it was estimated no trial actually reached the 50% rule for third instars, when the estimated number of larvae treated was corrected through instar checks, and summed across all trials, well over 30 000 third instars (actually closer to 60 000) were treated with no survivors.

Trial no.	Target life stage	Estimated number treated	% of life stages present at treatment % Egg	% First instar	% Second instar	% Third instar	Meets 50% rule requirement	Corrected number of insects treated Egg	First instar	Second instar	Third instar
1	Egg	11 463	100.0	–	–	–	Yes	11 463	0	0	0
	First instar	11 703	28.0	72.0	0.0	0.0	Yes	3 277	8 426	0	0
	Second instar	22 867	–	57.3	42.7	0.0	No	0	13 103	9 764	0
	Third instar	19 466	–	19.8	73.2	7.1	No	0	3 854	14 249	1 382
2	Egg	11 952	100.0	–	–	–	Yes	11 952	0	0	0
	First instar	9 996	20.3	79.7	0.0	0.0	Yes	2 029	7 967	0	0
	Second instar	24 595	–	24.2	75.8	0.0	Yes	0	5 952	18 643	0
	Third instar	20 056	–	12.3	65.3	22.4	No	0	2 467	13 097	4 493
3	Egg	12 297	100.0	–	–	–	Yes	12 297	0	0	0
	First instar	12 731	21.8	78.2	0.0	0.0	Yes	2 775	9 956	0	0
	Second instar	30 499	–	36.3	63.7	0.0	Yes	0	11 071	19 428	0
	Third instar	14 662	–	8.5	48.8	42.7	No	0	1 246	7 155	6 261
4	Third instar	49 171	–	3.6	62.7	33.7	No	0	1 770	30 830	16 571
5	Third instar	66 599	–	2.9	54.8	42.3	No	0	1 931	36 496	28 171
TOTAL								**43 793**	**67 743**	**149 662**	**56 878**

and third instars treated. Incorporating the instar results into all calculations dramatically changes the results and the total number of eggs, first, second and third instars treated with no survivors recorded were 43 793, 67 743, 149 662 and 56 878, respectively.

The '50% rule' has been written for commodities that are good hosts to fruit fly and have consistent development rates, but in this example development rates were highly variable. Third instar fruit in all five trials in Table 10.9 were handled identically and all placed in the cold room 12 days after infestation based on earlier life-history studies. Yet the percentage of third instars present in each of the five trials was 7.1%, 22.4%, 42.7%, 33.7% and 42.3%. One option may have been to allow more time for development (e.g. 14 days), but the risk with this strategy was that there would also then be a small percentage of late third instars (larvae that have jumped out of the fruit to pupate) or pupae present in the samples. Given the definition of efficacy in this trial was 'failure to pupate', holding the fruit for longer would increase the risk of failure. Another important factor in why we were reluctant to hold the fruit any longer was that the response of late third instars to sudden drops in temperature (i.e. being placed in a cold room) is that they leave the fruit. While the larvae will die before they can pupate it does mean the instar results are invalid, which then impacts the estimates of the number of insects treated and the trial would need to be repeated.

One alternative we could have tried was artificial infestation. However, we believe that adjusting our calculations to the estimated number of insects treated is more appropriate and less contentious than placing larvae into fruit or using *in vitro* infestation techniques. Again it is strongly recommended that you consult your NPPO before adopting this methodology. It is a new concept that has not been approved internationally.

10.5.9   Efficacy levels

Ultimately, the level of efficacy required to meet the appropriate level of protection for an importing country will depend on the risk that the introduction of a particular pest would impose. Setting the level of risk is the sovereign right of every nation. As an example, New Zealand has established three risk groups with associated measures for pre-export clearance, and corrective actions should live insects be intercepted (the greater the risk, the stronger the measure). Risk Group 1 (RG1) pests are low-risk pests that would have little impact but are not present in New Zealand. Risk group 2 (RG2) pests are pests that would have a negative impact on export markets, national production or environment, but only a small number of crops or commodities would be affected. RG3 pests (e.g. fruit fly) are pests that would cause major disruption in export markets for a number of significant commodities, significantly impact on national production of these commodities, and have potentially significant adverse effects on the environment. The New Zealand preclearance and on-arrival action varies for each risk group and ranges from inspections to regulated treatments. Interception of live insects for RG1 or RG2 pests may result in consignments being treated, re-sorted, returned or destroyed. In the case of RG3 pest interceptions, the consignment will be returned or destroyed, and trade can be suspended until the cause of the non-compliance is identified and corrected.

Internationally, the efficacy rates required for high-risk pests such as fruit fly have been debated and reviewed by numerous authors, and both Probit 8.7 and Probit 9 (95% confidence interval) have been approved worldwide. The difference between the two efficacy rates simply reflects differences between NPPOs and their assessments on what is an acceptable level of protection. One error that has been referenced repeatedly is that Probit 9 is the 'international standard' for efficacy trials. This is simply not correct. It is/was the standard for the USDA treatments and many countries, including Thailand and China, adopted USDA-approved treatment schedules. While this is an excellent example of the 'equivalence principal', it does not make Probit 9 the international standard for efficacy trials. Many countries in the ASEAN region, including Australia, Japan and New Zealand, use Probit 8.7 (95% CI) for regional trade and only utilised Probit 9 (95% CI) for access to the United States. Once again this is an example where expansion of international standards could reduce delays in negotiations and duplication of research.

To date the IPPC has published 32 annexes in ISPM 28 (see Section 11.3.1). All annexes clearly identify the minimum mortality rate expected from the treatment schedule, but this should not be confused as the IPPC endorsing a fixed minimum efficacy rate. Every treatment schedule in the annexes of ISPM 28 is reviewed and approved on a case-by-case basis and the approved mortality ranges run from 99.9872% to 99.9968% (95% CI).

Another issue with efficacy rates is how the results are interpreted. Technically, Probit 8.7 (95% CI) can range from zero survivors from 29 957 insects, one survivor from 47 431 insects, or 10 survivors from 169 630 insects, and so on. The unfortunate reality is that trade negotiations will be simpler if no survivors are recorded, as few countries will accept a traditional quarantine treatment where there have been failures. While the IPPC specifically states there is no such thing as zero risk, the reality is that zero survivors is the standard that researchers try to achieve for high-risk pests such as fruit fly. So if survivors are recorded, researchers normally tend to undertake further trials until such time as zero survivors are recorded, even if the 'failed' treatment technically met the minimum efficacy requirement. But alternatively, for a treatment to be commercially viable, it must result in minimal damage or reduction in shelf life of the commodity. Many treatments are close to the upper limit of fruit tolerance and there are very small margins to increase treatment dose/time/concentration.

Another issue with focusing on obtaining zero survivors is that many authors when they publish tend not to include the failed treatments. This is something that needs to change and publishing 'failed treatments' would provide invaluable information for future treatment development, especially if modelling of the thermal tolerance of tephritids is to be used to help develop generic treatments.

While phytosanitary regulators and researchers have focused on treatments with no survivors, the reality is that producers also need to meet the quality assurance standards in the markets they supply. In Australia many of the major retailers have zero tolerance for live insects. The presence of a single live insect can result in whole consignments being condemned. This is not because it is a breach of quarantine, as fruit grown in an endemic area and sold within an endemic area does not require treatment. It is simply because many consumers do not like live insects. Additionally, many consumers do not know the difference between a pest species and a beneficial species that may have been deliberately introduced in the production system as part of an integrated pest management system. Public awareness campaigns are needed to educate consumers and so hopefully decrease the need for unnecessary phytosanitary treatments.

While there is debate on the appropriate efficacy rate, there is also debate over assessments of mortality. Historically, mortality for heat and cold treatments has been classified as either chronic mortality or acute mortality. Chronic mortality is where eggs may hatch and larvae may survive for short periods, but no life stage survives long enough to pupate. Acute mortality, as the name suggests, does not allow for the presence of any live life stages. The argument for chronic mortality is that it better represents commercial conditions where inspections are undertaken for the presence of live larvae. However, inspections are notoriously inaccurate and because of this some researchers do not conduct inspections for eggs, first and second instars immediately. They store the fruit until the insects have had a chance to develop into third instars and inspections are easier to undertake. The issue of chronic and acute mortality is an area that needs to be reviewed, but chronic mortality has been used for decades with no commercial issues raised. Additionally, inspectors do not care what technique the researcher used. If they find live larvae the consignment will be condemned: as such both assessments are treated equally.

10.6 Future Work

Despite decades of research on fruit fly worldwide there are still many aspects of fruit fly biology and ecology that are not well understood. As a result, many of the international and regional standards developed for phytosanitary measures against fruit fly are very broad (rather than prescriptive) and based on expert opinion. But the important point is that disinfestation research is very conservative and treatments on a wide range of technologies have been developed and operated very successfully for decades. Many concerns about the use of laboratory-reared flies, infestation techniques, the use of chronic or acute mortality, or appropriate efficacy rates are valid concerns. They are definitely areas of research that need to be pursued, but it is important that research on these issues is undertaken using commercial treatment conditions. When a significant difference in tolerance is found at sub-lethal levels it does not mean that the treatment is failing. It simply highlights that more research needs to be undertaken.

One thing that can be improved immediately, and very simply, is the way results are reported. By providing detailed information on the methodology used and results recorded, it will be easier to undertake retrospective reviews of data sets if new research does identify weaknesses in current research methodology. So if there is a difference in the tolerance between wild flies and laboratory-reared flies it will be important to precisely record how the laboratory colonies were established, the media they were reared on, and how long colonies were maintained before being replaced or supplemented.

Similarly, it is very important to start reporting detailed information on the life stages present when a treatment was undertaken. Once again, if future research identifies that a particular life stage is more tolerant than previously thought then it will be easier to review older data sets. Additionally, reporting failed treatments or treatments that experienced problems (e.g. a temperature spike in a cold treatment) will provide invaluable information for future treatment development especially if modelling of the thermal tolerance of tephritids is to be used to develop generic treatments.

Phytosanitary research can be very labour and resource intensive, and so researchers may need to improvise on standard treatment techniques depending on the fly species/commodity combination and the resources available. This is acceptable and nearly any technique you consider appropriate can be used, as long as you can demonstrate the methods used do underestimate the treatment efficacy.

10.7   Further Reading and References Cited

Anderson, D.T. (1962) The embryology of *Dacus tryoni* (Frogg.) [Diptera, Tephritidae (=Trypetidae)], the Queensland fruit fly. *Journal of Embryology and Experimental Morphology*, 10, 248–292.

Anon. (1996) *Textbook of Vapour Heat Disinfestation Test Technicians* (Revised). Japan Fumigation Technology Association, Okinawa International centre, Japan International Cooperation Agency.

Anon. (2001) MAF Biosecurity Authority Standard 155.02.03 – Specification for the determination of fruit fly disinfestation treatment efficacy. New Zealand Ministry of Agriculture and Forestry.

De Lima, C.P.F., Mansfield, E.R. and Poogoda, S.R. (2017) International market access for Australian tablegrapes through cold treatment of fruit flies with a review of methods, model and data for fresh fruit disinfestation. *Australian Journal of Grape and Wine Research*, 23, 306–317.

Dohino, T., Hallman, G.J., Grout, T.G., Clarke, A.R., Follett, P.A., Cugala, D.R., Tu, D.M., Murdita, W., Hernandez, E., Pereira, R. and Myers, S.W. (2017) Phytosanitary treatments against *Bactrocera dorsalis* (Diptera: Tephritidae): Current situation and future prospects. *Journal of Economic Entomology*, 110, 67–79.

FAO/APPPC (2004) Regional Standards for Phytosanitary measures- Guidelines for the development of heat disinfestation treatments of fruit fly host commodities. The Asia and Pacific Plant Protection Commission, Bangkok. RAP Publication 2004/23. Available at http://www.fao.org/docrep/007/ad515e/ad515e00.htm (accessed 18 December 2018).

FAO/APPPC (2005) Guidelines for the confirmation of non-host status of fruit and vegetables to Tephritid fruit flies. Regional Standard for Phytosanitary Measures (RSPM) No. 4. Bangkok: The Asia and Pacific Plant Protection Commission (APPPC). Available at http://www.fao.org/docrep/008/ae942e/ae942e00.htm (accessed 18 December 2018).

FAO/IPPC (2017). Recommendation on: Replacement or reduction of the use of methyl bromide as a phytosanitary measure. Recommendation for the Implementation of the IPPC. Available at https://www.ippc.int/static/media/files/publication/en/2017/04/R_03_En_2017-04-26_Combined.pdf (accessed 18 December 2018).

FAO/North American Plant Protection Organization (2012) *Specifications for a Standard on Host Status (NAPPO_2012-02)*. Kentucky, USA: North American Plant Protection Organization. Available at https://www.nappo.org/files/7814/3753/7920/HostStatus_specification31-07-2013-e.pdf (accessed 18 December 2018)

Follett, P.A. (2004) Irradiation to control insects in fruits and vegetables for export from Hawaii. *Radiation Physics and Chemistry*, 71, 163–166.

Follett, P.A. (2014) Phytosanitary irradiation for fresh horticultural commodities: generic treatments, current issues and next steps. *Stewart Postharvest Reviews*, 3, 1–7.

Follett, P.A. and Armstrong, J.W. (2004) Revised irradiation doses to control melon fly, Mediterranean fruit fly, and Oriental fruit fly (Diptera: Tephritidae) and a generic dose for tephritid fruit flies. *Journal of Economic Entomology*, 97, 1254–1262.

Follett, P.A. and Hennessey, M.K. (2007) Confidence limits and sample size for determining nonhost status of fruits and vegetables to Tephritidae fruit flies as a quarantine measure. *Journal of Economic Entomology*, 100, 251–257.

Follett, P.A. and Neven, L.G. (2006) Current trends in quarantine entomology. *Annual Review of Entomology*, 51, 359–385.

Follett, P.A. and Weinert, E.D. (2012) Phytosanitary irradiation for tropical commodities in Hawaii: generic treatments, commercial adoption, and current issues. *Radiation Physics and Chemistry*, 81, 1064–1067.

Hallman, G.J., Myers, S.W., Jessup, A.J. and Islam, A. (2011) Comparison of *in vitro* heat and cold tolerance of the new invasive species *Bactrocera invadens* (Diptera: Tephritidae) with three known tephritids. *Journal of Economic Entomology*, 104, 21–25.

Hallman, G.J., Wang, L., Uzel, G.D., Cancio-Martinez, E., Caceres-Barrios, C.E., Myers, S.W. and Vreysen, M.J.B. (2018) Comparison of populations of *Ceratitis capitata* (Diptera: Tephritidae) from three continents for susceptibility to cold phytosanitary treatment and implications for generic cold treatments. *Journal of Economic Entomology*, 112, 127–133.

Hancock, D.L., Hamacek, E.L., Lloyd, A.C. and Elson-Harris, M.M. (2000) *The Distribution and Host Plants of Fruit Flies (Diptera: Tephritidae) in Australia*. DPI Publications, Brisbane, Australia.

Heather, N.W. and Hallman, G.J. (2008) *Pest Management and Phytosanitary Trade Barriers*. CAB International, Wallingford, UK, 257pp.

Jang, E.B., Nagata, J.T., Chan, H.T. and Laidlaw, W.G. (1999) Thermal death kinetics in eggs and larvae of *Bactrocera latifrons* (Diptera: Tephritidae) and comparative thermotolerance to three other tephritid fruit fly species in Hawaii. *Journal of Economic Entomology*, 92, 684–690.

Jessup, A.J. and McCarthy, D. (1993) Host status of some Australian-grown cucurbits to *Bactrocera tryoni* (Froggatt) (Diptera: Tephritidae) under laboratory conditions. *Journal of the Australian Entomological Society*, 32, 97–98.

Kancyuki, M., Kobashigawa, Y., Yamamoto, T., Kikukawa, K., Miyazaki, I., and Adachi, H. (2016) Effect of age and feeding on heat tolerance in each larval instar period of *Bactrocera dorsalis* and *Bactrocera cucurbitae* (Diptera: Tephritidae). *Research Bulletin of the Plant Protection Service*, 52, 29–36.

Li, L., T.Liu, B.Li, F.Zhang, S.Dong, and Y. Wang. (2014) Toxicity of phosphine fumigation against *Bactrocera tau* at low temperature. *Journal of Economic Entomology*, 107, 601–605.

Lloyd, A.C., Hamaceh, E.L., Smith, D., Kopittke, R.A. and Gu, H. (2013) Host fruit susceptibility of citrus cultivars to Queensland fruit fly (Diptera: Tephritidae). *Journal of Economic Entomology*, 106, 883–890.

Secretariat of the International Plant Protection Convention (2016) *ISPM 37: Determination of host status of fruit to fruit flies (Tephritidae)*. International Plant Protection Convention and FAO, Rome. 18pp. Available at https://www.ippc.int/static/media/files/publication/en/2016/05/ISPM_37_2016_En_HostStatus_2016-04-26.pdf (accessed 18 December 2018).

Shellie, K.C. and Mangan, R.L. (2000) Postharvest disinfestation heat treatments: response of fruit and fruit fly larvae to different heating media. *Postharvest Biology and Biotechnology*, 21, 51–60.

11 Regulatory Controls

11.1 General Introduction

Pre-harvest (Chapter 9) and post-harvest (Chapter 10) treatments, if applied properly, can ensure a commercial crop is available for harvest, and guarantee (to a given level of risk) that the picked commodities are free of fruit fly infestation. Nevertheless, an importing nation is still likely to be concerned about the biosecurity risks posed by fresh-commodity imports because it does not know what controls may have been applied to a crop. How then to ensure that the exporting nation has applied adequate pre- and/or post-harvest controls to minimise risk to a standard acceptable to the importing nation? In short, how to allow the free movement of fresh commodities around the world while minimising the biosecurity risk this poses? For the last 65 years, the global solution to this problem lies in the International Plant Protection Convention (IPPC) and its International Standards for Phytosanitary Measurements (ISPMs).

The IPPC and the ISPMs relevant to fruit fly are the topic of this chapter. While it may seem unusual to include an entirely regulatory pest management chapter in an otherwise 'biology' book, it is essential for gaining a full understanding of dacines as pests, and knowing how they are managed within a global free-trade environment. My experience is that many growers, academic researchers and field entomologists are largely ignorant of the regulatory requirements associated with international market access. Yet the economic viability of many fresh-commodity industries relies on access to international markets and, if *Bactrocera* are present in a growing region, then those markets are likely to remain closed without government-to-government negotiated agreements. Thus, understanding the role of the IPPC in facilitating fresh-commodity trade in the presence of fruit flies is just as important, and indeed may be more important, than any other pest management component.

11.2 The International Plant Protection Convention and ISPMs

The IPPC is an international, legally binding agreement on plant health to which 182 nations are signatories. Created in 1951, the IPPC aims to protect cultivated and wild plants by preventing the introduction and spread of plant pests. The IPPC is governed by the Commission on Phytosanitary Measures (CPM) and is operationally implemented by the IPPC Secretariat. The CPM, which meets once a year, consists of a member from each of the signatory countries (normally the head of the nation's plant protection organisation); while the Secretariat is provided by the United Nations Food and Agricultural Organisation (FAO) and is based at FAO headquarters in Rome.

So what does a high-level, multinational agreement have to do with day-to-day management of fruit flies? The answer is a great deal. Despite the fact that many growers and researchers do not know about the existence of the IPPC, their adopted standards – known as the International Standards for Phytosanitary Measures – directly impact on how fresh commodities produced in fruit fly areas can be traded, and hence how fruit flies must be managed if targeting international markets.

11.2.1 How the IPPC operates

The aim of the IPPC (https://www.ippc.int) is to facilitate international movement and trade of plant and plant products, while minimising the risk of spreading plant pests. The CPM represents the global community and, on advice from experts and with a comprehensive negotiation and commentary process, approves the ISPMs. The ISPMs are the basis for any phytosanitary conditions associated with plant commodity trade – for example ISPM 07 is *Phytosanitary certification system* and ISPM 11 *Pest risk analysis for quarantine pests*. Approval of ISPMs is not done lightly – in the 67 years since the signing of the convention, only 42 have been approved, with or without additional 'Annexes' (as at February 2019). All ISPMs and Annexes can be found at https://www.ippc.int/core-activities/standards-setting/ispms.

The ISPMs and the IPPC are important because they create a binding international policy and operational framework under which fresh-commodity trade can operate, while minimising global biosecurity risk. The IPPC represents the 'policy' of this framework, setting out 23 articles that are summarised in Table 11.1. While the articles are critical to outlining how global plant protection may exist along-side global commodity trade, with the exception of phytosanitary certification (Article V) the IPPC does not provide technical advice on how to meet its operational objectives. Instead this is the role of the ISPMs, the creation of which are mandated under IPPC Article X. The ISPMs are technically focused documents that outline operational steps and procedures to implement the objectives of the IPPC. The implementation and following of ISPMs forms the basis of bilateral negotiations between importing and exporting countries.

One problem of the ISPM system is that they are high-level documents that go through an extensive, iterative technical and political process before approval.

Table 11.1. Summary of the articles of the International Plant Protection Convention (1997). The reader should always and only use the full text of the Convention (available at https://www.ippc.int) for any official or planning purpose.

Article	Title	Summary of article
I	Purpose and responsibility	Identifies the purpose of the Convention ('*of securing common and effective action to prevent the spread and introduction of pests of plants and plant products*'); and identifies each contracting party (generally but not always an independent nation) as being responsible for taking action under the Convention to the best of their competency. This article also notes that the convention need not be restricted to plants and plant products, but may also deal with packing materials, storage containers, other organisms, etc., which may be capable of spreading plant pests.
II	Use of terms	Defines the terms used in the Convention (e.g. '*Pest*' – *any species, strain or biotype of plant, animal or pathogenic agent injurious to plants or plant products*.) (Note: a full glossary of phytosanitary terms is provided in ISPM No 5.)
III	Relationship with other international agreements	Simply states: '*Nothing in this Convention shall affect the rights and obligations of the contracting parties under relevant international agreements*'.
IV	General provisions relating to the organisational arrangements for national plant protection	Outlines the requirement for each contracting party to establish a National Plant Protection Organisation (NPPO) to oversee regulatory plant health arrangements including inspection and certification, surveillance and reporting of plant pests in their countries, conduct pest risk analysis, and make provision for plant protection research.
V	Phytosanitary certification	Identifies that contract parties must make arrangements for phytosanitary certification by qualified and authorised individuals. A model phytosanitary certificate is appended as an Annex to the Convention text.
VI	Regulated pests	Recognises that contracting parties may require phytosanitary measures for quarantine and regulated non-quarantine pests so long as those phytosanitary measures are technically justified and are not more than required to protect plant health.
VII	Requirements in relation to imports	In summary, this is the article that states the right of importing countries to request the imposition of phytosanitary measures (including denial of entry or destruction) on exporters so as to reduce risk of entry of quarantine and regulated pests. However, the aim of the article is to minimise interference with international trade and so any phytosanitary measures requested must be technically justified and are not more than required to protect plant health.
VIII	International cooperation	Identifies the requirement to share information on plant pests and cooperate in their control.

IX	Regional plant protection organisations	Identifies that NPPOs may cooperate with each other to form regional plant protection organisations for inter-regional coordination and cooperation.
X	Standards	A simple article that states that contracting parties will cooperate in the development and adoption of agreed international standards (these have become the ISPMs).
XI	Commission on Phytosanitary Measures	Describes the functions, membership and duties of the Commission on Phytosanitary Measures (CPM) that implements the objectives of the IPPC.
XII	Secretariat	Describes the functions and duties of the IPPC Secretariat that carries out the day-to-day implementation of the IPPC objectives as identified by the CPM.
XIII	Settlement of disputes	As the name suggests, describes the dispute settlement process between contracting parties if one party believes another party is in conflict with obligations under the IPPC (for example in allowing commodity trade). [Importantly, the IPPC has no legal mechanism to inforce compliance. If dispute settlement cannot be reached using the identified IPPC process, then the opportunity exists for the case to be escalated to the Dispute Settlement Body of the World Trade Organisation.]
XIV	Substitution of prior agreements	Identifies three earlier international agreements that the IPPC replaces.
XV	Territorial application	States that at any time, following notification, a contracting party may notify what parts of its territories (all, partial, none) shall be subject to obligations under the IPPC.
XVI	Supplementary agreements	States the contracting parties may enter supplementary agreements for meeting special problems of plant health (e.g. for dealing with a particular pest outbreak or other specific problem).
XVII	Ratification and adherence	A largely historical article, noting the intention of the Convention to be ratified as soon as possible after 1 May 1952.
XVIII	Non-contracting parties	States that contracting parties should encourage non-contracting parties to join.
XIX	Languages	That the authentic languages of the IPPC will be all official languages of FAO.
XX	Technical assistance	*'The contracting parties agree to promote the provision of technical assistance to contracting parties, especially those that are developing contracting parties.'*
XXI	Amendment	Identifies the formal process by which any amendments to the IPPC can be requested or made.
XXII	Entry into force	*'As soon as this Convention has been ratified by three signatory states it shall come in force.'*
XXIII	Denunciation	Any contracting party can, at any time, denounce the IPPC; with that denunciation taking effect 12 months later.

Thus the final wording is often highly legalised and indirect, such that operationally they can be difficult to apply to specific pest situations. Recognising this, a new generation of ISPMs have been created in recent years that are aiming to be much more pest specific – fruit flies have figured highly in this process and this directly impacts on the future of global fruit fly management.

11.3 Fruit Fly Specific ISPMs and Annexes

Of the 42 ISPMs and their Annexes, three ISPMs and 23 Annexes in two further ISPMs are specific to tephritid fruit flies (Table 11.2). Directly relevant to *Bactrocera* are the three ISPMs (ISPM 26, ISPM 35 and ISPM 37) and eight annexes in ISPM 28 that are about post-harvest disinfestation treatments. Descriptions of the Annexes and ISPMs follow, but the reader is always directed to the IPPC website to seek the authoritative versions.

11.3.1 Annexes to ISPM 28, Post-harvest treatments

The annexes for ISPM 28 are relatively simple, highly specific technical documents that provide statements on the efficacy of the given post-harvest treatments. As one example, ISPM 28 Annex 5 (Irradiation treatment for *B. tryoni*) states that an irradiation treatment with a minimum absorbed dose of 100 Gy has a treatment efficacy (for preventing emergence of *B. tryoni* adults) of $ED_{99.9978}$ at the 95% confidence level. It further notes that the irradiation may not cause direct mortality, and so inspectors should be aware that they may encounter live, but not viable, larvae or pupae. The acceptance of this Annex by the CPM means that this treatment, if used by an exporter as a single-step disinfestation treatment for a *B. tryoni* susceptible commodity, should not need to be further justified to the importing country: the annex is all the evidence that should be needed. Should the importer reject the treatment outright, then a case could be made before the World Trade Organization that a technical barrier to trade had been erected. Commodity negotiations rarely reach that level, but that is the power of accepted ISPMs and their annexes. Importantly, the subsequent ISPM 28 Annex 7 (irradiation treatment for fruit flies of the family Tephritidae (generic)) records that the minimum absorbed dose of 150 Gy will prevent the emergence of adults of all tephritid fruit flies with an efficacy and confidence level of the treatment at $ED_{99.9968}$ at the 95% confidence level. For any country exporting from a fruit fly affected area, which has the opportunity to use irradiation post-harvest treatments, then this becomes a universal treatment for fruit fly for export commodities. Unfortunately, irradiation treatment sites are still not yet routinely available, nor are consumers all accepting of irradiated produce.

11.3.2 ISPM 26, Establishment of pest-free areas for fruit flies

Pest-free areas (PFAs) are the ideal situation for growers requesting market access. If the area is declared free of the target pest, then market access from

Table 11.2. List of Commission on Phytosanitary Measures adopted by the International Standards for Phytosanitary Measures and Annexes directly relevant to tephritid fruit flies.

ISPM	Annex (if relevant)	Title
ISPM 26		Establishment of pest-free areas for fruit flies (Tephritidae)
	Annex 01	Guidelines on corrective action plans
	Annex 02	Control measures for an outbreak within a fruit fly pest-free area
ISPM 27		Diagnostic protocols for regulated pests
	Annex 09	DP C9: Genus *Anastrepha* Schiner
ISPM 28		Phytosanitary treatments for regulated pests
	Annex 01	PT 1 (2009): Irradiation treatment for *Anastrepha ludens*
	Annex 02	PT 2 (2009): Irradiation treatment for *Anastrepha obliqua*
	Annex 03	PT 3 (2009): Irradiation treatment for *Anastrepha serpentin*
	Annex 04	PT 4 (2009): Irradiation treatment for *Bactrocera jarvisi*
	Annex 05	PT 5 (2009): Irradiation treatment for *Bactrocera tryoni*
	Annex 07	PT 7 (2009): Irradiation treatment for fruit flies of the family Tephritidae (generic)
	Annex 08	PT 8 (2009): Irradiation treatment for *Rhagoletis pomonella*
	Annex 14	PT 14 (2011): Irradiation treatment for *Ceratitis capitata*
	Annex 15	PT 15: Vapour heat treatment for *Bactrocera cucurbitae* on *Cucumis melo* var. *reticulatus*
	Annex 16	PT 16: Cold treatment for *Bactrocera tryoni* on *Citrus sinensis*
	Annex 17	PT 17: Cold treatment for *Bactrocera tryoni* on *Citrus reticulata* × *C. sinensis*
	Annex 18	PT 18: Cold treatment for *Bactrocera tryoni* on *Citrus limon*
	Annex 21	PT 21: Vapour heat treatment for *Bactrocera melanotus* and *Bactrocera xanthodes* on *Carica papaya*
	Annex 24	PT 24: Cold treatment for *Ceratitis capitata* on *Citrus sinensis*
	Annex 25	PT 25: Cold treatment for *Ceratitis capitata* on *Citrus reticulata* × *C. sinensis*
	Annex 26	PT 26: Cold treatment for *Ceratitis capitata* on *Citrus limon*
	Annex 27	PT 27: Cold treatment for *Ceratitis capitata* on *Citrus paradisi*
	Annex 28	PT 28: Cold treatment for *Ceratitis capitata* on *Citrus reticulata*
	Annex 29	PT 29: Cold treatment for *Ceratitis capitata* on *Citrus clementina*
	Annex 30	PT 30: Vapour heat treatment for *Ceratitis capitata* on *Mangifera indica*
	Annex 31	PT 31: Vapour heat treatment for *Bactrocera tryoni* on *Mangifera indica*
	Annex 32	PT 32: Vapour heat treatment for *Bactrocera dorsalis* on *Carica papaya*
ISPM 35		Systems approach for pest risk management of fruit flies (Tephritidae)
ISPM 37		Determination of host status of fruit to fruit flies (Tephritidae)

the area can be negotiated with no further phytosanitary treatments required. However, establishing a fruit fly PFA for trade negotiation is not as simple as stating that flies are not present. Proof of absence needs to be demonstrated both in setting up the PFA and while the PFA is operating, and plans need to be put in place dealing with emergency responses if the PFA is compromised (i.e. flies are caught). ISPM 26 identifies the following requirements for a PFA.

A PFA is 'an area in which a specific pest does not occur as demonstrated by scientific evidence and in which, where appropriate, this condition is being officially maintained'. PFAs may be naturally free of pests due to the presence of barriers or climate conditions, and/or maintained free through movement restrictions and related measures, or may be made free by an eradication pro-gramme. Before establishing a fruit fly pest-free area (FF-PFA) the area needs to be delimited, buffer zones need to be established if physical isolation is not enough to stop entry of the fly, and surveillance (trapping and fruit sampling) needs to be carried out for a least 12 consecutive months to demonstrate absence of the pest. Importantly, collection of a single adult fly may not be enough to disqualify an area for FF-PF status, but there should not be detection of an immature specimen, two or more fertile females, or a single inseminated female. Fruit fly trapping procedures and fruit sampling protocols to demon-strate area freedom are described in detail in the ISPM.

Declaration of area freedom is made by the national plant protection organisation (NPPO), who also have responsibility for monitoring ongoing sur-veillance and control activities in order to continuously verify pest-free status. The NPPO also needs prepared plans in the event of a fruit fly outbreak in the FF-PFA, including corrective action plans to suspend the FF-PFA until it is reinstated. Reinstatement follows after the meeting of a pre-determined period of time during which no further flies are trapped or reared. If fruit fly outbreaks continue and cannot be managed, then area freedom will be lost or suspended. Guidelines for corrective action plans are supplied as an annex of the ISPM.

While growers, particularly, greatly benefit from a FF-PFA, they are not a universal panacea and significant thought needs to be considered before attempting to establish one. Firstly, and most importantly, FF-PFAs are not cheap. Significant costs (in Australia more than AU$10 million for one state in 1 year for the former FF-PFA in south-eastern Australia) are associated with maintaining and servicing the comprehensive trapping network required to demonstrate area freedom, and if a buffer zone is required then additional sur-veillance and fruit fly control costs are incurred. There are also direct and indi-rect societal costs associated with movement restriction of fruit fly susceptible commodities into the FF-PFA. Such costs are commonly born by governments and, in many countries, governments are becoming increasingly unwilling to meet them. A second important consideration of establishing a FF-PFA is that it can, surprisingly, be a risky strategy. If market access is being gained entirely through pest-free status, then a declared outbreak during peak harvest may cause huge economic loss as those markets are lost until pest-free status can be regained. If fruit picking occurs over a short seasonal window, as it does for most commercial crops, an entire year's market opportunity may be lost by one outbreak.

11.3.3 ISPM 35, Systems approach for pest risk management of fruit flies

Systems approaches for fruit fly management are straightforward in theory, but much less so in practice. A systems approach uses two or more *independent* risk-reduction steps to reduce the risk of a fruit fly establishing in a new country or region free of that pest. The risk-reduction steps can be applied anywhere from growing and harvest, packing, post-harvest and transport, and at the entry and distribution point within the receiving country. Systems are best suited when a single risk-reduction step (e.g. a post-harvest treatment) is not possible or effective enough. A system might include growing a fruit that is of low host status, in an area with low pest prevalence, and then cold storing for a given time. Any one of these treatments on their own may be considered by the importing country as insufficient to reduce the risk of spreading fruit fly, but when combined they may be sufficient. The US has to date been the largest approver of systems approaches for market access and one of these is provided as an illustrative case in Box 11.1. This example illustrates a number of points covered under the IPPC and the ISPMs. These include: the right of the importing country to set phytosanitary conditions so long as they are scientifically justified, the key role of the NPPO (in Box 11.1 the Brazilian Ministry of Agriculture), the integration of an area of low pest prevalence into a systems approach, the need for record keeping, and the need for actions if a protocol fails.

11.3.4 ISPM 37, Determination of host status of fruit to fruit flies (Tephritidae)

If a fruit is not a host to a fruit fly, then all other market access conditions become irrelevant with respect to reducing fruit fly risk – because there is no risk. However, determining what is and is not a fruit fly host fruit is not straightforward. For polyphagous flies, particularly, there are many biological reasons why a fruit may be rarely utilised by a fly, but a positive host record may still exist (see Chapter 6). Alternatively, old or incorrect taxonomy may mean a host record has been associated with the incorrect fly. An onus is therefore placed on the exporting nation to conclusively demonstrate non-host status using internationally agreed testing protocols.

ISPM 37, which was approved in 2016, deals with this tricky problem, and does so through a biologically logical and sequential process (Fig. 11.1). The first stage is to use existing information to determine if there is already evidence to confirm unambiguously if the fruit is a host or non-host. If this cannot be done, then the next step is to do field surveys of the crop being targeted for export, in the environment that the crop will be exported from, and at the physiological stage and quality of export. If infestation is found then the fruit is confirmed as a host. If no field infestation is found, but other information suggests it may still be a host (e.g. small cage laboratory trials or historical data), then trials must be conducted under semi-natural conditions. These conditions demand that tested fruit is attached naturally to the plant, the whole plant is being used (potted or in the ground) and work is done in shade house or glass house, or on bagged branches of a whole plant. The intent of the ISPM is to make the test

Box 11.1. An illustrative example of a systems approach for a fruit fly affected commodity. The document illustrates not only the requirements for systems approach to market access, but also the addition of an area of low pest prevalence into that system. The document was accessed from http://www-naweb.iaea.org/nafa/ipc/SA_Several_E_Miller.pdf on the 23rd May 2018.

A. Title: Papayas from Brazil to the United States.

B. Source: United States CFR 319.56-2w. Conditions governing the entry of papayas from Central America and Brazil.

C. Pest of concern: Mediterranean fruit fly (Medfly) and South American fruit flies (*Anastrepha fraterculus*).

D. Major mitigation measures: Poor host status, low prevalence area, specific cultivars, specific maturity stage and hot water dip.

E. Specific mitigation measures.

F. Notes: Since this pest is of major concern to the United States and since it can become established in the United Stated easily, these conditions are required for this secondary host.

319.56-2w Administrative instruction; conditions governing the entry of papayas from Central America and Brazil.

The Solo type of papaya may be imported into the continental United States, Alaska, Puerto Rico and the US Virgin Islands only under the following conditions:

(a) The papayas were grown and packed for shipment to the United States in the State of Espirito Santo, of Brazil.

(b) Beginning at least 30 days before the harvest began and continuing through the completion of harvest, all trees in the field where the papaya were grown were kept free of papayas that were ½ or more ripe (more than one-quarter of the shell surface yellow), and all culled and fallen fruits were buried, destroyed or removed from the farm at least twice a week.

(c) The papayas were treated with a hot water treatment consisting of 20 minutes in water at 40°C (120.2°F).

(d) When packed, the papayas were less than half ripe (the shell surface was no more than ¼ yellow, surrounded by light green), and appeared to be free of all injurious insect pests.

(e) The papayas were safeguarded from exposure to fruit flies from harvest to export, including being packaged so as to prevent access by fruit flies and other injurious pests.

(f) All cartons in which papayas are packed must be stamped 'Not for importation into or distribution in HI'.

(g) All activities described in paragraphs (a) through (f) of this section were carried out under the supervision and direction of plant health officials of the national Ministry of Agriculture.

(h) Beginning at least 1 year before the harvest begins and continuing through the completion of harvest, fruit fly traps were maintained in the field where the papayas were grown. The traps were placed at a rate of one trap per hectare and were checked for fruit flies at least once weekly by plant health officials of the national Ministry of Agriculture. Fifty per cent of the traps were of the McPhail type, and 50% were of the Jackson type. If the average Jackson trap catch was greater than seven Medflies per trap per week, measures were taken to control the Medfly population in the production area. The national Ministry of Agriculture kept records of fruit fly finds for each trap, updated the records each time traps

Continued

Box 11.1. Continued.

were checked, and made the records available to APHIS inspectors upon request. The records were maintained for at least 1 year.

(i) If the average Jackson trap catch exceeds 14 Medflies per trap per week, importations of papayas from the production area must be halted until the rate of capture drops to an average of seven or fewer Medflies per trap per week.

(j) In the State of Espirito Santo, Brazil, if the average McPhail trap catch was greater than seven South American fruit flies (*Anastrepha fraterculus*) per trap per week, measures were taken to control the South American fruit fly population in the production area. If the average McPhail trap catch exceeds 14 South American fruit flies per trap per week, importation of papayas from the production area must be halted until the rate of capture drops to an average of seven or fewer South American fruit flies per trap per week.

(k) All shipments must be accompanied by a phytosanitary certificate issued by the national Ministry of Agriculture stating that the papayas were grown, packed, and shipped in accordance with the provisions of this section.

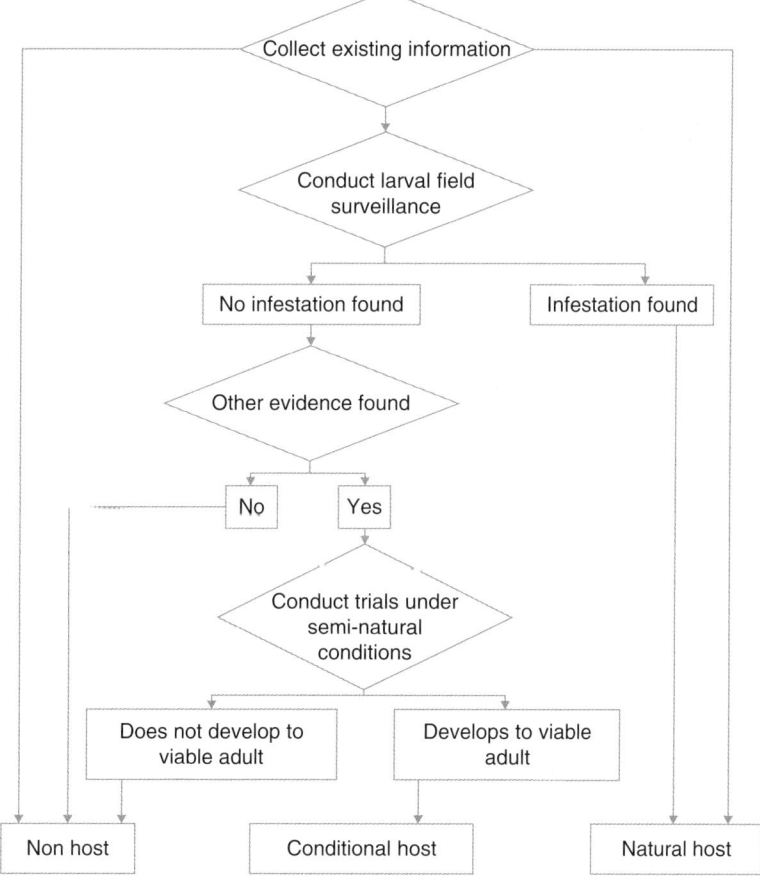

Fig. 11.1. The steps for determination of host status of fruit to fruit flies as determined by ISPM 37 (redrawn from Figure 1 from ISPM 37, FAO 2016).

as realistic as possible, given the inherent biological problems in such trials. If in such tests no viable adults are bred through from the test fruit (but they are in appropriate controls) then the ISPM states the plant or cultivar is a non-host. If flies are reared through then the plant becomes a conditional host and there will presumably be risk-reduction procedures required for export.

Experimentally demonstrating non-host status for polyphagous species that may egg-dump is notoriously difficult. Nevertheless, this ISPM provides an internationally agreed protocol that is a well balanced mix of biological science and political pragmatism.

11.3.5   Why can implementing certain ISPMs appear so difficult?

If internationally accepted guidelines exist for systems approaches for fruit flies, why do they seem so difficult to implement, which is the experience of many countries? The problem is the ISPMs, which while defining what these terms mean and what is needed to set them up, are largely silent on critical issues such as 'how few flies do you need to have to get an area of low pest prevalence', or 'how do you assess the risk reduction of different risk reduction steps'?

That the ISPMs are largely silent on these issues is not a fault of the authors of these documents, but the nature of trade negotiation and also problems of science. Under the agreements of the IPPC, determining the acceptable level of risk posed by an exported commodity is the sovereign right of the importing nation (while recognising that there is no such thing as no risk). Thus what might be deemed 'risky' to one nation, might be quite acceptable to another. If this concept is translated to systems approaches, it can be easily seen that different risk-reduction steps and amounts might be acceptable to different nations, and this is one reason why ISPMs are rarely prescriptive in the fine details.

There are also major issues related to fruit fly biology that impact on the issue of 'how low is low'. Not all fruit types, or even varieties within a fruit type, are equally susceptible to fruit fly damage. For example, for *B. tryoni* nearly all citrus are hosts, but some types (e.g. murcott mandarins) are more susceptible than others (e.g. Myer lemon); whereas all citrus are bad hosts compared to nearly all stone fruit. In the field this means the risk posed by a given number of flies to stone fruit is much greater than the risk posed by the same number of flies to citrus. This issue is again modified by what other potential hosts are around. If good hosts and bad hosts are both available at the same time the flies will choose the good host almost exclusively, but if only the bad host is available then that fruit may get quite heavily damaged. This situation occurs for all polyphagous *Bactrocera* pests. When such issues are combined with ripening effects (flies prefer ripe or over-ripe to under-ripe fruit), time of year effect (spring versus autumn), the number of generations the fly has per year, and so on, then calculating what is an acceptable threshold number of flies in a trap to reduce the commodity infestation risk to an acceptable level (which is in itself dependent on the importer) is difficult.

One other major technical problem, which has not yet been overcome, also impinges on systems approaches. Systems approaches rely, by definition, on two or more risk-reduction steps acting together to reduce risk to an acceptable (to the importing nation) level. For single-step post-harvest treatments (e.g. heat, cold, irradiation, etc.), calculating the efficacy of the treatment is technically challenging, but still relatively straightforward. Fruit are infested with a known number of eggs or maggots, the fruit is subjected to the treatment, and the number of survivors counted. When this is done with enough replication, over enough different treatment types, accurate estimates of efficacy can be calculated and presented to a trading partner. However, if you have a system that combines, for example, seasonal low pest prevalence (e.g. a winter crop), stage of ripeness (mature green rather than ripe), and picking and packing shed cull, calculating the risk-reduction step posed by each independent treatment varies from difficult to impossible. Rather, what tends to happen in practice is that a final point assessment is made (e.g. the infestation level of packed fruit) and the whole system is judged on that. While that works, it makes it difficult to work backwards to find the weak link if the system fails, and it makes it difficult to determine redundancy in the system.

11.4   Other ISPMs of Particular Relevance to *Bactrocera* and Other Fruit Flies

11.4.1   ISPM 10, Requirements for the establishment of pest-free places of production and pest-free production sites

A pest-free place of production is a 'place of production in which a specific pest does not occur as demonstrated by scientific evidence and in which, where appropriate, this condition is being officially maintained for a defined period'; whereas a pest-free production site is a management unit within a place of production. Pest-free places and sites of production as a market access tool operate on the same general principle as a pest-free area: that is, market access is greatly simplified by demonstrated absence of the pest. However, they differ from PFAs in three important aspects. First, there is the geographic size. A PFA should be at least a geographic region, and can be as big as a country, or several neighbouring countries. In stark contrast, a pest-free place of production can be individual or collected premises, or collection of fields, operated as a single production unit. A pest-free site of production can be a single unit (e.g. field or building) within a place of production. The second major difference with a PFA is that a pest-free place and pest-free site of production are managed individually by the producer, under supervision and responsibility of the NPPO. In contrast, a pest-free area is managed by the NPPO. The third difference lies in the duration of pest freedom. A PFA is generally maintained over many years without interruption, whereas the status of a pest-free place of production may be maintained for only one or a few growing seasons.

Pest-free production areas and pest-free sites of production have rarely been considered for *Bactrocera* and other species, as the mobility of adults and field growing of crops make them poorly suited to such strategies, which are better suited to glass-house production systems and similarly intensive production. Nevertheless, this may change in the future. Horticultural glass-house systems are getting larger and larger and offer the potential to become pest-free places of production if they can be appropriately screened and the external fly populations can be suppressed (i.e. to effectively become a buffer). Isolated production areas (e.g. irrigation sites well removed from 'normal' production areas) that constitute one or only a small number of collaborating growers could also argue for pest-free production site if local eradication and intervening buffer zone suppression could be demonstrated. As horticultural production intensifies globally, especially protected horticulture, pest-free production areas and pest-free sites of production may become more common for fruit flies.

11.4.2   ISPM 02, Framework for pest risk analysis; ISPM 11, Pest risk analysis for quarantine pests; ISPM 21, Pest risk analysis for regulated non-quarantine pests

Pest risk analysis (PRA) is a tool used for identifying appropriate phytosanitary measures. The PRA process may be used for organisms not previously recognised as pests, recognised pests, pathways and review. The process consists of three stages: stage 1, initiation; stage 2, pest risk assessment; and stage 3, pest risk management (Table 11.3).

The initiation phase (stage 1) studies pathways (e.g. a particular commodity type being exported from country A to country B), or a specific plant pest or disease (e.g. Oriental fruit fly) to determine if the proposed pathway may potentially move phytosanitary pest(s), or in case of a specific organism if that organism is of phytosanitary concern to the importing country (or region). The process continues to stage 2 (pest risk assessment) if a regulatory pest is identified, or is halted if a regulatory pest is not identified. Even when the PRA is initiated for a known pest, such as Oriental fruit fly, or for a pathway in which a known pest occurs (e.g. Oriental fruit fly in mango exports), it does not automatically follow that the PRA will proceed past stage 1. For international trade, if the pest identified is a non-regulated quarantine pest in the place of destination, then it is not a pest in terms of the formal PRA process. For example, *B. dorsalis* is established and not under regulated control in both Malaysia and Thailand. In this instance, if a PRA for mango commodity trade was being considered from Thailand to Malaysia then *B. dorsalis* is likely to be considered as part of a PRA stage 1, but it would not be considered beyond that stage because it already occurs as a non-regulated pest in Malaysia. In contrast, if the same commodity was being considered for export from Thailand to Australia, then *B. dorsalis* would be identified as a pest of regulatory concern, as it does not occur in Australia and is a quarantine pest of that country.

Stage 2 of the PRA involves determining the likely risk posed by the identified plant pests or diseases. As identified in Table 11.3, the risk posed considers both biological and economic factors. The biological risk includes the probability of the pest organism moving along the pathway, entering the focus area, establishing and then spreading (note: these are the classical phases of a biological invasion process). The economic risk includes not just an assessment of the direct crop damage, but also indirect effects (e.g. a new pest organism may cause loss of market access for the newly affected growers) and an analysis

Table 11.3. Stages of the pest risk analysis (PRA) process as taken from ISPM 11, *Pest risk analysis for quarantine pests*. (The most recent version of this ISPM can be downloaded from www.IPPC.int)

Stage 1: Initiation

> 1.1 Initiation points
> PRA initiated by the identification of a pathway
> PRA initiated by the identification of a pest
> PRA initiated by the review or revision of a policy
> 1.2 Identification of PRA Area
> 1.3 Information
> Previous PRA (yes/no)
> 1.4 Conclusion of Initiation (decision to stop or continue)

Stage 2: Pest Risk Assessment

Pest categorisation

> 2.1 Pest categorisation
> Identity of pest; presence or absence in PRA area; regulatory status; potential for establishment and spread in PRA area; potential for economic consequences in PRA area; conclusion of pest categorisation
> 2.2 Assessment of the probability of introduction and spread
> Identification of pathways for a PRA initiated by a pest; probability of the pest being associated with the pathway at origin; probability of survival during transport or storage; probability of pest surviving existing pest management procedures; probability of transfer to a suitable host
> 2.3 Probability of Establishment
> Availability of suitable hosts, alternate hosts and vectors in the PRA area; suitability of environment; cultural practices and control measures; other characteristics of the pest affecting the probability of establishment
> 2.4 Probability of spread after establishment
> 2.5 Conclusion on the probability of introduction and spread
> 2.6 Conclusion regarding endangered areas

Assessment of potential economic consequences

> 2.7 Pest effects
> Direct pest effects; indirect pest effects;
> Analysis of Economic Consequences
> Time and place factors; analysis of commercial consequences; analytical techniques; non-commercial and environmental consequences
> 2.9 Conclusion of the assessment of economic consequences
> 2.10 Degree of uncertainty
> 2.11 Conclusion of the pest risk assessment stage (decision to stop or continue)

Stage 3: Pest risk management

> 3.1 Level of risk
> 3.2 Technical information required
> 3.3 Acceptability of risk
> 3.4 Identification and selection of appropriate risk management options
> Options for consignments; options preventing or reducing infestation in the crop; options ensuring that the area, place or site of production or crop is free from the pest; options for other types of pathways; options within the importing country; prohibition of commodities
> 3.5 Phytosanitary certificates and other compliance measures
> 3.6 Conclusion of pest risk management
> Monitoring and review of phytosanitary measures

4 Documentation of Pest Risk Analysis, Documentation requirements

of the medium- to long-term economic consequences. If the pest is not considered to be a biological risk (either because its chances of entry, establishment and spread are too low), or because it is not going to cause any economic or environmental damage, then the process terminates at the end of stage 2: if there is potential risk then the process continues.

Stage 3 of the PRA process investigates strategies that can be applied to reduce risk to an acceptable standard so that commodities can be traded. Risk-reduction steps can be applied in-field, in the packing shed, post-harvest and during transport, and at the point of onward distribution at the port of entry. In event that risk reduction cannot be achieved to an acceptable level through phytosanitary treatments, commodities can be prohibited entry. ISPM 11, however, sees this as a measure of last resort.

While all of this is relatively technical and well established for fruit flies (see earlier chapters), what constitutes 'acceptable risk' is far from clear. The IPPC sets in its opening pre-amble the right of sovereignty; that is, an individual country's right to determine what is best for it and what is acceptable to it. However, international free-trade agreements also recognise that there is no such thing as 'no risk'. International arguments arise when in bilateral negotiations one country has set its acceptable risk limits at a level whereby the second country considers them too prohibitive to allow the free movement of trade.

The process of PRA is a complex one, requiring biological, mathematical risk analysis and economic modelling expertise. This introduction is purely to set the context if the reader becomes involved in discussions with growers or exporters that may involve *Bactrocera*-affected commodities. The NPPO has full responsibility for carrying out PRA and it is not done by unauthorised individuals.

11.4.3   ISPM 05, Glossary of phytosanitary terms

Biologists tend to use terms that we all think we know the meaning of, for example 'buffer zone', 'fruit and vegetables', or 'outbreak'. However, even common terms, such as 'pest' may have different meanings to different people, and certainly different meanings to quarantine and market regulators versus non-regulators. ISPM 5 is a glossary of phytosanitary terms and supplies official definitions for these and many other terms. If becoming involved in potentially trade sensitive research or discussions, knowing this glossary exists, and using it (!), is of critical importance.

11.5   Further Reading and References Cited

Aluja, M. and Mangan, R.L. (2008) Fruit fly (Diptera: Tephritidae) host status determination: critical conceptual, methodological, and regulatory considerations. *Annual Review of Entomology*, 53, 473–502.

Armstrong, J.W. (2001) Quarantine security of bananas at harvest maturity against Mediterranean and Oriental fruit flies (Diptera: Tephritidae) in Hawaii. *Journal of Economic Entomology*, 94, 302–314.

Dominiak, B.C., Wiseman, B., Anderson, C., Walsh, B., McMahon, M. and Duthie, R. (2015) Definition of and management strategies for areas of low pest prevalence for Queensland fruit fly *Bactrocera tryoni* Froggatt. *Crop Protection*, 72, 41–46.

Follett, P.A. and Vargas, R.I. (2010) A systems approach to mitigate Oriental fruit fly risk in 'sharwil' avocados exported from Hawaii. *Acta Horticulturae*, 880, 439–445.

Jang, E.B. (1996) Systems approach to quarantine security: postharvest application of sequential mortality in the Hawaiian grown 'Sharwil' avocado system. *Journal of Economic Entomology*, 89, 950–956.

Jang, E.B. and Moffitt, H.R. (1994) Systems approaches to achieving quarantine security. In: Sharp, L. and Hallman, G.J. (eds.) *Quarantine Treatments for Pests of Food Plants*. Westview, San Francisco, California, USA, pp. 225–237.

Orden, D. and Peterson E. (2007) Science, opportunity, traceability, persistence, and political will: necessary elements of opening the U.S. market to avocados from Mexico. In: Grote, U., Basu, A.K. and Chau, N.H. (eds.) *New Frontiers in Environmental and Social Labelling*. Physica-Verlag HD, Heidelberg, Germany, pp. 133–150.

Secretariat of the International Plant Protection Convention (2016) *ISPM 02: Framework for pest risk analysis*. International Plant Protection Convention and FAO, Rome, Italy, 20 pp. Available at https://www.ippc.int/en/publications/592/ (accessed 23 May 2018).

Secretariat of the International Plant Protection Convention (2016) *ISPM 05: Glossary of phytosanitary terms*. International Plant Protection Convention and FAO, Rome, Italy, 38 pp. Available at https://www.ippc.int/en/publications/622/ and https://www.ippc.int/en/publications/621/ (accessed 23 May 2018).

Secretariat of the International Plant Protection Convention (2016) *ISPM 10: Requirements for the establishment of pest free places of production and pest free production sites*. International Plant Protection Convention and FAO, Rome, Italy, 12 pp. Available at https://www.ippc.int/en/publications/610/ (accessed 23 May 2018).

Secretariat of the International Plant Protection Convention (2016) *ISPM 11: Pest risk analysis for quarantine pests*. International Plant Protection Convention and FAO, Rome, Italy, 40 pp. Available at https://www.ippc.int/en/publications/639/ (accessed 23 May 2018).

Secretariat of the International Plant Protection Convention (2016) *ISPM 26: Establishment of pest free areas for fruit flies (Tephritidae)*. International Plant Protection Convention and FAO, Rome, Italy, 60 pp. Available at https://www.ippc.int/en/publications/594/ (accessed 23 May 2018).

Secretariat of the International Plant Protection Convention (2016) *ISPM 27: Diagnostic protocols for regulated pests*. International Plant Protection Convention and FAO, Rome, Italy, 14 pp. Available at https://www.ippc.int/en/publications/593/ (accessed 23 May 2018).

Secretariat of the International Plant Protection Convention (2016) *ISPM 28: Phytosanitary treatments for regulated pests*. International Plant Protection Convention and FAO, Rome, Italy, 14 pp. Available at https://www.ippc.int/en/publications/591/ (accessed 23 May 2018).

Secretariat of the International Plant Protection Convention (2016) *ISPM 35: Systems approach for risk management of fruit flies (Tephritidae)*. International Plant Protection Convention and FAO, Rome, Italy, 12 pp. Available at https://www.ippc.int/en/publications/635/ (accessed 23 May 2018).

Secretariat of the International Plant Protection Convention (2016) *ISPM 37: Determination of host status of fruit to fruit flies (Tephritidae)*. International Plant Protection Convention and FAO, Rome, Italy, 18 pp. Available at https://www.ippc.int/en/publications/82520/ (accessed 23 May 2018).

12 Looking Forward

12.1 Introduction

Until only a few years ago the great majority of research on *Bactrocera* and the related genera fell into one of two main categories: taxonomy, or work directly related to pest management (including biological control). This is logical given the extreme pest status of several species, and the need for robust taxonomy to underpin pest management. However, it also means that we know amazingly little about the vast number of non-pest Dacini, and even little about the general biology and ecology of the pest species. However, a casual review of the current literature shows that this pattern is changing, driven by the emergence of new research tools, new centres of research excellence (notably in China), a changing global market place, and new environmental drivers. By way of a conclusion, this brief chapter speculates on the future of selected areas of Dacini research.

12.2 'Omics' and Whole-Systems Biology

For the last 20 years, entomologists have been utilising the ever increasing sophistication of genetic tools, and at the fore-front of this research have been tephritid workers. Although led initially by *C. capitata* workers, *Bactrocera* workers now have draft or partial genomes for *B. dorsalis*, *B. tryoni*, *B. neohumeralis* and *B. jarvisi*, with others (e.g. *B. carambolae*) under development. Transcriptomes are now publicly available for large numbers of *Bactrocera* species, as well as some *Zeugodacus* and *Dacus* species. The collection of transcriptomic data, particularly, has become routine in the last few years. Still far less common in *Bactrocera*, but starting to appear in a few papers and as 'work in progress' at conferences, are studies that also utilise metabolomics and proteomics in a whole-systems biology approach.

To date a lot of the published 'omics' research has simply been the collection and summation of large amounts of genetic data with, at best, correlative assessment made of genomic/transcriptomic differences between groups (e.g. transcriptome differences between immature and mature flies). However, researchers are increasingly utilising these resources to test specific hypotheses or to solve key technical problems. For example: population genomics is being used to explore local adaptation to selected environmental variables to develop better sterile insect technique (SIT) lines; transcriptomics coupled with qPCR and RNA interference (RNAi) is being used to test the role of individual genes in physiological pathways (e.g. vision and pesticide resistance); transcriptomic and metabolomic approaches are providing insights into cryptic phenotypic interactions between larvae and their host fruit; and genomic and transcriptomic datamining is being used to find unique diagnostic markers for cryptic species within complexes, or to identify source populations for an invasive species.

These new tools are clearly the future for most fields of *Bactrocera* work, from evolutionary phylogenetics through to post-harvest disinfestation. Because the tools, and particularly the bioinformatics, still require moderate to high levels of research funding and expertise the 'omics' revolution in the Dacini is not going to happen over-night, but it has already started and will continue to gain momentum as the tools and the data analysis become more commonplace and affordable.

12.3   A World Without Pesticides

Since the public release of DDT in the years immediately following World War II, applied entomology has had the safety net of synthetic insecticides to control pest populations. Even when biological controls have been the primary method of pest control, pesticides have still been there to manage outbreaks or seasonal population spikes. Increasingly though, the usefulness of the synthetic pesticides is diminishing. Older chemistries are being withdrawn or the manufacturers are simply ceasing their production, consumer markets are refusing to accept pesticide-treated products, and reliance on the small number of remaining chemistries is increasing the chance of resistance. For these reasons a future without synthetic pesticides is likely.

The loss of the pesticides will place a greater emphasis on non-pesticide-focused integrated pest management (IPM) systems for fruit fly management. Micro-SIT, as currently practised in Israel for *C. capitata* control, may increase in usage. As practised, micro-SIT focuses on geographically and temporally targeted releases of sterile flies, based on a local 'as-needs' basis as determined through regular field monitoring and predictive spatial models. Alternatives to components of, or replacements for, classical SIT are also on the horizon, or in some cases well above the horizon. These include the use of RNAi technology for male sterilisation, the release of insects with dominant lethal genes, and (cytoplasmic) incompatible insect technologies. All of these have technological and/or societal hurdles to overcome before they can become true replacements for classical SIT, but if they or alternatives do then this may also help to make SIT more widely adapted.

I am convinced that in the next decades parasitoids will be much more pro-actively utilised in fruit fly IPM. Largely relegated to usage as classical biolog-ical control agents, Hawaiian and French researchers have demonstrated that augmentoria can be used to get the benefits of crop hygiene while not breaking the parasitoid cycle; this approach should be adopted more widely. Recent research has demonstrated that opiines can be trained to forage more efficiently on certain crops, and if being bred for inundative release then this line of research should be followed further. I am also not aware of any active research for *Bactrocera* on orchard habitat manipulation to improve the environment for parasitoids and predators (excepting weaver ants), despite Mediterranean work demonstrating that more complex environments lead to greater natural enemy control of olive fruit fly. If we are to effectively manage fruit flies without pesticides, IPM packages that include the proactive manipulation of natural enemies need to be more widely researched, developed and extended.

My final prediction for growth research areas in a world without pesticides is host fruit resistance. Within any crop type there is huge variation in fruit fly susceptibility between cultivars and varieties. Yet incorporation of fruit fly resistance traits into traditional fruit breeding programmes is rarely done, nor does there appear to be any uptake of fruit fly resistance into new generation breeding using plant biotechnology approaches. This is in stark contrast to plant diseases, where resistance breeding is routine. By identifying the physiological mechanisms by which fruit respond to egg and larval infestation, the 'omic' approaches, particularly, have the potential to give us significant insights into why some fruits are resistant to fruit fly and others are not. *Anastrepha* research-ers in Mexico have begun such work, as have olive researchers (entomologists plus breeders); the rest of us need to follow suite.

12.4   Biodiversity Assessment

Biodiversity research tends to focus on charismatic vertebrates and flowering plants, yet the insects dominate the species level diversity of the super-diverse tropical rainforests. However, with the exception of some butterfly groups, insects are rarely the focus of detailed rainforest biodiversity assessment and conservation because their small size makes them difficult to collect and harder to identify, especially to the species level. By being unable to easily work with the insect fauna of rainforests, subtle changes in the environmental health of the forests may remain undetected, impeding opportunities for conservation.

Thanks predominantly to the career work of R.A.I. Drew, the Dacini of Asia and the Pacific represent one of the few super-diverse, rainforest-dominant insect groups for which the species level taxonomy is largely known. Further, in Papua New Guinea, Vojtech Novotny and colleagues have experimentally demonstrated that *Bactrocera* make ideal subjects for rainforest community ecology studies because they can be readily and repeatedly sampled using male lure traps. As novel male lures are developed, greater and greater percentages of the total local fauna will be trappable, giving high levels of regional species coverage. Unfortunately, the processing of large numbers of fruit fly trap catches is still time consuming and limited by taxonomic capacity. However, in the near future rapid diagnostic techniques will become available that will be able to overcome this impediment, by being able to handle bulk *Dacini* samples: identifying both the species present and their relative abundance. This step is not far away as bulk-sample diagnostic techniques already exist, and all that remains is to get comprehensive coverage of diagnostic genes (e.g. COI barcode or equivalent) for regional faunas. Australia is already close to this.

Because *Bactrocera* can be intimately associated with particular species of fruiting plants, replicated assessment of local *Bactrocera* faunas over space and time will give a direct measure of the stability or otherwise of not only a speciose insect group, but also the flowering plants upon which they depend. This information can be used to support studies around rainforest modification and clearing, habitat shifts with climate change, and broader community ecology. I thus envisage that as bulk diagnostic assessments become available, the Dacini will play a lead role in rainforest biodiversity assessment studies.

12.5 Conclusion

Throughout this book I have tried to stress the dual nature of the Dacini – major pests that need control, but also an incredibly rich dipteran clade worthy of their own ecological and evolutionary studies. However, the fields of Dacini pest management and evolutionary biology are increasingly merging as new diagnostic tools and sustainable control measures are sought. This is truly one group of insects for which there is no such thing as 'basic science' and 'applied science', but simply 'good science'.

Index

Page numbers in *italics* refer to Figures.

247

CABI – who we are and what we do

This book is published by **CABI**, an international not-for-profit organisation that improves people's lives worldwide by providing information and applying scientific expertise to solve problems in agriculture and the environment.

CABI is also a global publisher producing key scientific publications, including world renowned databases, as well as compendia, books, ebooks and full text electronic resources. We publish content in a wide range of subject areas including: agriculture and crop science / animal and veterinary sciences / ecology and conservation / environmental science / horticulture and plant sciences / human health, food science and nutrition / international development / leisure and tourism.

The profits from CABI's publishing activities enable us to work with farming communities around the world, supporting them as they battle with poor soil, invasive species and pests and diseases, to improve their livelihoods and help provide food for an ever growing population.

CABI is an international intergovernmental organisation, and we gratefully acknowledge the core financial support from our member countries (and lead agencies) including:

 UKaid from the British people

Ministry of Agriculture People's Republic of China

 Australian Government Australian Centre for International Agricultural Research

 Agriculture and Agri-Food Canada

 Ministry of Foreign Affairs of the Netherlands

 Schweizerische Eidgenossenschaft Confédération suisse Confederazione Svizzera Confederaziun svizra

Swiss Agency for Development and Cooperation SDC

Discover more

To read more about CABI's work, please visit: **www.cabi.org**

Browse our books at: **www.cabi.org/bookshop**, or explore our online products at: **www.cabi.org/publishing-products**

Interested in writing for CABI? Find our author guidelines here: **www.cabi.org/publishing-products/information-for-authors/**